Lecture Notes of the Institute for Computer Sciences, Social Informatics and Telecommunications Engineering   347

More information about this series at http://www.springer.com/series/8197

Shuai Liu · Liyun Xia (Eds.)

Advanced Hybrid Information Processing

4th EAI International Conference, ADHIP 2020
Binzhou, China, September 26–27, 2020
Proceedings, Part I

 Springer

Editors
Shuai Liu (ID)
Hunan Normal University
Changsha, China

Liyun Xia
Hunan Normal University
Changsha, China

ISSN 1867-8211 ISSN 1867-822X (electronic)
Lecture Notes of the Institute for Computer Sciences, Social Informatics
and Telecommunications Engineering
ISBN 978-3-030-67870-8 ISBN 978-3-030-67871-5 (eBook)
https://doi.org/10.1007/978-3-030-67871-5

This Springer imprint is published by the registered company Springer Nature Switzerland AG
The registered company address is: Gewerbestrasse 11, 6330 Cham, Switzerland

Preface

We are delighted to introduce the proceedings of the fourth edition of the European Alliance for Innovation (EAI) International Conference on Advanced Hybrid Information Processing (ADHIP 2020). This conference brought together researchers, developers and practitioners around the world who are leveraging and developing hybrid information processing technology for smarter and more effective research and applications. The theme of ADHIP 2020 was "Industrial applications of aspects with big data".

The technical program of ADHIP 2020 consisted of 190 full papers, with acceptance ratio about 46.8%. The conference tracks were: Track 1 –Industrial application of multi-modal information processing; Track 2 –Industrialized big data processing; Track 3 –Industrial automation and intelligent control; and Track 4 –Visual information processing. Aside from the high-quality technical paper presentations, the technical program also featured two keynote speeches. The two keynote speakers were Dr. Khan Muhammad from Sejong University, Republic of Korea, who is currently working as an Assistant Professor at the Department of Software and Lead Researcher of the Intelligent Media Laboratory, Sejong University, Seoul, Republic of Korea, and is an editorial board member of the Journal of Artificial Intelligence and Systems and Review Editor for the Section "Mathematics of Computation and Data Science" in the journal Frontiers in Applied Mathematics and Statistics; as well as Dr. Gautam Srivastava from Brandon University in the Canada, who has published a total of 143 papers in high-impact conferences in many countries and in high-status journals (SCI, SCIE) and has also delivered invited guest lectures on Big Data, Cloud Computing, Internet of Things and Cryptography at many Taiwanese and Czech universities. He is an Editor of several international scientific research journals.

Coordination with the steering chairs, Imrich Chlamtac, Guanglu Sun and Yun Lin, was essential for the success of the conference. We sincerely appreciate their constant support and guidance. It was also a great pleasure to work with such an excellent organizing committee team for their hard work in organizing and supporting the conference. In particular, the Technical Program Committee, led by our TPC Chair, Dr. Shuai Liu, completed the process of peer-review of technical papers and made a high-quality technical program. We are also grateful to the Conference Manager, Natasha Onofrei, for her support and to all the authors who submitted their papers to the ADHIP 2020 conference and workshops.

We strongly believe that the ADHIP conference provides a good forum for all researchers, developers and practitioners to discuss all scientific and technical aspects that are relevant to hybrid information processing. We also expect that future ADHIP conferences will be as successful and stimulating, as indicated by the contributions presented in this volume.

Shuai Liu

Conference Organization

Steering Committee

Imrich Chlamtac   University of Trento
Yun Lin   Harbin Engineering University
Guanglu Sun   Harbin University of Science and Technology

Organizing Committee

General Chairs

Shuai Liu   Hunan Normal University
Yun Lin   Harbin Engineering University

General Co-chair

Gautam Srivistava   Brandon University

TPC Chair and Co-chair

Xunli Zhang   Binzhou University

Sponsorship and Exhibit Chair

Zhaoyue Zhang   Civil Aviation University of China

Local Chairs

Ligang Chen   Binzhou University
Aixue Qi   Binzhou University

Workshops Chair

Gautam Srivastava   Brandon University

Publicity and Social Media Chair

Weina Fu   Hunan Normal University

Publications Chair

Khan Muhammad   Sejong University

Web Chair

Wei Wei   Xi'an University of Technology

Posters and PhD Track Chair

Liyun Xia   Hunan Normal University

Panels Chair

Xiaojun Deng   Hunan University of Technology

Demos Chair

Jie Gao   Hunan Normal University

Tutorials Chair

Qingxiang Wu   Yiyang Vocational and Technical College

Technical Program Committee

Hari M. Srivastava	University of Victoria
Guangjie Han	Hohai University
Amjad Mehmood	University of Valencia
Guanglu Sun	Harbin University of Science and Technology
Gautam Srivastava	Brandon University
Guan Gui	Nanjing University of Posts and Telecommunications
Yun Lin	Harbin Engineering University
Arun K Sangaiah	Vellore Institute of Technology
Carlo Cattani	University of Tuscia
Bing Jia	Inner Mongolia University
Houbing Song	Embry-Riddle Aeronautical University
Qingxiang Wu	Yiyang Vocational and Technical College
Xiaojun Deng	Hunan University of Technology
Zhaojun Li	Western New England University
Weina Fu	Hunan Normal University
Han Zou	University of California
Xiaochun Cheng	Middlesex University
Wuyungerile Li	Inner Mongolia University
Huiyu Zhou	University of Leicester
Weidong Liu	Inner Mongolia University
Juan Augusto	Middlesex University
Jianfeng Cui	Xiamen University of Technology
Xuanyue Tong	Nanyang Institute of Technology
Qiu Jing	Harbin University of Science and Technology
Mengye Lu	Inner Mongolia University
Heng Li	Henan Finance University
Lei Ma	Beijing Polytechnic
Mingcheng Peng	Jiangmen Vocational and Polytechnic College
Wenbo Fu	Datong university
Yafei Wang	Ping dingshan University

Yanning Zhang	Beijing Polytechnic
Guangzhou Yu	Guangdong Ocean University
Dan Zhang	Xinyang Vocational and Technical College
Fuguang Guo	Henan Vocational College of Industry and Information Technology
Weibo Yu	Changchun University of Technology
Dan Sui	Califoraia State Polytecnic University-Pomona
Juan Wang	Zhengzhou Institute of Technology
Xinchun Zhou	Baoji University of Arts and Sciences
Qingmei Lu	University of Louisville
Hong Tian	Baotou lron steel vocational technical college
Yuling Jin	Chizhou Vocational And Technical College
Yongjun Qin	Guilin Normal College
Wen da Xie	Jiangmen Polytechnic
Shuai Yang	Changchun University of Technology

Contents – Part I

Industrialized Big Data Processing

Industrial Automation and Intelligent Control

Contents – Part II

Visual Information Processing

Industrial Application of Multi-modal Information Processing

Design of Unmanned Aerial Vehicle Automatic Endurance System

Jiang Heng[1], Pan Di-zhao[1], Hou Xiaofeng[1], Tang Yujia[1],
Chen Ligang[1,2(✉)], and Ma Guoli[1,2]

[1] Institute of Aeronautical Engineering, Binzhou University,
Binzhou 256603, China
clgwlx@126.com
[2] Shandong Engineering Research Center of Aeronautical Materials
and Devices, Binzhou 256603, China

Abstract. With electricity as the main power energy, the UAV has been affected by the impact of battery storage on the flight time and flight time, which makes the UAV need to constantly replace the battery to ensure the long-term operation. The automatic endurance system takes the remaining power of the UAV as the independent variable, and the average value of the flight data of the UAV battery for nearly three times as the judgment basis. It realizes the automatic navigation in the UAV area, charges the base station after landing accurately, and improves the endurance time of the UAV.

Keywords: UAV · Automatic endurance · Wireless charging · Electromagnetic induction

1 Introduction

With the rapid development of technology, drone applications have become more and more extensive, but due to the bottleneck problem of drone battery technology, the flight time and flight distance limitations in practical applications are subject to [1]. The design of the subject's automatic endurance system has become one of the key technologies to realize the automatic endurance of the UAV area.

2 Overall Design of UAV Automatic Endurance System

The system design plan is intended to propose a design theory for an automatic drone endurance system. The automatic drone endurance system consists of a wireless charging base station, a drone endurance management system, and an automatic endurance system general controller (hereinafter referred to as the total controller) partly composed.

The charging base station controller monitors the working status and GPS position of the base station in real time, and sends these status information to the general controller at the same time. After receiving the screening of the status information of the base station, the general controller sends the location of the currently available base

S. Liu and L. Xia (Eds.): ADHIP 2020, LNICST 347, pp. 3–10, 2021.
https://doi.org/10.1007/978-3-030-67871-5_1

station to each drone. The UAV monitors and counts its own battery status information, calculates the distance that can be flown per unit of power in different power ranges, and uses it as reference data for the next flight; and sets a remaining power threshold θ based on the battery information when the battery monitoring module finds When the remaining power reaches the remaining power threshold θ, the UAV starts to receive the current available base station position coordinates, compares and calculates it with the current own position coordinates, selects the most suitable base station and sends a charging request to the general controller. After receiving the request, the master controller locks the base station to avoid repeated selection by other drones. The process is shown in Fig. 1.

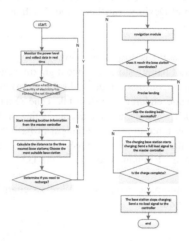

Fig. 1. Design flow chart of UAV automatic endurance

3   Wireless Charging Base Station

The wireless charging base station is one of the key points of system design. The basic principle adopts the electromagnetic induction method. The alternating current of a certain frequency in the primary coil generates a certain intensity of current in the secondary coil through electromagnetic induction, thereby transferring energy from the transmitting end to the receiving end. A transmitting coil is set in the base station, and a receiving coil is mounted on the bottom of the drone, and the electric energy can be transmitted from the transmitting end to the receiving end [2, 3]. It is composed of base station controller, charging circuit, GPS module, base station monitoring module, etc. The wireless charging base station is powered by external 220 V AC power, and is converted to DC power by switching power supply and buck regulator module all the way to power the controller and other functional modules; All the way is directly connected with the launch coil, controlled by the relay, and it is charged when the drone is landing. The overall structure of the wireless charging base station is shown in Fig. 2.

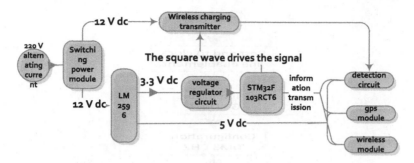

Fig. 2. Overall design structure diagram of wireless charging base station

Using STM32F103RCT6 as a wireless charging base station controller, cortex-M3 high-performance core, low power consumption, short delay, strong interrupt handling ability, very suitable for base station control and management. STM32F103RCT6 as a base station controller in addition to controlling the GPS module and wireless communication module to achieve positioning and information transmission, also has an important role is to use the TIM timer to generate PWM square wave after TPS28225 conversion, drive NMOS tube to achieve push-pull output, To the resonant circuit to achieve the transmission of electrical energy. The transmitter circuit of the wireless charging base station is shown in Fig. 3.

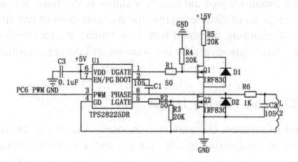

Fig. 3. Wireless charging transmitter circuit diagram

STM32F103RCT6 uses TIM3_ch2 to generate PWM with resonance frequency as a signal to drive the inverter circuit. The flow chart of PWM generation and configuration is shown in Fig. 4.

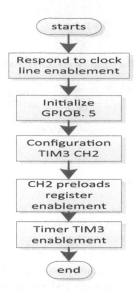

Fig. 4. PWM initialization program flow chart

4 UAV Endurance Management System

Accurate real-time monitoring of the battery's status is the basis for automatic drone battery life. The reception of electric energy, the measurement of remaining power, the setting of the UAV charging threshold based on battery characteristics, and the best choice of charging base stations are the innovations and design priorities of the system design.

4.1 Wireless Charging Receiver Design

The receiving end of wireless charging is composed of a coil at the receiving end, a high-frequency voltage stabilizing rectifier circuit, and a charging management circuit. The wireless charging receiving and rectifying circuit is shown in Fig. 5.

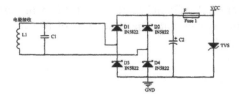

Fig. 5. Wireless charging receiving circuit diagram

The charging management circuit is built with the CN3703 integrated circuit as the core, indicating the charging status of the battery through the on and off of the LED light, and adjusting the charging current by changing the size of the resistor. The specific circuit design is shown in Fig. 6.

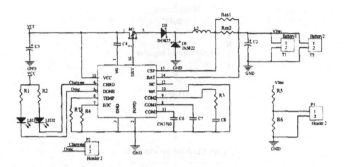

Fig. 6. Receiver charge management circuit diagram

4.2 Acquisition of Remaining Power

The microcontroller sends a voltage detection command to the DS2762 every 88 ms. If the pressure difference is positive, the battery is in the charging state; when the pressure difference is negative, the battery is in the discharging state. The single chip microcomputer can use the control command to control the DS2762 to monitor the battery voltage, current, charge and discharge status and remaining power and other parameters in real time, and automatically store the data into the corresponding register, which is read and used by the single chip microcomputer [4–6].

To obtain the remaining power, the microcontroller can directly read the value in the current accumulation register. The DS2762 automatically monitors the current value in real time and stores it in the current accumulation register. During the reading process, you need to check the status of the EEC flag first. The specific procedure The process is shown in Fig. 7.

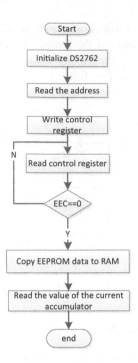

Fig. 7. Flow chart of the remaining battery reading program

4.3 Setting Method of Charging Threshold

After obtaining the remaining power of the drone, collect the relationship between the flying distance of the drone and the remaining power [5–9]. The specific methods are as follows:

Taking each w% of electricity as a detection unit, each time the drone's electricity drops by w%, record the flying distance L of the drone during this process, so as to obtain $L_1, L_2, ..., L_n$, corresponding to the flying distance The power range is 0-w%, w%-2w%, ..., (100-w)%-100%; after recording 3 sets of $L_1, L_2, ..., L_n$ data, calculate the The average value of the flight distance A_k in the same detection unit, and then each time a new set of $L_1, L_2, ..., L_n$ data is entered, replacing it with the first set of data entered; after the collected flight distance A_k corresponds to the power range After the relationship, according to the downward trend of the flight distance corresponding to each w% of power with the decrease of the power range, the flight distance of the aircraft in one or more power ranges with lower power is obtained, and the obtained flight distance is subtracted by a preset value as the final prediction.

4.4 The Choice of Charging base station

When the remaining power reaches the set power threshold, the UAV starts to receive the location information of available wireless charging base stations, and determines the most suitable charging base station based on its location information and remaining power, and requests charging from the general controller [10–12]. The location selection scheme of the specific wireless charging base station is as follows:

When the power of the drone reaches the preset threshold θ, the drone controller reads the average value of the last three flight records in its own battery information memory, roughly estimates the current power can fly the longest distance, and records it as Q_0; the drone starts Receive the location information of the available charging base stations transmitted by the general control, and calculate and filter out the available base stations whose horizontal and vertical coordinate distances are not greater than the UAV's own GPS module, combined with the UAV's own vertical height information, calculate And select a base station with the minimum distance in the direction of the flight target available, the distance between the two is recorded as Q_1, and a minimum distance opposite to the flight target can be used as the location information of the base station, the distance is recorded as Q_2.

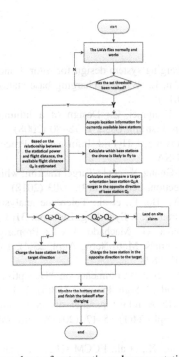

Fig. 8. Flow chart of automatic endurance station selection

The general rules for the selection of the automatic endurance base station for the drone are as follows: when $Q_0 < Q_1$ and $Q_0 < Q_2$, the drone landed in situ, and sent the location information to the general controller to alarm, waiting for the staff to replace the battery or other processing. When $Q_0 < Q_1$ and $Q_0 > Q_2$, the available base station with the smallest distance to the target is charged. When $Q_0 > Q_1$, the available base station with the smallest distance to the direction close to the target is charged. When the battery management module detects that its own battery has been charged, the drone continues to take off and fly to the target location. The process is shown in Fig. 8.

5 Conclusion

Automatic drone battery life is a problem that must be solved in the development of regional drones in the future, and it is also a basis for the realization of intelligent drone. Therefore, in the absence of innovations in battery technology, the automatic drone battery life system is One of the effective ways to solve the problem of drone power energy. The solution to this problem allows the drone to save landing time, and it can work continuously and uninterruptedly in a certain area with high efficiency, such as express delivery and delivery.

Acknowledgements. This work was supported by the National Natural Science Foundation of China (NSFC) (U1731121), by a Project of Shandong Province Key R&D Program Project (2019GSF109105) and Shandong Province Higher Educational Science and Technology Program (J18KB108).

References

1. Zhen, H.: Autonomous charging system design for rotor drones (2019)
2. Zhiying, Z., Shu, N., Linlin, L.: Wireless charging base stations for drones. Sci. Technol. Innov. Appl. **3**, 41–42 (2019)
3. Xiaojie, T., Cungen, L., Dongmei, Z.: Design of a lithium battery monitoring circuit composed of DS2762 chip. Instrum. Users **1**, 78–80 (2008)
4. Shuai, L., Gelan, Y.: Advanced Hybrid Information Processing, pp. 1–594. Springer International Publishing, USA
5. Fei, W., Xin-Bo, C., Chen-Guang, S., et al.: Target tracking while jamming by airborne radar for low probability of detection. Sensors **18**(9), 2903 (2018)
6. Tianpeng, L., Xizhang, W., Bo, P., et al.: Tolerance analysis of multiple-element linear retrodirective cross-eye jamming. J. Syst. Eng. Electron. **31**(3), 460–469 (2020)
7. Liu, S., Bai, W., Srivastava, G., Machado, J.A.T.: Property of self-similarity between baseband and modulated signals. Mob. Netw. Appl. **25**(4), 1537–1547 (2019). https://doi.org/10.1007/s11036-019-01358-9
8. Shi, J., Liu, X., Yang, Y., et al.: Comments on "Deceptive jamming suppression with frequency diverse MIMO radar". Sign. Process. **158**(5), 1–3 (2018)
9. Liu, S., Liu, G., Zhou, H.: A robust parallel object tracking method for illumination variations. Mob. Netw. Appl. **24**(1), 5–17 (2018). https://doi.org/10.1007/s11036-018-1134-8
10. Wang, X., Zhang, G., Wang, X., et al.: ECCM schemes against deception jamming using OFDM radar with low global PAPR. Sensors **20**(7), 2071 (2020)
11. Zhou, F., Tian, T., Zhao, B., et al.: Deception against near-field synthetic aperture radar using networked jammers. IEEE Trans. Aerosp. Electron. Syst. **55**(6), 3365–3377 (2019)
12. Liu, S., Liu, D., Srivastava, G., et al.: Overview and methods of correlation filter algorithms in object tracking. Complex Intell. Syst. (2020). https://doi.org/10.1007/s40747-020-00161-4

Research and Design of UAV Environmental Monitoring System

Shen Xiaoyu[1], Liu Yuanhang[1], Chen Ligang[1,2(✉)], Zhang Xin[1,2],
and Ma Guoli[1,2]

[1] Institute of Aeronautical Engineering, Binzhou University,
Binzhou 256603, China
clgwlx@126.com
[2] Shandong Engineering Research Center of Aeronautical Materials
and Devices, Binzhou 256603, China

Abstract. With the rapid development of China's economy and the increasing level of urbanization, more and more attention has been paid to urban air pollution. Accurate monitoring of air quality at high altitude is always a problem. It is difficult to collect air quality data at high altitude, and the high cost leads to the lack of data collection of air quality at vertical height. It is imperative to combine UAV and traditional environmental monitoring devices to learn from each other when it is necessary to periodically test and compare air quality data at different altitudes. Environment monitoring system based on UAV, environment monitoring device and Internet. The system includes four rotor UAV, environmental monitoring terminal and computer environmental monitoring platform. This paper mainly studies and designs from these three parts.

Keywords: UAV · Environmental monitoring · 4G module · Monitoring management system

1 Introduction

At present, many cities in China are conducting environmental air quality assessment and prediction. The basic monitoring sample data for the assessment and prediction of ambient air quality come from one automatic monitoring system of ambient air quality. In China, multi-rotor UAV has become a new favorite of all walks of life in recent years. With the opening of China's multi-space and the rapid development of consumer drones, multi-rotor UAV has come into our life. At present, air quality monitoring stations still mainly detect air quality in horizontal direction, and there are few data related to vertical height. Therefore, in the process of investigating airborne particulate matter, in order to better detect the numerical difference of suspended particulate matter at different altitudes and stratify the data vertically, it is imperative to apply uav technology to detect air quality.

2   Correlation Technique

2.1   The Overall Structure of the UAV Environmental Monitoring System

We will STM32F103C8T6 as control core, MPU6050 VR technology, cooperate with the external circuit controller, realize the operator to air of beauty appreciation, camera on the unmanned aerial vehicle (UAV) figure back to the images obtained with VR technology, make the line of sight of the operator feel on the unmanned aerial vehicle (UAV), through MPU6050 module, track fast and slow or quick, indeed make The cradle head of the drone follow the operator on the line of sight change and change, make the operator feel "immersive" lifelike [1] (Fig. 1).

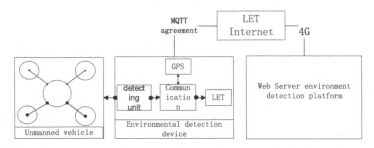

Fig. 1. General structure diagram of UAV environmental monitoring system

UAV environmental monitoring device is a set of independent equipment that needs the assistance of UAV. It is independently powered and can be connected to temperature and humidity sensors, particulate matter sensors and other sensors to accurately detect air quality. The device has its own barometer, GPS, 4G module, and memory for accurately recording air quality on the route. When flying at low altitude, the data can be transmitted to the environment monitoring and management platform in real time through 4G module, and the data can be viewed and managed online through Web client on the mobile terminal and computer terminal. At high altitude, due to the limited coverage of 4G signal, the data will be stored in the device's memory, and the Excel file can be exported after the flight. The data includes altitude, latitude and longitude, air quality index, number of particulate matter, temperature and humidity, etc., and sensors can be added if necessary. The environmental monitoring management platform can aggregate, process and display the data, conveniently, effectively and intuitively provide the air particulate data at different altitudes, and solve the problem of high difficulty and high cost for the monitoring of high-altitude particulate matter caused by the traditional environmental monitoring system.

2.2   Overall Structure of Environmental Monitoring Unit

This paper studies the design scheme and hardware structure design of the environmental monitoring device. The overall structure of the UAV environmental monitoring device is shown in Fig. 2. The environmental monitoring device can connect to the

Web server in real time and upload data through 4G signal and MQTT protocol. The primary task of the environmental monitoring device is to obtain the monitoring data of air particulate matter sensor and barometer, as well as the geographical coordinates and position data obtained by GPS, and then the master MCU Uploates the data to the Web server through THE STANDARD of MQTT protocol of LTE module, and provides the staff for management and view through the interface design [2, 3].

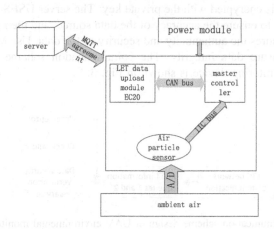

Fig. 2. The overall structure of the UAV environmental monitoring device

The internet-based UAV environmental monitoring device is mainly divided into parts, which are described as follows:

(1) Master MCU: STM32F103ZET6 chip is used for master control. STM32 series is based on armCortex-M3 kernel specially designed for embedded applications requiring high performance, low cost and low power consumption. Low power mode: STM32F103RCT6 supports three low power modes to achieve the best balance between low power, short startup time and available wake-up sources. Mainly responsible for reading data from air monitoring sensor, barometer and GPS module through IIC serial port bus, processing and processing the respective data, filtering noise data caused by sensor or circuit problems. The data is uploaded to the Web server in MQTT protocol via LTE module [4–6].

(2) Power supply module: Provide stable power supply for each functional module.

LET module/GPS module: LET and GPS module adopt EC20 chip, which encapsulates 4G communication LET chip and GPS chip. It is mainly used to provide geographic coordinate location and upload data to the Web server.

(3) Air particulate acquisition module: The acquisition module adopts sensors to obtain accurate original particulate data.

2.3   Communication Scheme Design of UAV Environmental Monitoring Device

Data transmission scheme between environment monitoring device and Web server: Internet connection is required to establish connection with the IP specified port of the server public network [7–9]. Therefore, LTE module is used to connect 4G network and upload to the server through MQTT protocol. The server establishes MQTT service, and the data is encrypted with the private key. The server USES the public key to decrypt and verify to ensure the accuracy of the data source, and then puts the data into storage, which ensures the authenticity and security of the data. Use MQTT messaging middleware to ensure data integrity. The communication scheme design of UAV environmental monitoring device is shown in Fig. 3.

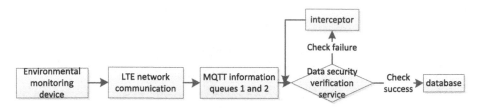

Fig. 3. Communication scheme design of UAV environmental monitoring device

2.4   Monitor Platform Development Process Design

The environmental monitoring platform is used for the staff to check and manage the data of atmospheric particulate monitoring of the UAV, and the data stored in the SD card can be uploaded to the server with one click through the system program (Explanation: When the UAV is in the upper air without 4G signal or the airspace with poor signal, the data will be stored in the SD card. When the 4G signal is good, the data will be directly uploaded to the server.) To save labor, and the server will analyze and store the data. Staff are also supported to export UAV monitoring reports through the system. More convenient and fast report generation, easy to work. It also supports real-time viewing of air quality data at different elevations of streets in different cities and regions. Based on the above requirements, the environment monitoring platform will include the following six systems: authority menu configuration system, personnel authority configuration system, module management system, large-screen data display system, report output system, and data upload and input system. The architecture of environment monitoring platform requirements is shown in Fig. 4.

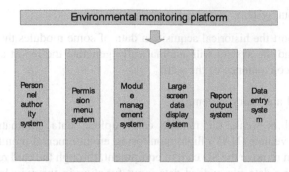

Fig. 4. Environment monitoring platform requirements architecture diagram

This paper will analyze and study the specific requirements of these six systems:

(1) Personnel authority system:

The basic functions of the management system can be added, deleted, frozen, unfrozen personnel account in the system. Can be configured to modify edit account permissions. When logging into the system, the staff shall verify through the staff permission system to determine whether the user is a white list user, the account permission level, what function points the current level contains, and the dynamic rendering function menu.

(2) Configure the system with the permission menu:

The basic functions of the management system, can be configured in the system function items and menu directory corresponding relationship, can be configured in the system of different authority accounts contained function points.

(3) Module management system:

The module (environmental monitoring device) in the imported system can then be connected to the system and upload data. Each module system will automatically generate a unique module code, which is the certificate for the module to upload data. Administrator can add, edit and delete modules through the module management system.

(4) Large-screen data display system:

The visual display module of EChat diagram is used to collect data, and the collected data of the module is presented on the page in real time for the convenience of staff. Historical data can also be displayed to show the dynamic changes of air quality particles in the same area at different times.

(5) Report output system:

The staff can export the historical acquisition data of some modules through the report output system, and the system will automatically generate the report according to the template for the convenience of the staff.

(6) Data upload and entry system:

When 4G signal is weak, the module can't upload data automatically, such as unmanned aerial vehicle (UAV) flight monitoring environmental monitoring device is complete, you can remove the SD card of equipment, through the card reader computer, copy of monitoring data file, upload data input function in the recycling system, will copy the monitoring data file upload data collection [10–12].

3   Conclusion

This paper focuses on the design of UAV environmental monitoring device and environmental monitoring platform. The data collection of users and staff on environmental status and air particles at different altitudes was realized, and the data was uploaded to the server in real time through LTE module. The server verified, stored and displayed the data.

Acknowledgements. This work was supported by the National Natural Science Foundation of China (NSFC) (U1731121), by a Project of Shandong Province Key R&D Program Project (2019GSF109105) and Shandong Province Higher Educational Science and Technology Program (J18KB108).

References

1. Song, Z., Fang, W.: Overall design of multi-UAV collaborative air quality monitoring system. Comput. Knowl. Technol. **15**(26), 92–93+100 (2019)
2. Li, G., Shen, Y., Wu, W., Liu, H., Niu, Z.: Design and implementation of indoor air quality monitoring device. Fujian Comput. **35**(04), 86–87 (2019)
3. Yang, G., Zhang, J., Han, F., Liu, T., Yue, Z., He, J.: Integrated design of air quality monitoring system based on BDS. Exp. Technol. Manag. **35**(06), 73–76+82 (2008)
4. Lin, K., Shi, J., Deng, Y., Zeng, Q.: Design and implementation of wearable air quality monitoring products. Guangxi Phys. **39**(01), 28–30 (2008)
5. Qin, C.: Discussion on monitoring methods of ambient air quality in public places. Sci. Technol. Innov. Bull. **18**, 127–128 (2009)
6. Jian, W., Jiansheng, S., Yaxiong, L., et al.: UAV 3D route planning based on artificial immune clone selection algorithm. Syst. Eng. Electron. **40**(1), 86–90 (2018)
7. Gao, Z., Liu, B., Cheng, Z., et al.: Marine mobile wireless channel modeling based on improved spatial partitioning ray tracing. China Commun. **17**(3), 1–1 (2020)
8. Zexuan, J., Kaihua, W., Shuo, W., et al.: Path planning algorithm for obstacle-bearing irregular areas of fully automatic plant protection unmanned aerial vehicles. Jiangsu Agric. Sci. **46**(05), 220–223 (2018)

9. Shuai, L., Liu, G., Zhou, H.: A robust parallel object tracking method for illumination variations. Mob. Netw. Appl. **24**(1), 5–17 (2019)
10. Shuai, L., Gelan, Y.: Advanced Hybrid Information Processing, pp. 1–594. Springer, New York (2019)
11. Liu, S., Lu, M., Li, H., et al.: Prediction of gene expression patterns with generalized linear regression model. Front. Genet. **10**, 120 (2019)
12. Lu, M., Liu, S.: Nucleosome positioning based on generalized relative entropy. Soft. Comput. **23**(19), 9175–9188 (2018). https://doi.org/10.1007/s00500-018-3602-2

Design of Temperature Measurement and Control System of Chemical Instrument Based on Internet of Things

Xiu-hong Meng, Shui Cao, You-hua Zhang, and Lin-hai Duan[✉]

Guangdong University of Petrochemical Technology, Maoming 525000, China
lhduan@126.com

Abstract. The traditional temperature measurement and control system of chemical instruments can not accurately grasp the basic condition of temperature data, resulting in low efficiency of measurement and control. Therefore, a temperature measurement and control system of chemical instruments based on the Internet of things is designed. According to the relevant performance of the hardware components of the system, the system information of the control center is studied, and the correlation of the internal system is studied. Based on this, the command of hardware mode transformation is executed. After the hardware design is realized, the system software design is realized on the premise of hardware data. Combined with the measurement and control algorithm, the measurement and control mode of the center is continuously studied, the data difference between the systems is adjusted, and the contradiction between the measurement and control data is avoided Complete the overall system design operation. The experimental results show that the design of chemical instrument temperature measurement and control system based on Internet of things has higher measurement and control efficiency, shorter measurement and control time, and improves the accuracy of measurement and control.

Keywords: Internet of things · Temperature measurement and control of chemical instruments · Temperature measurement and control system · Design of measurement and control system

1 Introduction

Chemical instruments will have a certain degree of danger during use. If improper operation will cause physical harm to users, for this reason, the use of chemical instruments must be investigated and supervised to improve their safety. Most researchers investigate the temperature status of chemical instrument, and use analysis algorithm to study the measurement and control information, so as to achieve the purpose of designing the temperature measurement and control system of chemical instrument.

Current research at home and abroad selects the required control data according to the authenticity of the system operation, and strengthens the internal data monitoring

S. Liu and L. Xia (Eds.): ADHIP 2020, LNICST 347, pp. 18–29, 2021.
https://doi.org/10.1007/978-3-030-67871-5_3

performance to ensure that the collected data is within the scope of the system operation to ensure the security of the control data [1].

Traditional chemical instrument temperature measurement and control system based on data mining technology uses Internet of things information skills to transform network data into system control data to achieve research and operation [2].

The traditional temperature measurement and control system for chemical instruments based on cloud computing achieves the purpose of measurement and control system design through the measurement and control calculation of the internal network system and the algorithm research. However, the current research does not select more accurate reference data, the degree of mastering the measurement and control system is low, which can not meet the needs of system operation [3]. Therefore, in view of the above problems, this paper designs a temperature measurement and control system of chemical instruments based on the Internet of things, and analyzes and solves the above problems. In this paper, a comprehensive data control method is designed, which improves the control accuracy of the control system, reduces the control time, and can better carry out the experimental research operation.

2 Hardware Design of Temperature Measurement and Control System of Chemical Instrument Based on Internet of Things

As a powerful data operation mode, the Internet of things has been widely used in data operation. In this paper, the status information of the Internet of Things is used to find the internal temperature data of chemical instruments, and the system hardware design operation is performed, and different operations are constructed according to different hardware properties. In order to strengthen the control accuracy of the control system, this paper sets up a data acquisition module, connects the channel between the data and the system, controls the collection strength of the data, and constructs the data collection diagram (Fig. 1):

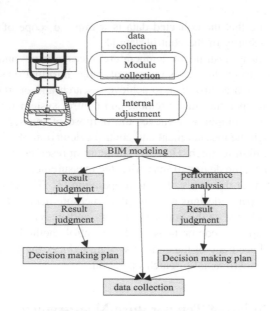

Fig. 1. Data collection diagram

The data collector in this paper adopts dczl23-wfet600s-i type collector, which adopts high-performance and low-power microcontroller hardware platform and embedded operating system software platform, with flexible system upgrading ability, supporting various communication modes such as power line carrier, micro power wireless, RS485 [4].

Uplink communication adopts a modular design, manages the operation information of the data system, and can transmit data information to the central control system while managing, and connects to the USB interface to ensure the security of data information transmission, and use different transmission channels to weaken the system data. Differential information promotes the unified development of data, eliminates the network signal data that does not conform to the operation of data system in time, and ensures the purity of data transmission.

According to the performance of data system control, further analysis of its necessary conditions, set up the corresponding data control module, standardize the processing of control data. In this paper, the data control module selects the control system mark controller, which has the function of serial port data acquisition, integrating resistance signal assistance, relay control and 485 communication. It can be applied to single coil SPI serial port encoder, displaying angle change and displacement change, serial port SSI three wire signal input, external power supply is 12 V or 5 V. It has absolute encoder with display serial port, meters of accumulated multi turn length, 0–10 k or 0 external access- 2.5 V. The voltage signal potentiometer, which is used in combination with the absolute value of single loop parallel port, can produce the use effect of multi loop absolute value, increase the success rate of data control, and study the main control chip status of the control system according to the process of central data analysis, so as to complete the design of data measurement and control module [5].

In this paper, the control chip is an IC that integrates a controller and a driver. It can independently control two robot gait data systems at the same time. It is a powerful system-on-a-chip. It integrates complex in a miniature 7 mm × 7 mm QFN package. The Six Point ramp generator, as well as industry-leading diagnostic and protection features.

With the addition of steadthchop silent driving technology, spread cycle anti jitter technology, coolstep current dynamic adjustment technology, it can save 75% energy. With dcstep torque adjustment technology and stallgard locked rotor detection function, it can promote the optimization of the system. With the expansion of n-channel MOSFETs, the motor current of each coil can reach 20A, and the voltage is 60 V DC, easy to operate, only need to find the target parameters, all the instrument temperature data system logic are running in the tmc5160. When NEMA 17 to NEMA 34 and larger systems are driven, no software operation is required, saving operation time, and improving control efficiency [6]. Connected to the host microcontroller via industry standard SPI or step/direction interface, TMC5160 performs all real-time position and speed step motion calculations. At the same time, TMC5160 also supports ABN encoder input to optimize input port data information and facilitate internal structure Sexual operation can provide relevant information on the basis of joint conversion, and strengthen the information analysis of internal control data.

According to the above steps, adjust the hardware structure of the system, and realize the research and operation of the hardware of the multi-point control system of the network engineering virtual unit [7].

The MSU processor in the node has the function of timed sleep, which can reduce the cost of power consumption. It is mainly responsible for the screening and transmission of analog quantities. The A/D conversion function in the processor can reduce the complexity of analog quantity screening, considering the power supply voltage. Differently add a capacitor in the MSU processor to reduce the impact of the voltage difference.

Each node has a set of analog interface circuit, which can timely transmit the real-time environment information collected from the orchard to the next data layer through the interface. In this system, the node analog interface circuit adopts the analog front-end with current signal of 2ma–10ma, and converts the current into voltage signal through the front-end.

The high-definition vision node is mainly composed of the grid high-definition electronic eye as the main body of the visual node. It is reliably powered by the JW48V–6A power adapter. A color processing chip is installed in the gate high-definition electronic eye, which improves the sensitivity and resolution of the electronic lens, and enhances the color processing ability, automatic light filling ability and automatic focusing ability of the electronic eye [8]. This can realize the high-definition vision node, which can provide 24-h security protection for the entire orchard range. The high-definition vision node is also equipped with a data computing platform, which can accurately measure the terrain position. The computing platform uses the D-2015 J model motor as the The drive provides a stable and sensitive environment for data operation.

3 Software Design of Temperature Measurement and Control System for Chemical Instruments Based on Internet of Things

After realizing the hardware design of the system, based on the operation data of the hardware system, this paper changes the data operation conditions, and stores the collected data in the network structure space in combination with the big data calculation mode. The operation flow chart of the relevant system software is set as follows (Fig. 2):

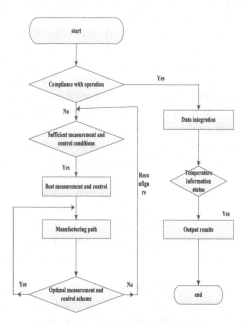

Fig. 2. Software operation flow chart

At the beginning of the program, the transmission node communicates commands to other nodes in the form of numbers. After receiving the command, the other nodes enter the Internet of Things. After the other nodes are successfully connected, they start to send real-time environmental information of the orchard to the transmission node. The node applies for joining the network and transmits information. After receiving the request, the transmitting node needs to check whether the node has the conditions to join the network in the network address. If it does not meet the conditions, it refuses to join; if it meets the conditions, it can join.

According to the structure sequence, the multi-point data of network engineering virtual unit is allocated and the data training operation is carried out. The big data calculation and analysis data and the multi-point data of the virtual unit are transmitted to the same operation channel, and the corresponding data transmission formula is set:

$$C = \sqrt{A + P} - \frac{S + U}{N} \tag{1}$$

In the above formula, C represents data transmission parameters, A represents operating channel environment data, P represents network structure space data, S represents virtual unit multi-point data storage location parameters, U represents data training sample parameters, and N represents the total number of data involved in data transmission. After the above research and operation, the transmitted data will be placed in the control system. After the system control of data is realized, the data location of the control will be allocated, and the data control area will be divided by referring to the multi-point data control content of the network engineering virtual unit, and the control parameters will always be within the adjustable range of the system. Further set the data parameter range control equation as follows:

$$C = \sum P + \frac{q - m}{T} \tag{2}$$

In the above formula, C represents the data control range data, P represents the basic parameters of the control area, q represents the allocation data location parameters, m represents the control content parameters, and T represents the system control space data information. Therefore, obtain the required range data, set the data operation range, standardize the management of internal system information, and transfer all the information in the system operation range to the control space, monitor the data position at all times, and there is a certain operation relationship between the data. The software in this paper alleviates certain data contradictions by cracking its relationship [9].First look for relational data, determine the possibility of operation between the data through the corresponding content of the relational data, and set the data determination formula to adjust the difference between the data:

$$P = \frac{k - l}{z} \cdot (u + d) \tag{3}$$

In the above formula, P is the data determination parameter, K is the data contradiction possibility parameter, l is the relational data information, Z is the software system operation function, u is the operation relationship index between data, and D is the internal space operation index [10–12].

According to the above operation, enhance the adjustment performance of internal data information, and combine the research content of the control system to convert the control data into control research data, maintain the operating distance between the data, improve the function of the information control system, complete the temperature measurement and control system of the chemical instrument Design research [13–15].

4 Experiment and Research

In view of the complexity of the operation data of the Internet of things and the difficulty in measuring the instrument temperature data of the chemical instrument temperature measurement and control system, it is necessary to screen the data of the experimental environment, adjust the original system state, continuously optimize the internal structure of the system, change the performance of information collection and analysis, implement the data measurement and control operation according to the relevant measurement and control principles, and strengthen the connection between information,

After each module of the node establishes the route, the MSU processing module will enter the low loss mode by using the timing sleep gap of the node. At this time, the whole system will enter the low loss stage. The trigger timer can use the timing sensor to collect and transmit the data. The basic function of the trigger timer is as follows:

$$S = \sum_{i=0} L + \frac{L_2}{C} \tag{4}$$

In the function, S represents the trigger time, L represents the starting module working time, represents the module enters low-loss time, and C represents the module's operating efficiency.

When the system node under low loss receives the command information outside the system, the normal operation of the entire system is resumed by changing the time address of the timer for the first time. The peripheral circuit of the MSU processor is shown in Fig. 3:

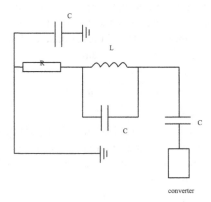

Fig. 3. Peripheral circuit of MSU processor

In this experiment, according to two different experimental parameters for experimental comparison, to further improve the overall comparison effect, and set the corresponding experimental parameter table as shown below (Table 1):

Table 1. Experimental parameters 1

Project	Parameter
Voltage	10 V
Network signal status	Stable
Program status	Offline
Run the model	IoT operation model
Analysis method	Platform computing analysis

In order to verify the effectiveness of the system in this paper, under the condition of experimental parameter 1, the efficiency of the temperature measurement and control of the chemical instrument temperature measurement and control system in this paper, the chemical instrument temperature measurement and control system based on data mining technology and the chemical instrument temperature measurement and control system based on data mining technology Analysis, the comparison results are shown in Fig. 4.

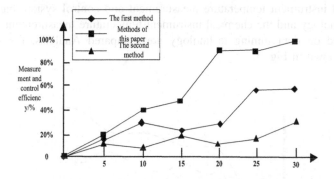

Fig. 4. Comparison of measurement and control efficiency

According to the above diagram, it can be analyzed that the traditional measurement and control system based on the data mining technology of chemical instrument temperature measurement and control is more efficient. Because this method centrally controls the operation performance of acquiring data, and studies the transformation direction of the internal system structure, the measurement and control according to the data. The standardized standard of data manipulation to improve the efficiency of measurement and control.

The traditional temperature measurement and control system of chemical instruments based on cloud computing has a low efficiency. Because the method has a small monitoring force on the internal data of the system, it reduces the effectiveness of data conversion, and the control concentration is weak, resulting in a low efficiency of measurement and control.

The measurement and control efficiency of the temperature measurement and control system of the chemical instrument based on the Internet of Things in this article is higher than that of the other two traditional methods. Contradiction with system control, to avoid data measurement and control errors during operation, and improve the efficiency of measurement and control.

Set the experimental parameter 2 for the secondary data performance test (Table 2):

Table 2. Experimental parameters 2

Project	Parameter
Monitoring program status	Connection Status
Network structure	Complete
Risk factor	0
Frame status	Basic framework
Interface	USB interface

In order to further verify the effectiveness of the system in this paper, under the condition of experimental parameter 2, the temperature measurement and control time of the chemical instrument temperature measurement and control system in this paper, the chemical instrument temperature measurement and control system based on data mining technology and the chemical instrument temperature measurement and control system based on data mining technology are compared Analysis, the comparison results are shown in Fig. 5.

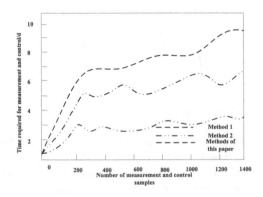

Fig. 5. Experimental comparison

According to the above diagram, it can be analyzed that the traditional design of chemical instrument temperature measurement and control system based on data mining technology takes a long time, while the traditional design of chemical instrument temperature measurement and control system based on cloud computing takes a short time. In this paper, the measurement and control time of chemical instrument temperature measurement and control system based on Internet of things is shorter than the other two traditional system designs. The reason for this difference lies in the system design and research of the internal information situation of the system, and find out the reasons for the state of the system, increase the data collection efforts, and store the data according to the relevant spatial data comparison system storage mode.

To a large extent, it ensures the accuracy of data storage, reduces unnecessary troubles, facilitates control operation, improves the efficiency of measurement and control, and reduces the time required for measurement and control operation. The traditional chemical instrument temperature measurement and control system based on data mining technology has designed and studied the integration mechanism between the data. Based on the management system center data, the temperature data is reviewed and analyzed to better control the performance of the system's control center and has a relatively stable Measurement and control route, but the operation steps are more complicated, with more repetitive steps, resulting in a certain amount of operation waste, resulting in a longer time for measurement and control.

The traditional design of temperature measurement and control system for chemical instruments based on cloud computing can control the flow direction and flow state of gait data while ensuring the effective transmission of information, which shortens the time required for measurement and control. However, because of this, the mastery of data is low, which can not meet the operation requirements of the system, and there are still deficiencies in control functions, unstable control routes, and efficient measurement and control Lower.

In summary, the design of the temperature measurement and control system for chemical instruments based on the Internet of Things in this paper can better adjust the contradictions between the systems, solve the system data information in time, control the effective circulation of temperature data, simplify the operation process, and avoid unnecessary waste of time. Has a broader space for development.

In order to further verify the effectiveness of the system in this paper, the temperature measurement and control accuracy of the traditional measurement and control system and the measurement and control system in this paper are compared and analyzed. The comparison results are shown in Fig. 6.

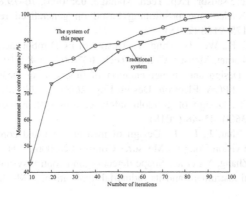

Fig. 6. Comparison of measurement and control precision results

According to Fig. 6, the temperature measurement and control accuracy of chemical instrument in this system can reach 100%, while that of traditional system is only 93%. The accuracy of temperature measurement and control of chemical instrument in this system is higher than that of traditional system.

5 Conclusion

Based on the traditional instrument temperature measurement and control system, this paper designs a new type of chemical instrument temperature measurement and control system based on the Internet of Things. The experimental results show that the design effect of the system is significantly better than that of the traditional system. This paper focuses on the operation characteristics of hardware and software design, reduces the contradiction rate in the design, and constantly adjusts the system space structure, optimizes the internal space state, simplifies the operation process, reduces unnecessary operation waste, can improve the information of the central components of the control system to a higher extent, has a higher operational feasibility, reasonably distributes the system operation principles, and strengthens the The data control performance of the control system can improve its control accuracy, which provides a solid theoretical basis for the follow-up research. The design effect is good and has a high research value.

Acknowledgments. This project was financially supported by the National Natural Science Foundation of Guangdong Province (2018A030307058).

References

1. Fu, Y., Feng, G., Tian, H., et al.: Design of multipoint thermal infrared mobile object temperature measurement system. Video Eng. **043**(004), 91–93, 108 (2019)
2. Zhu, X., Li, Z., Ge, Z., et al.: Design and implementation of digital temperature measurement system in vacuum thermal test. Comput. Measur. Control **026**(005), 21–24 (2018)
3. Zhang, Z., Guo, D., Lü, W., Zhang, L.: Design of temperature measurement system based on DS18B20 temperature sensor. Exp. Tech. Manage. **035**(005), 76–79,88 (2018)
4. Lu, M., Liu, S.: Nucleosome positioning based on generalized relative entropy. Soft. Comput. **23**(19), 9175–9188 (2018)
5. Li, H., Song, A., Li, H., Wei, H., Ding, T.: Design of EOD robot measurement and control system based on ethernet. Measur. Control Tech. **037**(007), 5–8 (2018)
6. Chen, Z., Liu, Y.: Design and implementation of time-sharing measurement and control system based on the FPGA. Electron. Design Eng. **26**(8), 75–78 (2018)
7. Wang, T., Wang, T.: Design of gondola safety control system based on digital sensor. Machin. Electron. **36**(3), 45–48 (2018)
8. Wang, Y., Xu, S., Zhai, J., Li, T.: Design of measurement and control system based on embedded machine vision.Comput. Measure. Control **026**(006), 104–106 (2018)
9. Yang, L., Yu, H., Zhang, Y., et al.: Shape detection and control system of cold rolling strip based on the virtual instrument and its industrial application. J. Mech. Eng. **054**(014), 1–7 (2018)

10. Zheng, P., Shuai, L., Arun, S., Khan, M.: Visual attention feature (VAF): a novel strategy for visual tracking based on cloud platform in intelligent surveillance systems. J. Parallel Distrib. Comput. **120**, 182–194 (2018)

11. Fu, W., Liu, S., Srivastava, G.: Optimization of big data scheduling in social networks. Entropy **21**(9), 902 (2019)

12. Vlahakis, E., Halikias, G.: Temperature and concentration control of exothermic chemical processes in continuous stirred tank reactors. Trans. Inst. Measure. Control **41**(15), 4274–4284 (2019)

13. Palle, D.V., Kanchi, R.R.: Cloud-based monitoring and measurement of pressure and temperature using CC3200. In: 11th International Conference on Intelligent Systems & Control, pp. 393–397. IEEE (2017)

14. Biswas, P., Wang, Y., Attoui, M.: Sub-2 nm particle measurement in high-temperature aerosol reactors: a review. Curr. Opin. Chem. Eng. **21**, 60–66 (2018)

15. Liu, S., Liu, D., Srivastava, G., et al.: Overview and methods of correlation filter algorithms in object tracking. Complex Intelligent Systems (2020). https://doi.org/10.1007/s40747-020-00161-4

Location and Path Planning of Cross-Border E-Commerce Logistics Distribution Center in Cloud Computing Environment

Yi-huo Jiang[✉]

Fuzhou University of International Studies and Trade, Fuzhou 350202, China
hbgv96012@126.com

Abstract. In order to solve the problem of high transportation cost in the range of transportation distance from 3.5 km to 7.5 km, this paper proposes a method of location and path planning for cross-border e-commerce logistics distribution center in cloud computing environment. Cross-border e-commerce logistics distribution center location model is established under the cloud computing environment to achieve cross-border e-commerce logistics distribution center location. Cross-border e-commerce logistics path planning is realized by constructing cross-border e-commerce logistics path planning model in cloud computing environment. The logistic path planning model of cross-border e-commerce includes time and event sequence sub-model, state variable quantum model, external information sub-model, state transfer sub-model, objective function and optimal strategy sub-model. In order to prove that the transportation cost of this method is lower in the range of 3.5 km to 7.5 km, two original methods are compared with this method. The experimental results show that the transportation cost of this method is much lower than that of the other two methods, and the cost is reduced successfully.

Keywords: Cloud computing environment · Cross-border E-commerce · Logistics distribution · Distribution center location · Path planning

1 Introduction

Cross-border e-commerce logistics distribution center location and path planning method can help cross-border e-commerce reasonable, scientific planning of the distribution center address and distribution path. In recent years, the rapid development of cross-border e-commerce has changed from an economic phenomenon to an advanced business model, and will become the mainstream business model in the future international trade. At present, the Central Government is focusing on promoting the development strategy of "going global" of Chinese enterprises, encouraging enterprises to actively participate in the competition in the international market, and striving to become world-class cross-border e-commerce enterprises [1, 2]. In order to simplify the processing of cross-border e-commerce orders, shorten the time of logistics transportation and improve the efficiency of logistics transportation, the research on the location and path planning of cross-border e-commerce logistics distribution center has been carried out for a long time. In 2012, M.M. Nassar proposed a new distribution

S. Liu and L. Xia (Eds.): ADHIP 2020, LNICST 347, pp. 30–40, 2021.
https://doi.org/10.1007/978-3-030-67871-5_4

model based on maximum likelihood method and moment method, namely beta logistics distribution model. The model is a parameter estimation model. At the same time, N. Balakrishnan put forward the delivery model of order recurrence relation based on it. The scholar Hsiaw-Chan Yeh puts forward a method of location and path planning of cross-border E-commerce logistics distribution center based on order recursive relation distribution model. However, the research in China is relatively late. Some scholars put forward a method of location and path planning of cross-border E-commerce logistics distribution center based on fuzzy decision analysis, mainly through fuzzy decision analysis of cross-border E-commerce logistics to achieve its location and path planning [3]. In view of the fact that the location and path planning of cross-border e-commerce logistics distribution centers by using the above methods are unable to be transferred due to the difficulties in cross-border logistics control, and the transportation cost is high within the range of transportation distance from 3.5 km to 7.5 km, a method for location and path planning of cross-border e-commerce logistics distribution centers under cloud computing environment is proposed.

2 Location of Cross Border E-Commerce Logistics Distribution Center in Cloud Computing Environment

2.1 Location of Cross Border E-Commerce Logistics Distribution Center

Cross-border e-commerce logistics distribution center site selection is realized by building a cross-border e-commerce logistics distribution center location model under the cloud computing environment [4].

2.1.1 Model Assumptions and Analysis

First of all, the model assumptions need to be made as follows: all the macro-factors and meso-factors of the alternative addresses of cross-border e-commerce logistics distribution centers are the same, so only the impact of micro-factors on the location of logistics distribution centers is taken into account; and the construction costs of each logistics distribution center are fixed within a certain period [5].

According to the operation mode of cross-border e commerce logistics distribution, this paper analyzes the factors of cross-border e-commerce logistics distribution center location, and constructs cross-border e-commerce logistics center location model. The factors for the location of cross-border e-commerce logistics distribution centers analyzed are as follows:

Cross-border e-commerce logistics and distribution center costs include transportation costs, warehousing costs, warehousing costs, reverse logistics (costs of returning and exchanging goods), costs of handling unmarketable goods, warehousing insurance costs, total customs duties and [6].

The transportation cost is the sum of the transportation cost from domestic seller to domestic port city, from port city to logistics distribution center, and from logistics distribution center to overseas buyer.

Each transport cost shall be the product of transport distance, transport volume and unit freight.

The warehousing cost of a logistics distribution center shall be the product of the storage volume (total transportation volume), storage time and unit storage fee.

Build a position cost to give total cost directly can.

The reverse logistics cost is the product of the rate of return and exchange, transportation distance, total transportation volume, and unit freight.

The total tariff is the product of the total value of the goods multiplied by the tariff rate of the importing country.

Customer satisfaction is the ratio of the quantity of goods delivered through a logistics distribution center to the total quantity of goods delivered within the standard time for customer satisfaction [7].

2.1.2 Establishing the Location Model of Cross-Border E-Commerce Logistics Distribution Center

The location model of the cross-border e-commerce logistics distribution center shall be established as follows:

Formula (1) is an overseas warehouse cost model;

$$\min Z_1 | Z_1 = L_{sr} V_m C_{sr} + \sum_{w=1}^{k} D_{rw} L_{rw} Q C_{rw} \tag{1}$$

Type (2) is a satisfaction model using overseas warehouses;

$$\min Z_2 | Z_2 = -\sum_{w=1}^{k} H_{bw} \frac{V_{wb}}{Q} \tag{2}$$

Type (3) means that the traffic volume of overseas warehouses to overseas buyers is less than the total traffic volume;

$$\begin{cases} \sum_{w=1}^{k} \sum_{b=1}^{n} X_{wb} Q_{wb} \leq Q \\ w \in \{1, 2, \ldots, k\} \\ b \in \{1, 2, \ldots, n\} \end{cases} \tag{3}$$

Formula (4) indicates that the first trip freight is composed of the transportation cost of the goods from the domestic seller to the domestic port city and the transportation cost of the goods from the domestic port city to the overseas warehouse;

$$F = L_{sr} V_m C_{sr} + \sum_{w=1}^{k} D_{rw} L_{rw} Q C_{rw} \tag{4}$$

Formula (5) indicates that the total tariff of the importing country consists of the product of the total value of the goods and the tariff rate;

$$R = P\theta \tag{5}$$

Formula (6) indicates the selection scope of customer satisfaction.

$$\begin{cases} H_{bw} = \begin{cases} 1, \frac{T_{wb}}{G_{wb}} \leq 1 \\ 0, \text{else} \end{cases} \\ w \in \{1, 2, \ldots, k\} \\ b \in \{1, 2, \ldots, n\} \end{cases} \tag{6}$$

Where, s represents the domestic seller; b represents the overseas buyer; n represents the number of overseas buyers; r represents the domestic port city; w represents the logistics distribution center; k represents the number of overseas warehouses; v represents the total transportation volume of domestic sellers to the logistics distribution center; V_{wb} represents the transportation volume from the logistics distribution center to the buyer; V_w represents the warehouse capacity of the logistics distribution center; L_{sr} represents the distance from the domestic seller to the domestic port city r; L_{rw} represents the distance from the domestic port city to the logistics distribution center; L_{wb} represents the distance from the logistics distribution center to the overseas buyer; C_{sr} is the freight for the domestic seller to deliver each ton of goods to the domestic port city; C_{rw} is the freight for the domestic port city to deliver each ton of goods to the logistics distribution center; C_{wb} is the freight of each unit of goods delivered by the logistics distribution center to overseas buyers; C_{wt} is the storage fee of each unit of goods stored in the logistics distribution center every day; C_w is the annual warehouse building cost of the logistics distribution center; T_{sw} is the time for domestic sellers to store goods in the logistics distribution center; G_{wb} and T_{wb} are the standard time and actual time for domestic sellers to send goods from the logistics distribution center to overseas buyers; a is the insurance cost of the logistics distribution center; θ is the import tariff rate of the country where the logistics distribution center is located; F is the first route transportation; P is the total value of the goods; R is the total tariff of the country where the goods are imported [8].

In the cross-border e-commerce logistics distribution center location model, it is composed of transportation cost, storage cost, warehouse building cost, reverse logistics (return and exchange cost), processing cost, insurance cost and total tariff; the dual objective problem of the cross-border e-commerce logistics distribution center location model is transformed into a single objective problem to solve the model, so as to achieve the location of cross-border e-commerce logistics distribution center.

2.2 Cross-Border E-Commerce Logistics Path Planning in Cloud Computing Environment

Cross-border e-commerce logistics path planning is realized by constructing cross-border e-commerce logistics path planning model in cloud computing environment. Cross-border e-commerce logistics path planning model consists of time and event sequence sub-model, state variable quantum model, external information sub-model, state migration sub-model, objective function and optimal strategy sub-model [9, 10].

2.2.1 Time and Event Sequence

The time and event sequencing submodels are modeled as shown in Fig. 1.

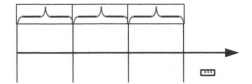

Fig. 1. Time vs. event sequencing submodel

In the time and event sequence sub model, the identifier on the timeline represents the time point, and the interval between the identifiers represents the time period. The part between time $t - 1$ and time t is called time period t state S and action A are known at time t, so it is also called decision time. External information W arrives in time period t. In cross-border e-commerce logistics path planning, decision-making is triggered when one or more vehicles arrive at customers or distribution centers at the same time. The sequence of events from t to $t + 1$ is as follows:

(1) When the system arrives at decision-making time t, it senses state S_t and obtains the real demand information W_t of all customers who have just arrived;

(2) Select action A_t according to current state S_t and random information W_t;

(3) Serve all newly arrived customers;

(4) Perform actions;

(5) The decision-making time of $t + 1$ is triggered by an arrival event, and the vehicle is transferred to state S_{t+1} to obtain the real demand information W_{t+1} of all customers who have just arrived.

2.2.2 State Variable

When planning cross-border e-commerce logistics path, the state variables need to provide enough information for decision-making, so the state variable sub model is usually composed of vehicle state and customer state.

Set vehicle set as M, quantity as m, customer set as Z and quantity as z. For any vehicle $i \in M$, its status can be expressed as:

$$F = (l_i, q_i, k_i) \tag{7}$$

Where $l_i \in Z$ represents the current location (before service) or the next destination (after service) of vehicle i; $q_i \in [0, Q]$ refers to the current remaining on-board cargo volume of vehicle i; $k_i \in [0, L]$ refers to the time when the vehicle arrives at the next destination; L refers to the service period limit. Let $(l, q, k) = (l_i, q_i, k_i)_{i \in M}$ indicate the status of the vehicle.

For any customer $j \in N/\{0\}$, its demand status can be expressed as (d_j, x_j), where d_j represents the unmet demand of customer j; x_j represents the real value of the observed customer j demand.

2.2.3 External Information

Because there is only one kind of external stochastic information, namely customer demand, for the multi vehicle routing problem with stochastic demand, the external information sub model is expressed as follows:

$$W_t = \{\hat{x}_{l_i}; i \notin M\} \tag{8}$$

Where, W_t represents external random information; $\hat{x}_{l_i}$ represents the real needs of the target customer l_i of vehicle i. When external information W_t is triggered at decision time t, the system will know immediately when it reaches state S_t.

2.2.4 State Migration

The state migration function can be expressed as:

$$S_{t+1} = S^M(S_t, A, W_{t+1}) \tag{9}$$

The superscript M refers to the model, and the independent variables of the function are state S_t, action A and external information W_{t+1}.

State transition is divided into two steps: the deterministic transition to the post decision state and the stochastic transition to the pre decision state. When action A is executed in state S_t, the state will definitely migrate to post decision state S_t^A.

2.2.5 Objective Function and Optimal Strategy

In the cross-border e-commerce logistics path planning with stochastic demand and service period, the goal is to maximize the customer demand, that is, the objective function is $\max \sum_{t=0}^{T} \rho(S_t, a)$. $\pi \in \Pi$ strategy a defines a mapping from state space to action space: $\pi : S \mapsto A(S)$. Let the value function of the state under strategy π be the long-term cumulative reward, and the evaluation standard of the strategy be the value function $V^\pi(S_0)$ of the initial state S_0, Then find the optimal strategy π^*, so that $V^{\pi^*}(S_0) \geq V^\pi(S_0), \forall \pi \in \Pi$, cross-border e-commerce logistics path planning [11, 12].

3 Experimental Test

3.1 Experimental Case

The design of cross-border e-commerce logistics distribution center location and path planning method in cloud computing environment was tested. The experimental case is as follows: Enterprise B is a third party logistics enterprise of cross-border e-commerce logistics and distribution. With the rise of export cross-border e-commerce in recent years, enterprise B's cross-border e-commerce business customers are mainly located in Quanzhou of Fujian Province, Hangzhou of Zhejiang Province, Hefei of Anhui Province, Kunming of Yunnan Province and Suzhou of Jiangsu Province. The overseas buyers of cross-border e-commerce enterprises are mainly located in France, Germany, Italy, the Netherlands and other European countries. With the increasing of business volume, B enterprises consider to add some logistics distribution centers and make path planning for them. The operating model of enterprise B is shown in Fig. 2.

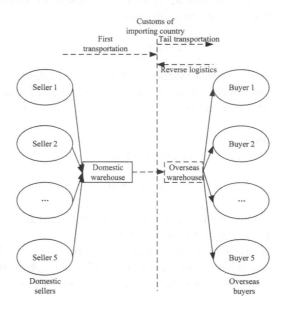

Fig. 2. B Business model

As shown in Fig. 2, enterprise B first selects a domestic port city to build a warehouse, assembles the goods of five cross-border e-commerce sellers, and then transports them by sea to the port city where the overseas warehouse is located. Finally, when the foreign buyer places an order on the cross-border e-commerce platform, the goods are directly delivered to the buyer through the overseas warehouse.

B enterprise has 5 domestic sellers concentration points, 7 overseas buyers concentration points, selected 4 domestic port cities as the warehouse selection points, 7 overseas warehouse selection points, the entire logistics network of a total of 23 nodes.

In this paper, the location and path planning of cross-border e-commerce logistics distribution center in B enterprise are tested by using the method of cloud computing. The transportation cost data in the range of 3.5 km to 7.5 km is used as the experimental data. In order to avoid the unitary experimental results, the two methods are used as comparative experimental methods, including cross-border e-commerce logistics distribution center location and path planning based on order-recursive distribution model and cross-border e-commerce logistics distribution center location and path planning based on fuzzy decision analysis. Similarly, the two methods are used to choose the location and path planning of B enterprise's cross-border E-commerce logistics distribution center, and the transportation cost data are used as the comparative experimental data. The experimental data of several experimental methods were compared.

3.2 Analysis of Experimental Results

Within the range of transportation distance from 3.5 km to 5.5 km, the comparison results of transportation cost experimental data between cross-border e-commerce logistics distribution center location and path planning under cloud computing environment and distribution model based on order recurrence relation and cross-border e-commerce logistics distribution center location and path planning based on fuzzy decision analysis are as shown in Table 1.

Table 1. Comparison of transport cost experimental data

Transport distance	Transport costs (yuan)		
	Approaches in a cloud computing environment	Method of distribution model based on order recurrence relation	Method based on fuzzy decision analysis
3.5 km	5163.32	6112.25	6932.65
3.6 km	5233.25	6115.29	7124.20
3.7 km	5263.32	6154.25	7132.36
3.8 km	5361.02	6235.55	7102.03
3.9 km	5401.02	6220.02	7004.09
4.0 km	5431.08	6323.05	7102.29
4.1 km	5441.01	6302.02	6932.30
4.2 km	5533.14	6324.25	7221.58
4.3 km	5563.32	6321.25	6962.32
4.4 km	5624.07	6421.01	7248.01
4.5 km	5713.20	6551.20	7120.06
4.6 km	5763.21	6513.20	7335.04
4.7 km	5801.02	6596.36	7469.68
4.8 km	5860.27	6563.24	7442.21
4.9 km	5820.20	6631.20	7541.20
5.0 km	5810.07	6632.04	7521.02
5.1 km	5863.20	6630.01	7620.1
5.2 km	5912.08	6725.36	7720.25
5.3 km	5934.20	6765.32	7712.23
5.4 km	5932.04	6831.20	7820.27
5.5 km	5978.25	6923.32	7921.30

According to the comparison of transportation cost experimental data in the range of transportation distance from 3.5 km to 5.5 km in Table 1, it can be concluded that the cost of cross-border e-commerce logistics distribution center location and path planning method in cloud computing environment is much lower than that in cross-border e-commerce logistics distribution center location and path planning method based on order recurrence relationship distribution model and fuzzy decision analysis.

Within the range of transportation distance from 5.5 km to 7.5 km, the comparison results of transportation cost experimental data between cross-border e-commerce logistics distribution center location and path planning under cloud computing environment and distribution model based on order recurrence relation and cross-border e-commerce logistics distribution center location and path planning based on fuzzy decision analysis are as shown in Table 2.

Table 2. Comparison of transport cost experimental data

Transport distance	Transport costs (yuan)		
	Approaches in a cloud computing environment	Method of distribution model based on order recurrence relation	Method based on fuzzy decision analysis
5.6 km	5123.36	6132.20	6759.36
5.7 km	5123.25	6132.20	6785.20
5.8 km	5163.20	6113.32	6798.62
5.9 km	5120.04	6223.21	6798.98
6.0 km	5263.32	6112.20	6895.32
6.1 km	5287.62	6298.96	6912.03
6.2 km	5221.20	6324.20	6943.21
6.3 km	5232.04	6321.20	6987.25
6.4 km	5363.24	6452.30	7021.20
6.5 km	5362.30	6425.30	7046.32
6.6 km	5462.32	6521.03	7059.32
6.7 km	5414.02	6564.20	7084.32
6.8 km	5420.21	6578.96	7125.32
6.9 km	5563.32	6584.21	7201.32
7.0 km	5524.21	6632.01	7698.32
7.1 km	5623.20	6698.32	7712.03
7.2 km	5652.32	6712.05	7785.20
7.3 km	5721.02	6789.32	7798.32
7.4 km	5813.20	6875.30	7821.02
7.5 km	5820.32	6897.21	7962.30

According to the transportation cost experimental data in the range of 5.6 km to 7.5 km in Table 2, the results show that the cost of cross-border E-commerce logistics distribution center location and path planning method in cloud computing environment is much lower than that in cross-border E-commerce logistics distribution center location and path planning method based on order recurrence relationship distribution model and fuzzy decision analysis. The cost and capacity constraints of e-commerce logistics distribution center are shown in Table 3.

Table 3. Cost and capacity constraints of e-commerce logistics distribution center

Distribution center candidate	Construction cost of candidate distribution center/ten thousand	Unit product operation cost/ten thousand	Capacity limitation/ten thousand
1	60	85	800
2	60	85	800
3	70	85	850
4	70	90	860
5	75	90	880
6	75	95	900
7	80	100	920

It can be seen from Fig. 2 that the location problem of e-commerce logistics distribution center involves not only the cost of construction, operation and distribution, but also the cost and capacity constraints. Therefore, it has practical application value to construct an optimization model of e-commerce logistics distribution center location, which comprehensively considers the factors of cost, time window constraint, cargo damage rate, objective function and optimal strategy [13, 14].

In conclusion, the transportation cost of location and route planning of cross-border e-commerce logistics distribution center in cloud computing environment is lower than the other two methods, which provides an important reference for practical application.

4   Conclusions

At present, the construction level of China's cross-border logistics and transportation system is far lower than that of developed countries in Europe and America, and the distribution service level cannot meet the needs of the rapid development of the market. We should actively promote the layout and construction of logistics hubs to promote and improve the quality and efficiency of national economic operation. Major cross-border logistics solutions all have some problems that cannot be solved by themselves, which are the main factors affecting the further development of cross-border e-commerce. Therefore, the site selection and path planning of cross-border e-commerce logistics distribution center must be carried out through the method of site selection and path planning.

The location and path planning of cross-border e-commerce logistics distribution centers under cloud computing environment can reduce transportation costs, but this method only considers the influence of micro factors. The two original methods are compared with this method, which has lower transportation costs within the range of 3.5 to 7.5 km. The experimental results show that the transportation cost of this method is much lower than the other two methods, and the cost is reduced successfully.

References

1. Yin, A., Wang, Y., Shu, X., et al.: Grain boundary distribution evolution of 00Cr12Ti FSS during annealing. J. Wuhan Univ. Technol. Mater. Sci. Ed. **34**(4), 932–939 (2019)
2. Tolchkov, Y.N., Mikhaleva, Z.A., Sldozian, R.D.A., et al.: Effect of surfactant stabilizers on the distribution of carbon nanotubes in aqueous media. J. Wuhan Univ. Technol. Mater. Sci. Ed. **33**(3), 533–536 (2018)
3. Pinto, L.C., De Mello, C.R., Norton, L.D., et al.: A hydropedological approach to a mountainous clayey humic dystrudept. Sci. Agric. **75**(1), 60–69 (2018)
4. Yan, Z., Song, L., Tang, M., et al.: Oxygen evolution efficiency and chlorine evolution efficiency for electrocatalytic properties of MnO2-based electrodes in seawater. J. Wuhan Univ. Technol. Mater. Sci. Ed. **34**(1), 69–74 (2019)
5. Wang, X., Ma, J., Zhang, L., et al.: Radioactive element distribution characteristics of red mud based field road cement before and after hydration. J. Wuhan Univ. Technol. Mater. Sci. Ed. **33**(2), 452–458 (2018)
6. Okamiya, H., Kusano, T.: Evaluating movement patterns and microhabitat selection of the Japanese common toad (*Bufo japonicus formosus*) using fluorescent powder tracking. Zoologicalence **35**(2), 153–160 (2018)
7. Kono, Y., Shibazaki, Y., Benson, C.K., et al.: Pressure-induced structural change in MgSiO3 glass at pressures near the Earth's core–mantle boundary. Proc Natlacad Sci U S A **115**(8), 1742–1747 (2018)
8. Liu, S., Li, Z., Zhang, Y., et al.: Introduction of key problems in long-distance learning and training. Mobile Netw. Appl. **24**(1), 1–4 (2019)
9. Fan, G., Xu, X., Wang, X., Lu, W., Liang, F., Wang, K.: Preparation of Ni/PZT core-shell nanoparticles and their electromagnetic properties. J. Wuhan Univ. Technol. Mater. Sci. Ed. **33**(1), 9–14 (2018)
10. Zheng, P., Shuai, L., Arun, S., Khan, M.: Visual attention feature (VAF): a novel strategy for visual tracking based on cloud platform in intelligent surveillance systems. J. Parallel Distrib. Comput. **120**, 182–194 (2018)
11. Dai, X., Yu, Y., Zu, F., et al.: Effect of chemical plating with Ni content on thermoelectric and mechanical properties of P-type Bi0.5Sb0.15Te3 bulk alloys. J. Wuhan Univ. Technol. Mater. Sci. Ed. **33**(4), 797–801 (2018)
12. Shuai, L., Weiling, B., Nianyin, Z., et al.: A fast fractal based compression for MRI images. IEEE Access **7**, 62412–62420 (2019)
13. Vincenti, Giovanni., Bucciero, Alberto., Helfert, Markus, Glowatz, Matthias (eds.): E-Learning, E-Education, and Online Training. LNICST, vol. 180. Springer, Cham (2017). https://doi.org/10.1007/978-3-319-49625-2
14. Ma, X., Gao, X., Sun, B., et al.: An in vitro investigation of acid etching treatment prior to post and core cementation and teeth fracture resistance. J. Wuhan Univ. Technol. Mater. Sci. Ed. **34**(5), 1233–1237 (2019)

Signal Collection Method of Wireless Radio Frequency Gas Sensor Array Based on Virtual Instrument

Li Ya-ping[⊠] and Zhao Dan

College of Mechatronic Engineering, Beijing Polytechnic, Beijing 100176, China
liyaping_mail29@163.com

Abstract. The traditional multi-dimensional array tactile sensor research and application can not have the characteristics of flexibility and multi-dimensional force measurement, so there is a big gap in the acquisition of sensor array signal, it is difficult to extract relevant information and make corresponding actions in complex environment. Based on this, a method of wireless RF gas sensor array signal acquisition based on virtual instrument is proposed. By mining the flexible three-dimensional force and temperature composite sensor array numerical characteristics of virtual instrument, the flexible sensitive value of sensor array is improved, and the flexible three-dimensional force sensor array signal acquisition and temperature compensation are designed, so as to effectively reduce the temperature to three-dimensional force detection The influence of measurement can improve the signal acquisition performance of wireless radio frequency gas sensor array. The experiment proves that compared with the traditional dedicated array acquisition method, the wireless RF gas sensor array signal acquisition method based on virtual instruments is easy to implement, flexible to use, and cost-effective, which can be used by researchers for reference.

Keywords: Virtual instrument · Radio frequency gas · Sensor · Signal acquisition

1 Introduction

With the development of sensor technology and signal processing methods, the functional requirements of sensor data acquisition device are also changing. The collection, analysis, processing and display of sensor array signal is one of the important research trends of current social development. Traditional sensor data collection devices are generally designed for a single sensor. Even multi-channel collection devices are often only for data collection of different sensors, and the requirements for synchronous signal collection are not very high [1]. Usually the data processed by the sensors are array signals. Therefore, this paper analyzes the structure of sensor synchronous acquisition module from the functional point of view, describes the construction process of array acquisition and bidirectional safe grasping of auxiliary target, so as to achieve the purpose of human-computer interaction.

© ICST Institute for Computer Sciences, Social Informatics and Telecommunications Engineering 2021
Published by Springer Nature Switzerland AG 2021. All Rights Reserved
S. Liu and L. Xia (Eds.): ADHIP 2020, LNICST 347, pp. 41–52, 2021.
https://doi.org/10.1007/978-3-030-67871-5_5

2 Signal Collection Method of Wireless Radio Frequency Gas Sensor Array

2.1 Multi Card Cascade Construction of Wireless RF Gas Sensor

With the development of signal processing and computers, especially with the intro-duction of virtual instruments, users are more willing to accept graphical representations and use computer technology to process the entire process of information. In the research of acquisition and display, the same kind of sensors are often arranged in a certain geometric shape, using the spatial characteristics of the signal to enhance the signal and effectively extract the spatial information of the signal [2]. Array sensing signal has a high requirement for synchronous acquisition. Generally, the acquisition card based on single channel design is not suitable for array sensing signal acquisition. According to the structural characteristics of the flexible three-dimensional force sensor array, the A/D conversion accuracy, sampling rate and other related values are converted and standardized. The 32-bit high-performance and low-power microcontroller stm32f103vet6 based on the armcortex-m3 core is selected. Its biggest advantage is fast acquisition and Processing data [3]. Multi channel A/D conversion is integrated in the system, with 12 bit accuracy and the fastest conversion speed of 1 μ s. It is equipped with more than 2 independent ADC controllers, which can quickly collect multiple analog signals at the same time. And build a DMA module that is independent of the Cortex-M3 core and parallel with the CPU to connect the peripheral mapped registers and access them at high speed, and its transmission is not controlled by the CPU [4]. ADC module is initialized to continuous conversion mode, ADC channel 0–8 is selected for DMA transmission, and software filtering is added to ensure the stability and reliability of data transmission. The specific frame structure is shown in the Fig. 1 below.

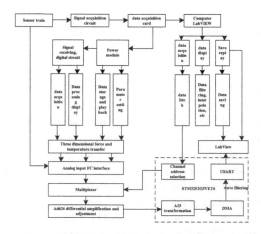

Fig. 1. Multi-card cascade construction framework of sensor signals

Based on the above frame structure, the flexible three-dimensional force value of the radio frequency gas sensor is analyzed. The composite sensor array is composed of

4×4 three-dimensional force sensor and 3×3 temperature sensor, which are arranged in a concave convex structure. The three-dimensional force sensor is arranged in a cross shaped four fork structure with the specification of 10 mm $\times$ 10 mm [5]. Under normal circumstances, the radio frequency gas sensor is composed of a 5 mm $\times$ 3.5 mm size interdigitated electrode. During operation, a temperature sensitive film is prone to appear. The temperature sensor made by the interdigitated electrode has a resistance that is linear with the temperature increase in the temperature measurement range. The added features require synchronous acquisition and can be triggered using the conversion start command of the AD chip. However, the A/D conversion chips are all working in time-series beats, which may cause a beat time error [6]. A more reliable way is to use timing control, which can ensure the simultaneous conversion of A/D signals. The specific conversion mode is as follows (Fig. 2).

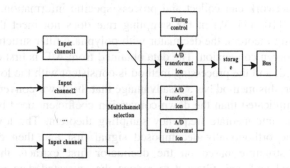

Fig. 2. Sensor signal A/D conversion mode

Generally, the composite array sensor consists of 16 three-dimensional force sensitive units and 9 temperature sensitive units, in which each three-dimensional force sensitive unit outputs four resistance signals. In order to reduce the mutual interference between sensitive units and give full play to the performance of MCU, it is necessary to design appropriate array acquisition circuit [7]. Row and column scanning is an effective way to save I/O port, but there is mutual coupling interference noise between sensitive units. Voltage mirror method is a common method to eliminate the coupling interference between resistive array units.

In the 3×3 array, the unselected row signals are grounded to reduce coupling interference of sensitive units. Single point acquisition is the easiest way to eliminate mutual coupling of sensitive units. With multi-channel analog switches, multi-channel data collection can be completed. Among them, 8 pieces of CD4051 are used to select the sensitive unit of the three-dimensional force sensor, and 1 piece of CD74HCT4067 is used as the temperature sensor acquisition channel [8, 9]. Due to the use of tactile array sensor, only through multi-channel selection circuit can the output signal of sensor be obtained orderly. CD4051 chip is used to control the acquisition sequence of sensor array. CD4051 is a single 8-channel digital control analog electronic switch, which has a series of advantages such as low-pass impedance, low cut-off leakage current, etc. among them, A, B, C are binary control input and INH input. Using the method of row and column scanning, the signal of the sensor array is connected to the

CD4051, and the signal acquisition of each sensor unit in the sensor array is controlled by the control chip ATMAGEI6. The ATmegal6 chip is based on the enhanced AVRRISC structure with high performance and low power consumption. AVR microcontroller, because the chip has more advanced instruction set and single clock cycle instruction execution time, the data throughput rate of ATmegal6 reaches 1MIPS/MHz, which can alleviate the contradiction between power consumption and processing speed of sensor signal acquisition.

2.2 Beamforming Algorithm of Sensor Array Signal

In the application environment of wireless sensor networks, network nodes complete the network topology through self-organization. Network nodes collaborate to perceive, collect, and process information of interest in the network coverage area. The wireless sensor network can collect and process specific information of the covered area at any time [10, 11]. When the sampling rate does not meet the intermediate frequency sampling theorem, the decimator with polyphase filter structure can be used to realize digital down conversion. The data obtained from A/D is first decomposed by orthogonal method, and the processing method is consistent with the low-pass filtering method. However, this method has a disadvantage that the down-conversion coefficient used is more complicated than the down-conversion coefficient used by the sampling rate to satisfy the intermediate frequency sampling theorem. The low-pass filtering method filters the orthogonally decomposed signal first and then extracts it. The polyphase filter structure based on the decimator first extracts the orthogonally decomposed signal and then filters it to reduce the extracted data rate. Filtering is achieved through multiplexing [12, 13]. In order to improve the accuracy of information acquisition, a typical depth structure beamforming algorithm is proposed. If there are m hidden layers in a deep belief network, and a^n represents the nth hidden layer vector, the deep belief network model can be expressed as follows:

$$\Delta p = \mathrm{m}\left(m|a^1\right)\left(a^1|a^2\right)...\mathrm{m}\left(a^{n-1}|a^n\right) \tag{1}$$

The conditional probability $\mathrm{p}\left(a^n|a^{n+1}\right)$ can be expressed as:

$$\mathrm{p}\left(a^n|a^{n+1}\right) = \lambda\left(-l_m^n - \sum_{k=1}^{n+1} w_{km}^n h_k^{n+1}\right) \tag{2}$$

In the formula, l_m^n represents the m th node of the n th layer, w^n represents the weight matrix of the n th layer, and λ can be expressed as:

$$\lambda(\mathrm{q}) = \frac{\Delta p - w_{km}^n}{1 + \exp \mathrm{p}\left(a^n|a^{n+1}\right)} \tag{3}$$

In order to obtain depth information, you need to obtain a set of data layer by layer, and in order to obtain the final depth data and conduct supervised training, at this time the signal sampling feature frequency is greater than the information problem under normal circumstances, there will be signal deviation, which can be expressed for:

$$W(a)_{\min} = \sum_{i=1}^{n} \lambda(q) + h_k^{n+1} \tag{4}$$

The function of fault information collection is further standardized, which can be expressed as follows:

$$W(b)_{\min} = \Delta p \sum_{i=1}^{m} W(a)_{\min} - \lambda(q) \tag{5}$$

If the array signal in the sensor can be recorded as a vector of $k_1, \cdots, k_m \in 2^m$ information; $\lambda_1, \cdots, \lambda_m$ is the amount of information loss; y represents the coefficient of the function; δ represents the conjugate function; a represents the original information variable; b represents the The variable of the fault information function; $t \geq 0$ is the function parameter. Then each array information variable c_i corresponds to an D_i algorithm that stores variables as:

$$D_i = \frac{k_m \left(c_i^t a - D_i\right)^2 \cdot y(a)}{2\lambda_m[W(b)_{\min} + W(a)_{\min}] + t} - |b|_i \tag{6}$$

Furthermore, the constrained array function of signal feature acquisition is calculated:

$$f(a)_{\min} = t\|D_i - Fa\|_3^3 + n\|b\|_2 \tag{7}$$

Among them: F belongs to the matrix of n fault information; $t \succ 0$ is the parameter of information constraint. A linear equation algorithm is used to describe the information collection situation, and the upper or lower limit of the data in the interval [0,5] is selected as the collection quantity, which is 5 or 10. Based on the above algorithm, the sensor array signal characteristics can be effectively collected.

2.3 The Realization of Sensor Array Signal Acquisition

Acess2003 database is used for operation. Create a data table through DB Tool Create Table.vi in the LabVIEW database toolkit, so that users can directly create and delete files in the database in Lab VIEW software. Through this VI, a table named user is established in the database to store the data of all users in the module, such as number, user name, user password, user authority and last login time. A table named pressure is established to store the collected pressure signals, including test time, tester and collected data value. A table named temperature is established to store the collected data the collected signal includes test time, tester and collected data value, so as to realize signal conditioning. The specific steps are shown in the Fig. 3.

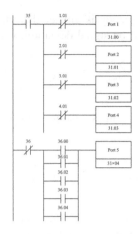

Fig. 3. Sensor array signal conditioning steps

Based on the above steps of signal conditioning, analyzing the signal characteristics of sensor array, a unique signal conditioning circuit is designed according to the different characteristics of acoustic, magnetic, vibration and infrared signals of sensor array, and the conditioned sensor signal is output to the subsequent acquisition module. Because the signal to be measured is a weak signal, the subsequent voltage signal is also very small and the amplitude is uncertain. In order to make the signal amplitude moderate, the signal needs to be amplified. The general-purpose operational amplifier cannot directly amplify the weak signal. Instrumentation amplifiers and instrumentation amplifiers must be used. It has the characteristics of high input impedance, low output impedance, strong resistance to common mode interference, low temperature drift, low offset voltage and high stable gain; in addition, before sampling, the signal must be processed by anti aliasing filter to remove the high-frequency noise, and the signal is in the low-frequency range. Therefore, the second-order Butterworth low-pass filter is used, which has flat amplitude frequency characteristics and good attenuation characteristics, so it is used in many filter circuit designs. Considering the background noise, frequency resolution and storage burden of the host, in the high-speed sampling mode, the number of sampling points of a single trigger is set to 20480, and the frequency resolution reaches 45.8 KHz. Next, this paper analyzes the performance of the high-speed array signal acquisition method in high sampling rate mode from two angles of signal measurement and error analysis. And record the correlation coefficient, as follows (Table 1):

Table 1. High sampling rate array signal data acquisition

Data index	Sampling points (SNR/DB)	Resolution
Array signal media value	11.15102	0.05481
Arithmetic mean	8.01813	0.18101
Weighted average	11.48411	0.04841

Further use DSN to access the output of the database sensor array associated with it through the analog switch through the signal amplification, adjustment circuit and sent to the MCU for A/D conversion. Select AD626 to zero-adjust the sensor unit, select the appropriate gain and send it to the microprocessor through the low-pass filter. The processed data is sent to the upper computer through the serial port to display the three-dimensional force and temperature test results. In the design of data acquisition software, the user information and equipment information are managed on the macro level and the data sampling time is designed on the micro level. It is designed in strict accordance with the real business process of data collection under the safe environment, and its sequence is: equipment calibration, parameter setting, data storage, data calculation display and export. Combined with virtual instrument technology, a set of VXI based data acquisition needs to be analyzed and processed, so as to view the processing results on site. According to different test applications to write different test assemblies, scalable functionality and versatility. Signal analysis dynamically displays the data processing process, which is convenient for users to analyze the collected data and finally give the results of signal analysis.

(1) Data playback: the array signal acquisition channels are many and the amount of data is large. The data playback function provides collected information, such as sampling rate, acquisition time and channel vector, etc.; it is convenient for users to view the acquisition data at any time and channel, and provides time-domain data display or power spectrum display.

(2) Beamforming: In order to extend the versatility, linear array and planar array beamforming are provided. The user can set various weighting methods or pour in custom weights, and view the configured beam pattern. According to the parameters set by the user, beamforming displays the output signals of each beam in real time, so that the user can compare the output results of each beam to analyze the collected data. Corresponding to the output of each beam, the detection output of each beam is given to facilitate the user to view the detection target.

(3) Azimuth estimation: azimuth estimation is one of the main purposes of array signal processing, which provides conventional and high-resolution azimuth estimation methods. In the application of underwater acoustic equipment testing, it is required to be able to distinguish coherent source and target. MUSIC algorithm is used for high-resolution azimuth estimation and spatial smoothing technology is used for coherent source and target decoherence. Based on this, the data collection process is regulated, as shown in the Fig. 4.

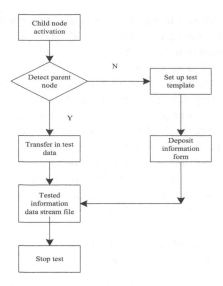

Fig. 4. Data collection process.

Array signal processing is an important branch of digital signal processing. The construction of a set of transducer array signal acquisition and analysis is of great significance in the field test, the design and development of the self-guided signal processor, and the study of signal processing theory and algorithms. Based on the above steps, accurate collection of signals from the wireless radio frequency gas sensor array can be effectively achieved.

3   Analysis of Experimental Results

3.1   Experimental Parameter Setting

In order to verify the actual application effect of the signal acquisition method of the wireless radio frequency gas sensor array based on the virtual instrument, the experimental detection was carried out, using the tactile sensor array, data acquisition circuit, USB2080 multi-channel data acquisition card and Lab VIEW software made by the laboratory Designed program. The designed acquisition circuit is used to accurately collect the signals of sensor array. The real-time and effective acquisition and storage of sampling data are realized by the Lab VIEW programmable software developed by Ni company, and the signals measured by sensor array are effectively, intuitively and vividly displayed by a large number of 2D, 3D and other controls of Lab View software. Further set the experimental environment and parameters in a standard way, as shown in the Table 2 below:

Table 2. Experimental parameter settings

Name parameter	Name parameter
Host	≤ 315 mm $\times$ 255 mm $\times$ 50 mm
The total amount	5.5 MB
Operating speed	$\leq 1 \times 10^{-5}$ (Input power ≥ -157.6 dBW)
Number of receiving channels	6 aisle
6 channels	1 ms $\pm$ 10 ns
Command function	Number of dependents that can be sent at the same time ≤ 100
Antenna cable length	≤ 20 m
Counterweight	≤ 3.5 Kg
Operating range	9–32

Place the sensor array in a BE-TH-80M8 constant temperature and humidity box, from 20 °C to 100°C, record the resistance value of any sensitive unit in the temperature sensor array every 10 °C, repeat 5 times, found in 20 ∼ the resistance value of the temperature-sensitive unit within 80 °C increases gently with increasing temperature. After 80 °C, the resistance value increases sharply. Considering the actual temperature measurement range and accuracy of the robot sensitive skin, 20–80 °C is selected as the effective range of the temperature sensor, the relationship between resistance and temperature in the measurement range can be expressed as $\Delta R = R_0(1 + aT)*RT$ is the temperature coefficient of resistance of the temperature sensor, R_0 is the initial resistance value, RT is the resistance value at temperature T, the transformation is: $\Delta R/R_0 = aT$.

3.2 Error Analysis

Furthermore, we use labsql to access the database, and build a free, multi database, cross platform LabVIEW database access toolkit. Before using labsql toolkit to access the database, first create a computer database with data source name DSN in ODBC data source of windows operation, select performance maintenance command management tool command data source component, call ODBC data source Manager dialog box, and create a DSN named acquisitionsystem. After that, you can create a DSN in ODBC data source manager. In order to ensure the security of data collection and prevent illegal operations by unauthorized users, we have designed a login interface, that is, users can only log in after entering the correct user name and password, and different permissions are given to different users. Demand, convenient operation requirements and protection of different users, improve the security of signal acquisition and data storage.

Under the condition of cloud computing application coefficient of 0.18, the total amount of recorded signal data is 0.5×10^9 T, 1.0×10^9 T, 1.5×10^9 T, 2.0×10^9 T, 2.5×10^9 T, 3.0×10^9 T, 3.5×10^9 T, 4.0×10^9 T, 4.5×10^9 T, 5.0×10^9 T, 5.5×10^9 T, the error value of the experimental group and the control group changes. The detailed experimental comparison results are shown in the following Table 3:

Table 3. Error value change value record

Set value	The method of this paper the traditional method			The method of this paper the traditional method		
	Acquisition volume	Difference	Number of repetitions	Collection	Error value	Repeat times
0.5×10^9 T	25.13	1.35	1	22.64	2.65	1
1.0×10^9 T	26.45	1.46	1	22.95	2.32	2
1.5×10^9 T	25.49	1.63	0	23.48	2.15	2
2.0×10^9 T	24.16	1.28	1	24.86	2.22	1
2.5×10^9 T	26.61	1.05	1	25.48	2.26	2
3.0×10^9 T	25.05	1.66	0	22.46	2.45	2
3.5×10^9 T	25.46	1.54	1	25.84	2.32	1
4.0×10^9 T	26.84	1.95	1	23.15	2.22	2
4.5×10^9 T	24.65	1.54	0	21.06	2.02	2
5.0×10^9 T	26.84	1.35	1	20.48	2.32	2
5.5×10^9 T	25.02	1.54	1	26.46	2.22	1

Based on the information in the above table, it is not difficult to see that, compared with the traditional signal acquisition method, the signal acquisition method of the wireless RF gas sensor array based on the virtual instrument proposed in this paper has significantly lower error rate value change compared with the traditional method, thus confirming that the error of the signal acquisition method of the wireless RF gas sensor array based on the virtual instrument is lower in the actual application process, the stability is relatively higher.

3.3 Acquisition Time

Further, the signal acquisition and detection efficiency of the wireless RF gas sensor array signal acquisition method based on virtual instrument in the operation process is compared and detected, and the detection results are recorded, as follows (Fig. 5):

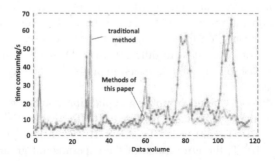

Fig. 5. Comparison test results

Based on the above test results, it can be seen that compared with the traditional methods, the time-consuming of the signal acquisition method based on the virtual instrument proposed in this paper is significantly lower in the practical application process, which confirms that the practical application effect of the signal acquisition method based on the virtual instrument proposed in this paper is relatively better and fully satisfied Research requirements.

3.4 Acquisition Accuracy

In order to further verify the effectiveness of the method in this paper, the traditional method and the method of wireless radio frequency gas sensor array signal acquisition accuracy are compared and analyzed, the comparison results are shown in Fig. 6.

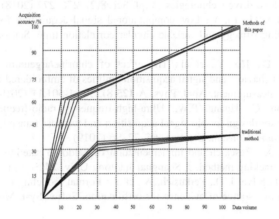

Fig. 6. Comparison of acquisition accuracy of two methods

According to Fig. 6, the highest signal acquisition accuracy rate of wireless radio frequency gas sensor array in this method can reach 100%, while that of traditional method is only 36%. It shows that the signal acquisition accuracy rate of wireless radio frequency gas sensor array in this method is higher than that of traditional method.

4 Conclusion

In this paper, a method of signal acquisition of wireless RF gas sensor array based on virtual instrument is proposed. Aiming at the problem of temperature interference in detection, a flexible sensor array signal acquisition and temperature compensation algorithm is designed. By temperature compensation of three-dimensional force sensor, the influence of temperature is effectively reduced, and the reliability of signal acquisition of wireless RF gas sensor array is improved Sex. In the later stage, the signal acquisition and processing can be integrated into the sensor array, and the data can be transmitted wirelessly to improve the practicality of the sensor array. The experimental results show that the wireless RF gas sensor array signal collection method can effectively improve the accuracy of signal collection.

5 Fund Projects

Research on temperature compensation method of silicon sapphire high temperature pressure sensor (CJGX2016-KY-YZK032).

References

1. Jiang, Y., Samuel, O.W., Liu, X., et al.: Effective biopotential signal acquisition: comparison of different shielded drive technologies. Appl. Sci. **8**(2), 276–277 (2018)
2. Zhang, Y., Wang, M., Li, Y.: Low computational signal acquisition for GNSS receivers using a resampling strategy and variable circular correlation time. Sensors **18**(3), 678–680 (2018)
3. Liu, H., Tang, D., Hu, Y., et al.: The effect of chlorine/argentum atomic ratios on electrochemical behaviors and signal acquisition abilities of embroidered electrodes for bio-potential signal measurement. Appl. Phys. A **125**(8), 501.1-501.11 (2019)
4. Lee, S.Y., Tsou, C., Huang, P.W.: Ultra-high-frequency radio-frequency-identification baseband processor design for bio-signal acquisition and wireless transmission in healthcare system. IEEE Trans. Consum. Electron. **66**, 77–86 (2019)
5. Chen, S., Mao, X.: Research and implementation of beidou-3 satellite multi-band signal acquisition and tracking method. J. Shanghai Jiaotong Univ. **24**(5), 571–578 (2019)
6. Sobrinho, A., Da Silva, L.D., Perkusich, A., et al.: Formal modeling of biomedical signal acquisition systems: source of evidence for certification. Softw. Syst. Model. **18**(2), 1467–1485 (2019)
7. Hao, Z., Jiang, L., Wang, W.: Impacts of sequential acquisition, market competition mode, and confidentiality on information flow. Naval Res. Logsit. **65**(2), 135–159 (2018)
8. Cao, Z., Guo, N., Li, M., et al.: Back propagation neutral network based signal acquisition for Brillouin distributed optical fiber sensors. Opt. Express **27**(4), 45–49 (2019)
9. Fu, W., Liu, S., Srivastava, G.: Optimization of big data scheduling in social networks. Entropy **21**(9), 902 (2019)
10. Liu, S., Bai, W., Srivastava, G., Machado, J.A.T.: Property of self-similarity between baseband and modulated signals. Mob. Netw. Appl. **25**(4), 1537–1547 (2019). https://doi.org/10.1007/s11036-019-01358-9
11. Lu, M., Liu, S.: Nucleosome positioning based on generalized relative entropy. Soft. Comput. **23**(19), 9175–9188 (2018). https://doi.org/10.1007/s00500-018-3602-2
12. Bermúdez Ordoez, J., Arnaldo Valdés, R., Gómez Comendador, F.: Energy efficient GNSS signal acquisition using singular value decomposition (SVD). Sensors **18**(5), 1586–1588 (2018)
13. Shuai, L., Weiling, B., Nianyin, Z., et al.: A fast fractal based compression for MRI images. IEEE Access **7**, 62412–62420 (2019)

Artificial Intelligence-Based Wireless Sensor Network Radio Frequency Signal Positioning Method

Zhao Dan[✉] and Qu Ming-fei

College of Mechatronic Engineering, Beijing Polytechnic, Beijing 100176, China
bishe16@163.com

Abstract. Aiming at the problem of low positioning accuracy of existing wireless sensor network node positioning methods, a distributed node positioning method based on radio frequency interference is proposed. Analyze the structure of the wireless sensor network, use two anchor nodes to form a radio frequency interference field, and use the movement of one of the anchor nodes to generate the Doppler effect, so that each node can obtain the instantaneous frequency indicated by its low frequency received signal field strength the angle information with the mobile anchor node, combined with the geographic location of the anchor node, the node merges multiple sets of positioning angle information to obtain the optimal position estimate. The simulation results show that, compared with other localization methods, the positioning accuracy of this method is significantly improved, and the localization time of radio frequency signal in wireless sensor networks is shortened.

Keywords: Artificial intelligence · Wireless sensor network · Radio frequency signal · Doppler effect · Node location

1 Introduction

In recent years, wireless sensor networks have received more and more attention. With the rapid development of microelectromechanical technology, the node terminals of wireless sensor networks are more intelligent, lightweight, and low energy consumption, and have the functions of measuring, sensing, and collecting external environmental information. Wireless sensor network has great practical value in the fields of national defense technology, smart home, sharing economy, fire disaster relief and other fields. It is considered as one of the most important technologies in the 21st century after the Internet [1].

In wireless sensor network (WSN), data without geographic location information is often meaningless, so wireless sensor node positioning plays an active role in data routing. There are many types of conventional positioning algorithms, but they are generally divided into two categories: one is based on non-ranging (Range-free) algorithms, such as the centroid method, the DV-Hop algorithm, and the APIT algorithm. These algorithms use inter-node Information exchange obtains information such as packet hops and distance per hop to estimate the location of the node to be measured.

S. Liu and L. Xia (Eds.): ADHIP 2020, LNICST 347, pp. 53–65, 2021.
https://doi.org/10.1007/978-3-030-67871-5_6

The required equipment is simple, but the positioning accuracy is poor; the other is a range-based algorithm, such as RSSI Algorithm 5), AOA algorithm, TOA algorithm, TDOA, etc., this type of algorithm uses the characteristics of radio signal phase information or radio signal strength (RSSI) with distance attenuation to perform ranging positioning. However, the positioning accuracy is poor, and the algorithm using phase has better accuracy, but there is often phase ambiguity, so it is necessary to introduce complex number to correct the algorithm. In addition, when ultrasonic signal and infrared signal are used for positioning, additional measurement equipment is needed, which will increase the complexity of the system, high energy consumption and short propagation distance [2].

In view of the problems of the above methods, this paper proposes a distributed node positioning method based on radio frequency interference. The node can obtain the angle information with the mobile anchor node by measuring the Doppler frequency deviation of the radio frequency interference signal, and then estimate the position. Compared with the RIPS method, this method has the following advantages: the node can complete the positioning only by performing signal reception measurement, and does not require centralized processing. It is a distributed location method. The nodes measure the signal frequency, and there is no phase ambiguity. The complexity of positioning is greatly reduced.

2   Radio Frequency Signal Location Method in Wireless Sensor Network

Wireless sensor network (WSN) is a self-organized network composed of a large number of sensor nodes deployed in the monitoring area through wireless communication technology. The emergence of wireless sensor network has profound social background and technical conditions, which not only comes from the actual needs, but also is the inevitable result of technological development. At present, information collection and acquisition methods have fallen far behind the development of computer computing speed, storage capacity and network bandwidth, so a wireless sensor network that combines sensor technology, network communication technology and information processing technology with information collection as the main purpose is ready And was born [3].

In the application of sensor network, it is almost useless to collect data without location information. It is a very important problem to determine the location of the event after detecting the event. For example, in the application of environmental monitoring, it is necessary to know the corresponding position of environmental data sensed by sensor nodes; for emergencies such as forest fire, it is necessary to know the specific location of fire. For these applications, sensor nodes must first know their geographic location, which is also an inherent requirement of wireless sensor networks [2].

This paper proposes a node positioning method combining RF interference and Doppler frequency offset measurement. A fixed anchor node and a mobile anchor node (such as an aircraft) are used to form an RF interference field. The mobile anchor node moves horizontally and uniformly to produce the Doppler effect. The frequency of the

RF signal transmitted by the two transmitting nodes is very close, and the superimposed signal received by the receiving node has a low frequency envelope; at the same time, due to the presence of the Doppler effect, the frequency of the low frequency envelope signal will change with the movement of the mobile anchor node [4]. Each node can obtain the relative position information of the unknown node and the flight path by measuring the change law of the instantaneous frequency of its own low-frequency RSSI (received signal field strength indication) signal. By moving the cross movement of the anchor node and fusing multiple relative position information, the unknown node can obtain its own absolute position information.

2.1 Wireless Sensor Network Structure

First, the network structure of the positioning system is introduced. The positioning system consists of a fixed error node, a mobile shop node and a large number of unknown nodes. The fixed anchor node and the mobile anchor node transmit single-frequency signals with similar frequencies, and the signal strength is sufficient to cover the entire network. The mobile anchor node is equipped with GPS equipment and broadcasts its own GPS information in real time. Mobile anchor nodes (such as helicopters or drones) do low-altitude intersections and fly straight in the positioning area. The network structure is shown in Fig. 1.

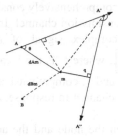

Fig. 1. Network structure of positioning system.

In Fig. 1, A is a mobile shop node, B is a fixed anchor node, and m is a wireless sensor node to be located. AA′, A′A″ are the two flight paths of the mobile anchor node, and the arrow points to the direction of movement [5]. P and q are the vertical feet of the vertical line made by the crossing point m on the flight path, the mobile anchor node Movement distance is long enough.

The positioning method adopted here is two-dimensional positioning, and the following parts are the mapping of flight path and unknown nodes to two-dimensional plane analysis. The unknown node receives the radio frequency interference signal, because the mobile anchor node moves, the frequency of the radio frequency interference signal measured at the unknown node contains Doppler frequency shift, so it is the core content of this algorithm to analyze the frequency change, so as to realize the positioning.

2.2 Doppler Effect

Doppler effect refers to that if there is relative movement between the wave source and the observer, the receiving frequency of the observer will be different from that of the vibration source. When the wave source moves to the observer, the receiving frequency will be higher, while when the wave source is far away from the observer, the receiving frequency will be lower [6], as shown in Fig. 2.

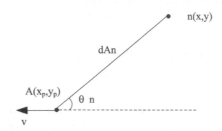

Fig. 2. Schematic diagram of the Peller effect

In Fig. 2, the mobile anchor node A transmits the tone signal $\exp[j(2\pi f_A t) + H_A]$, and moves at the instant (x_p, y_p) at the instantaneous velocity v along the negative direction of the z axis, where H_A comprehensively considers the response of the A local oscillator initial phase and the transmission channel. In a line-of-sight (LOS) environment, the received baseband complex signal of node n can be expressed as $\exp\left[j(2\pi(f_A - f_n)t) + H_A - \frac{2\pi dAn}{\lambda - s_n}\right]$, where f_n is the local oscillator frequency of n, λ is the wavelength of the RF signal, and s_n comprehensively considers the initial phase of n local oscillator and the receive channel In response, dAn is the distance between the sending and receiving nodes.

We call the angle θ_n between the node and the anchor node A and the negative direction of A movement at a certain moment as the node's Doppler angle at that moment. According to the simple geometric relationship, the instantaneous change rate $v\cos\theta_n$ of the distance dAn between the node and the anchor node A at this moment can be obtained. Due to the influence of the Doppler effect, the baseband complex signal of the node generates a Doppler frequency offset, so the instantaneous frequency of the received signal at this moment is $(f_A - f_n) - \frac{v\cos\theta_n}{\lambda}$, where $-\frac{v\cos\theta_n}{\lambda}$ is the Doppler frequency offset.

If the Doppler frequency offset can be accurately measured, the nodes can get the estimation of the Doppler angle at that time. However, for general communication systems, it is almost impossible to accurately measure the small frequency change caused by Doppler effect based on the local frequency deviation of transmitter and receiver nodes, and the local frequency drift will seriously affect the instantaneous frequency change. The above-mentioned problems that can not accurately measure the Doppler frequency deviation can be solved by using the radio frequency interference signal. In the RF interference field formed by the two transmitting anchor nodes, the mobile anchor node generates a Doppler effect, which makes the low-frequency RSSI signal of each node produce a Doppler frequency offset that is easy to measure. Due to the influence of vibration drift, the node can obtain accurate measurement of its own Doppler frequency deviation under low complexity [7, 8].

2.3 RF Interference

The mobile anchor node A and the stationary anchor node B respectively emit single-frequency signals $\cos(2\pi f_A t + H_A)$ and $\cos(2\pi f_B t + H_B)$ with similar frequencies (for example, a difference of about 1 kHz). Suppose that at some point in the LOS environment, the initial distance between unknown node m and A and B is dAm and dBm, respectively. The radio frequency interference signal received by node m can be expressed as:

$$G_n(t) = a_A \cos\left(2\pi f_A\left(t - \frac{dAm(t)}{c}\right) + H_A\right) + a_B \cos\left(2\pi f_B\left(t - \frac{dBm(t)}{c}\right) + H_B\right)$$

(1)

The baseband complex signal obtained by down converting the RF interference signal is as follows:

$$
\begin{aligned}
r_n(t) = &\ a_A \cos\left(2\pi(f_A - f_n)t - \left(2\pi(f_A - f_n)\frac{dAm(t)}{c} - H_A\right)\right) \\
&+ a_B \cos\left(2\pi(f_B - f_n)t - \left(2\pi(f_B - f_n)\frac{dBm(t)}{c} - H_B\right)\right)
\end{aligned}
$$

(2)

In the formula, a_A and a_B are the amplitude of the transmitted signal of the anchor node to reach node n; c is the speed of light; H_A and H_B comprehensively consider the initial phase of the anchor node local oscillator and the response of the transmission channel.

Therefore, the received field strength indication (RSSI) signal obtained by node M can be expressed as:

$$2\pi(f_A - f_n)t + H_A - H_B + 2\pi f_B \frac{dBm(t)}{c} - 2\pi f_A \frac{dAm(t)}{c}$$

(3)

From Eq. (3), we can see that the frequency of the RSSI signal is $\|f_A - f_B\|$, f_A and f_B are very close (such as a difference of 1kHz), the node can use its own simple equipment to achieve the measurement of the low-frequency interference signal frequency. The advantage of this method is that many chips have RSSI output, the use of low-speed AD can solve the problem of RSSI signal acquisition, without the need to use additional transceiver equipment [9, 10].

2.4 Puhler's Angle Estimation

At a certain moment t, the anchor node A moves at the instantaneous velocity at the (x_p, y_p) axis in the negative direction of the z axis. At this time, the Doppler angle of the node row is the angle θ_n between the line connecting the bamboo and A and the positive direction of the z axis. Due to the influence of the Doppler effect, the node's RSSI signal produces a Doppler frequency offset. At this moment, the instantaneous frequency of the RSSI signal is $f_{RSSI,n}$. This instantaneous frequency is derived from the interference between anchor nodes A and B on the one hand, It is due to the movement of the anchor node A. The former is the same at each receiving node in the whole network, while the latter is proportional to the cosine of the Doppler angle of the node. The schematic diagram of the Doppler angle of the node is shown in Fig. 3.

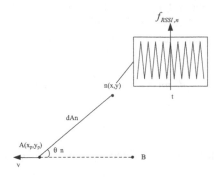

Fig. 3. Doppler included angle diagram

If the frequency of the interference signal at this time and the instantaneous moving speed of the anchor node A are known, then the Doppler angle estimation $\cos \theta_n$ at this time can be obtained by measuring the instantaneous frequency of the RSSI signal. However, because the interference signals formed by the anchor nodes A and B have a certain frequency drift, and the instantaneous movement speed of A has a large instability, the node cannot directly obtain its own Doppler angle by the above formula, which requires additional Two monitoring anchor nodes C and D are added to assist the Doppler angle estimation of the node [11, 12]. At time t, anchor nodes C and D respectively measure the instantaneous frequency of their RSSI signals as

$$f_{RSSI,C} = (f_A - f_B) - \frac{v \cos \theta_C}{\lambda} \tag{4}$$

$$f_{RSSI,D} = (f_A - f_B) - \frac{v \cos \theta_D}{\lambda} \tag{5}$$

Since the positions of the anchor nodes C and D are known, their respective Doppler angles θ_C and θ_D can also be calculated at this time. Therefore, combining the instantaneous frequencies of the RSSI signals of the anchor nodes C and D, an estimate of the interference frequency and the instantaneous moving speed of the anchor node can be obtained.

$$f_A - f_B = \frac{f_{RSSI,C} \cos \theta_D - f_{RSSI,D} \cos \theta_C}{\cos \theta_D - \cos \theta_C} \tag{6}$$

$$v = \frac{\lambda (f_{RSSI,C} - f_{RSSI,D})}{\cos \theta_D - \cos \theta_C} \tag{7}$$

Substituting $f_A - f_B$ and v in the above formulas (4) to (7) into the formula for estimating the angle of the node Doppler can be obtained:

$$\cos \theta_n = \frac{\cos \theta_D (f_{RSSI,C} - f_{RSSI,D}) - \cos \theta_C (f_{RSSI,D} - f_{RSSI,C})}{(f_{RSSI,C} - f_{RSSI,D})} \tag{8}$$

To sum up, with the help of anchor nodes C and D, each node can estimate its Doppler angle by measuring the instantaneous frequency of RSSI signal.

2.5 Node Positioning

Since the unknown node receives the GPS broadcast information of the mobile anchor node in real time, the CPS information at the vertical foot (P, Q in Fig. 1) of the mobile path and the vertical line of the unknown node and the moving direction of the mobile node can be obtained. The mobile anchor node moves along the route shown in Fig. 1. The GPS information of perpendicular P and Q can be obtained by the method introduced in part 3. The intersection of two perpendicular lines is the position of unknown node. Let the unknown node coordinates (x, y), P-point coordinates (PX, py), q-point coordinates (QX, QY), M1 and M2 be the slopes of the two flight paths respectively, and then we can get formula (9), and then we can get the node position by solving formula (9).

$$\begin{bmatrix} 1 & m1 \\ 1 & m2 \end{bmatrix} \begin{bmatrix} x \\ y \end{bmatrix} = \begin{bmatrix} m1p_y, p_x \\ m2q_y, q_x \end{bmatrix} \rightarrow a \times D = b \tag{9}$$

In the formula, a and b are called positioning matrix.

In order to improve the positioning accuracy of the network, the least squares method can be used to estimate the multiple results after making multiple intersections and movements in the unknown node area. $A = [a_1...a_n...a_N]^T$ and $B = [b_1...b_n...b_N]^T$ are the positioning matrices determined by the intersection and movement. This method uses interference to measure frequency, and the nodes only need to measure low-frequency interference signal, which reduces the complexity of the equipment hardware; through the difference of the received frequency, it can simply and effectively eliminate the influence of frequency offset; and it uses multiple flight positioning to improve the accuracy of positioning. In the whole positioning process, the unknown node only receives the high-power broadcast signal of the anchor node from a long distance without sending, and the energy consumption is low, which is suitable for positioning in a large-scale network.

3 Experimental Analysis

3.1 Experimental Environment

Due to the lack of experimental conditions in real sensor network environment, this paper chooses to use simulation software to simulate the node positioning process. The experimental environment is MATLAB, version 7.0. Matlab is an integrated experimental environment developed by Math Works of the United States for conceptual design, algorithm development, modeling and simulation and other purposes. It has a wide range of applications, including aerospace, medicine, finance, education and other fields. It has become the first choice for scientific computing, modeling and simulation, and information engineering system design and development.

3.2 Node Distribution and Initialization

Matlab environment can produce a random distribution topology in a specific area. In this paper, we specify a square with boundary value of 100. The nodes are randomly distributed in this square area. We can regard this area and the nodes in it as the practical application environment of wireless sensor network. For each node, it needs to be initialized. This information is indispensable for positioning using the CDLI algorithm. Because the randomly generated nodes themselves have coordinates, these coordinates need to be stored and correspond to the serial numbers of the nodes. This experiment uses a Sxy matrix to store this information. The coordinates of beacon nodes and unknown nodes need to be distinguished when the algorithm is implemented, so the coordinates of these two types of nodes are stored in two different matrices respectively. Of course, the coordinates of unknown nodes are not allowed to be used except for calculating the positioning error in the last step. Because the nodes are generated by MATLAB in the experiment, the coordinates of unknown nodes can be obtained, but In the practical application environment, we need to determine the coordinates of unknown nodes, which are unknown in advance. In the experiment, the final use of the coordinates of unknown nodes is to verify the positioning accuracy of the algorithm. When calculating the average distance per hop of the beacon node, the algorithm needs to use the distance information between the beacon nodes. Because the position of the beacon node has been determined, the distance can also be calculated accurately. Distance information. In addition, because matlab does not have the function to describe the communication signal transmission, the hop number is used to measure the distance between nodes, and the matrix is used to store the hop information between nodes. In the initial calculation of the hop number, the node coordinate information stored in the front, including the unknown node coordinates, needs to be used. In practical application, the sending and arriving of the communication signal can be segmented in the MATLAB experiment, the judgment is based on the distance between the unknown node and other nodes. If the distance is less than the communication radius, the corresponding node is in the range of the unknown node Inside. After initialization, the shortest path algorithm is needed to calculate the number of hops between nodes.

3.3 Storage of Node Information

According to the characteristics of MATLAB experimental environment, node information is stored by matrix, and the data operation of node information is also based on matrix. The key of storage is to change the row and column labels of matrix correctly according to the specific requirements. For example, for a matrix storing the distance between beacon nodes, assuming that the number of beacon nodes is the number of beacons, the number of rows and columns of the matrix is scalar; for the matrix that stores the coordinates of beacon nodes, the number of rows and columns is 2 And Beacon Amount.

3.4 Upgrading Secondary Beacons of Unknown Nodes

In the location algorithm, the unknown nodes that have been located are regarded as beacon nodes, and continue to participate in the location of other unknown nodes, which are called secondary beacon nodes. In the experiment in this paper, a matrix Beacon1 is set, which initially stores the coordinates of all beacon nodes. As the positioning progresses, some unknown nodes determine their own coordinates. These coordinates are added to the Beacon1 matrix. The centroid algorithm is used for positioning The beacon node in this matrix, of course, includes the original beacon node and the secondary beacon node. The unknown nodes upgrading to secondary beacons need to broadcast packets with their own location information to other unknown nodes, which will consume a certain amount of energy. However, the cost of common nodes is far lower than that of beacon nodes, and this paper focuses on the energy saving of beacon nodes, so the increase of energy consumption of common nodes is acceptable.

3.5 Evaluation Performance Index of Positioning Algorithm

The location accuracy is an important performance index for evaluating the location algorithm. The high location accuracy of an algorithm determines that the algorithm has at least some practical application value. As for whether it can be put into use on a large scale, it is necessary to analyze the network characteristics and other performance of the algorithm. However, if the positioning accuracy of a positioning algorithm is not satisfactory, then no matter how good the other performance is, it is impossible to apply it to actual positioning alone. Generally speaking, when investigating a positioning algorithm, we should first pay attention to its positioning accuracy. Only when the positioning accuracy is guaranteed, will we evaluate other performance indicators of this algorithm, and then consider applying it to practice.

The positioning accuracy is judged by the positioning error. The smaller the positioning error is, the higher the positioning accuracy is. The calculation formula of positioning error of unknown node i is as follows:

$$E(i) = \frac{\sqrt{(x_1 - x_2)^2 + (y_1 - y_2)^2}}{R} \tag{10}$$

Among them, (x_1, y_1) is the actual position of node i; (x_2, y_2) is the estimated position of node i; R is the communication radius of node i.

Assuming that there are M nodes in the network, N of which are anchor nodes ($i = 1$ to $i = N$ are anchor nodes and $i = N + 1$ to $i = M$ are unknown nodes), then the average positioning error of all unknown nodes is as follows:

$$E' = \frac{\sum\limits_{i=N+1}^{M} E(i)}{M - N} \tag{11}$$

The following is a comparison of the performance of the DV-Hop algorithm, AOA algorithm, and the algorithm in this paper from the perspective of positioning accuracy and network coverage.

3.6   Result Analysis

200 nodes randomly distributed in a 100 m $\times$ 100 m $\times$ 100 m cube network are used for simulation experiments, of which 20 nodes are anchor nodes and the rest are unknown nodes. The positioning results are shown in Table 1.

Table 1. Location result analysis (r = 15).

Positioning error	Number of nodes
≤ 0.1	15
0.1–0.5	20
0.5–1.0	25
1.0–1.5	30
≥ 1.5	40

It can be seen from Table 1 that the algorithm in this paper has high positioning accuracy, all unknown nodes can be located and the positioning error is small, so the algorithm has ideal effect in optimizing positioning.

(1) Positioning accuracy under different number of anchor nodes.

As shown in Table 2, when the node communication radius R = 16, and the number of anchor nodes N gradually increases from 1 to 50, the positioning accuracy

Table 2. Average positioning error under different number of anchor nodes.

Project	Number of anchor nodes 20	Number of anchor nodes 30	Number of anchor nodes 50
Algorithm in this paper	0.465	0.431	0.237
DV hop algorithm	0.604	0.442	0.324
AOA algorithm	0.623	0.601	0.452

of the three algorithms is improved. As the number of reference nodes and the number of test triangular pyramids increase, the estimation range of unknown node positioning is reduced, and the positioning accuracy is naturally improved.In addition, anchor nodes have a direct impact on the possibility of node location. When the number of anchor nodes is small, the positioning accuracy of this algorithm is obviously better than DV hop algorithm and AOA algorithm. This algorithm improves the division of positioning modules, reduces the estimation range of positioning, and speeds up the positioning speed. The estimated range of positioning can be reduced to half of other algorithms. More importantly, the algorithm in this paper can solve the situation where it cannot be located, so the positioning accuracy has been greatly improved.

Table 3. Average positioning error under different communication radius

Project	Communication radius 10	Communication radius 20	Communication radius 30
Algorithm	0.812	0.475	0.263
DV-Hop algorithm	1.402	0.604	0.512
AOA algorithm	1.493	0.925	0.553

(2) Positioning accuracy under different communication radius.

As shown in Table 3, when the number of anchor nodes n = 18, and the communication radius r of nodes gradually increases from 1 to 30, the average positioning errors of the three algorithms are rapidly reduced. This is because the increase in R allows unknown nodes to communicate with more anchor nodes, so the number of test pyramids continues to increase and the estimated range of nodes gradually decreases. As the size of the communication radius of the node has a significant impact on the positioning accuracy, it can be concluded that the positioning accuracy of this algorithm is significantly better than DV hop algorithm and AOA algorithm. With the increase of R, the number of anchor nodes within the one-hop communication range will also be greater, and the number of test pyramids obtained by the algorithm in this paper will also be greater.

In order to further verify the effectiveness of the method in this paper, the positioning time of radio frequency signal of wireless sensor network based on the proposed method and the traditional method is compared and analyzed. The comparison results are shown in Fig. 4.

According to Fig. 4, the positioning time of radio frequency signal in wireless sensor network of this method is within 30 s, while that of traditional method is within 60 s, which shows that the positioning time of radio frequency signal in wireless sensor network can be shortened by using this method.

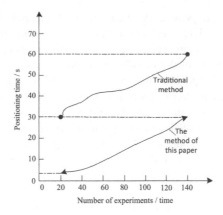

Fig. 4. Positioning time comparison results

4 Conclusion

In summary, this article elaborated on the principles and methods of interferometric positioning, and simulated and analyzed the interferometric positioning algorithm to verify the feasibility of the algorithm and analyze the factors that affect the accuracy. The location method of WSN based on interference effect can realize distributed location only by measuring the received signal phase at each receiving node. It has the advantages of simple equipment, high positioning accuracy and low power consumption. It is a better method. Of course, compared with other positioning algorithms, this algorithm has higher time complexity and increased network energy consumption. Future research directions should focus on reducing the energy consumption required for positioning while improving positioning accuracy and network coverage.

Acknowledgements. The research is supported by Research and design of equipment management system based on RFID(CJGX2016-KY-YZK041).

References

1. Wu, D., Zhu, D., Liu, Y., et al.: Location verification assisted by a moving obstacle for wireless sensor networks. IEEE Internet Things J. **5**(1), 322–335 (2018)
2. Sabale, K., Mini, S.: Anchor node path planning for localization in wireless sensor networks. Wirel. Netw. **25**(1), 49–61 (2019)
3. Qin, N., Chen, K.: A wireless sensor network location algorithm based on insufficient fingerprint information. Mod. Phys. Lett. B **32**(4), 1840093 (2018)
4. Kulkarni, V.R., Desai, V., Kulkarni, R.V.: A comparative investigation of deterministic and metaheuristic algorithms for node localization in wireless sensor networks. Wireless Netw. **25**(5), 2789–2803 (2019)
5. Yarinezhad, R., Hashemi, S.N.: Distributed faulty node detection and recovery scheme for wireless sensor networks using cellular learning automata. Wirel. Netw. **25**(5), 2901–2917 (2019)

6. Khan, I., Singh, D.: Energy-balance node-selection algorithm for heterogeneous wireless sensor networks. ETRI J. **40**(5), 604–612 (2018)
7. Liu, S., Li, Z., Zhang, Y., et al.: Introduction of key problems in long-distance learning and training. Mob. Netw. Appl. **24**(1), 1–4 (2019)
8. Lee, W.K., Schubert, M.J.W., Ooi, B.Y., et al.: Multi-source energy harvesting and storage for floating wireless sensor network nodes with long range communication capability. IEEE Trans. Ind. Appl. **54**(3), 2606–2615 (2018)
9. Shuai, L., Gelan, Y.: Advanced Hybrid Information Processing, pp. 1–594. Springer International Publishing, Heidelberg (2019)
10. Kashniyal, J., Verma, S., Singh, K.P.: A new patch and stitch algorithm for localization in wireless sensor networks. Wirel. Netw. **25**(6), 3251–3264 (2019)
11. Liu, S., Bai, W., Srivastava, G., Machado, J.A.T.: Property of self-similarity between baseband and modulated signals. Mob. Netw. Appl. **25**(4), 1537–1547 (2019). https://doi.org/10.1007/s11036-019-01358-9
12. Liu, S., Glowatz, M., Zappatore, M., et al. (eds.): e-Learning, e-Education, and Online Training, pp. 1–374. Springer, Heidelberg (2018)

Design of Big Data Control System for Electrical Automation

Lin-ze Gao[✉]

Guilin University of Electronic Technology, Guilin 536000, China
gaolz20@126.com

Abstract. The traditional big data control system is limited by hardware and software resources, which leads to low efficiency of data access and calculation. Therefore, a big data control system for electrical automation is designed. The system is mainly designed from two aspects of hardware and software. The hardware mainly designs a digital acquisition circuit, selects the model of each part in the overall working module of the acquisition board, and designs a pseudo dual-port RAM to store data, which is convenient for the processing of massive data and Control; in software design, according to the needs of the system, light MySQL and distributed HBase are used to jointly design the ER diagram and database table of the database user database to make the database performance more optimized. In the control algorithm of the system, the butterfly operation is used to optimized calculate the speed, obtain the operation error through the operation and correct it to ensure the efficiency of the system operation. The experimental results show that the performance of the designed system is better than the traditional system in data insertion speed, access speed and calculation speed, which fully verifies the application value of the system.

Keywords: Electrical automation · Big data · Database · Butterfly algorithm · Digital acquisition circuit

1 Introduction

The invention of electric energy is known as one of the greatest achievements since the 18th century. Its wide application has set off the second climax of industrialization, and made the human society march into the electrical age with its head held high. Since its development, electric energy has entered every aspect of people's life, which is not only related to the quality of people's life, but also of great significance to national economic development, national defense security and social stability [1]. Many countries in the world have used their degree of application as an important indicator for judging the national development level. Industrial production is inseparable from electricity, and many fields in the industry have put forward the requirements of high-speed real-time computing. Although traditional industrial field data collection systems have good reliability, accuracy and compatibility, they are large and costly high. Although the cost of the system based on single chip microcomputer is low, there are many devices, the system is complex, the development time is relatively long, and the

© ICST Institute for Computer Sciences, Social Informatics and Telecommunications Engineering 2021
Published by Springer Nature Switzerland AG 2021. All Rights Reserved
S. Liu and L. Xia (Eds.): ADHIP 2020, LNICST 347, pp. 66–76, 2021.
https://doi.org/10.1007/978-3-030-67871-5_7

reliability of the system is poor [2, 3], which has a certain impact on the processing efficiency of massive data.

In view of the above problems, a big data control system for electrical automation is designed. The hardware of the system mainly includes digital acquisition circuit and database, which is convenient for processing massive data. On the basis of hardware design, the system software is mainly designed. The user block E-R diagram and database table of database are designed by using lightweight MySQL and distributed HBase. Butterfly operation is used to improve the calculation speed of the system, and the calculation error is corrected. The experimental results show that the designed system has the advantages of data insertion speed, access speed and calculation speed, and has high utilization value.

2 Hardware Design of Electric Automation Big Data Control System

2.1 Design Digital Acquisition Circuit

The data acquisition board adopts PC/104 bus standard, 16-bit data transmission controlled by I/O commands, and has program trigger and external trigger functions. The external analog signal is converted into a digital signal by an AD converter. The conversion accuracy is 120 digital signals, and then the number of digits is expanded to 16 bits, which is sent to the FPGA for FFT conversion [4, 5]. The control signals required for the entire collection and subsequent data transmission are completed by the FPGA. The FPGA interface control module is mainly composed of address decoding, address generation, and FFT processing unit. The address decoding part translates the corresponding port through the 10 bit address from the bus, and generates the corresponding control signal; the address generation part is mainly responsible for generating the address of on-chip dual port RAM and off chip SRAM. The designed dual port RAM adopts ping-pong time sharing mode and has two sets of independent data, address and control line. Two data ports are read-only and write only (also known as pseudo dual port RAM).

The A/D conversion model is AD 1672. A belongs to a monolithic analog-to-digital converter (ADC). The chip contains 4 high-performance sample-and-hold amplifiers (SHA), 4 flashing ADCs and a voltage reference [6]. The output is equipped with an error correction logic circuit to ensure 12-bit accuracy at a sampling speed of 3MSPS and no missing codes over the entire operating temperature range. The sampling voltage is −2.5 V to +2.5 V. The data latch 74HC374 is an 8-bit tristate buffer. At the rising edge of the clock, the data is sent from the d end to the Q end to complete the cache. The schematic diagram of the overall working module of the acquisition board is shown in Fig. 1:

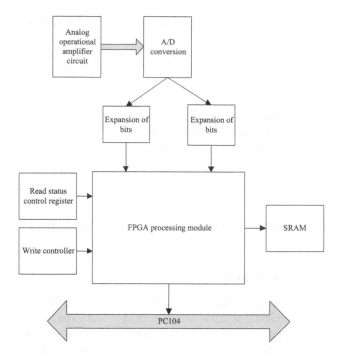

Fig. 1. Schematic diagram of the overall working module of the acquisition board

In the figure above, the rising edge of a clock AD 1672 completes the transfer of a data. In this design, AD 1672 and 74HC374 share a clock INCLK to maintain synchronization, INCLK is provided by the address generation unit inside the FPGA. In the design, two 74HC374s are used for bit number expansion, and the 12-bit signed number after A/D sampling is expanded to 16 bits. The control register model is 74HC373. Two circuit control registers are used in the circuit design. One is used to record the status signal on the acquisition board for query, and one is used to write the control word to control the working state of the entire acquisition board [7, 8].

The data memory model is IS61C1024, which is an 8-bit 128 KB high-speed CMOS static RAM with a minimum storage time of 12 ns. It can meet the requirements of high-speed data reading and writing, and is also suitable for the temporary storage of large capacity data. This design uses four pieces of IS61C1024 to cache FFT processing results. The bidirectional data bus transceiver model is HC245HC245, which is an eight bidirectional bus transceiver used for bidirectional asynchronous communication between data buses. The output allows the control terminal/G to be effective when the level is low, and when the level is high, the two ends are high impedance. In the circuit, MegaWizard designed a 2048 × 32 bits dual-port RAM to store the intermediate data of FFT operation.

The internal storage unit is divided into two parts. When the first part is used to provide operation data to the current level, the second part is used to store the operation results of the previous level. When the 1024 point operation of the current level is completed, the functions of the two parts are exchanged.

2.2 Design Database

Because the system database not only needs to store massive unstructured power data, but also involves the association operation of structured data such as user information, job information, system block information, etc. Therefore, plans to use lightweight MySQL database to process structured data and relational computing, and distributed HBase database to access massive unstructured data. MySQL is a small relational database management system, which has the characteristics of small size, high speed and low cost. User information, system block information and Hadoop job information in the system will involve query methods such as condition and association [9], so MySQL database is selected for storage. Among them, the E-R diagram of user board is shown in the Fig. 2, and there is a many to many relationship between them.

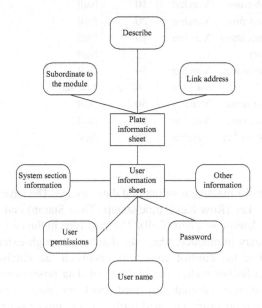

Fig. 2. E-R diagram of user sector

In the system designed in this paper, the data table structure of MySQL database is shown in Table 1:

Table 1. Database table

Field name	Field type	Field size	Allow to be empty
Uid	Int	4	Not null
User name	Varchar	30	Not null
Pass word	Varchar	20	Not null
Authority	Int	3	Not null
Other	Varchar	50	Null
Mid	Int	4	Not null
Module	Varchar	10	Null
Depict	Varchar	40	Null
Address	Varchar	100	Not null
Authority	Int	3	Not null
Jid	Int	4	Not null
Job name	Varchar	10	Null
Start time	Varchar	20	Null
Percentage	Varchar	10	Null
State	Int	2	Null
Job maker	Varchar	30	Null
Fid	Int	4	Not null
File name	Varchar	30	Null
Start time	Varchar	20	Null
Job maker	Varchar	30	Null

HBase is suitable for massive unstructured data storage. The structure of the table is determined by a row key (Row Key), time stamp (Time Stamp) and column (Column) three dimensions to determine a cell (Cell). The data type in the cell is not limited, and it is converted to binary in practice. Open the database through external commands of revit, so as to realize the control connection between the database and the user. However, if want to further realize the problem of data processing efficiency on the system, the large database automatically feeds back preventive measures, resolution measures, occurrence probability, etc., and further programming is needed to achieve it. At this point, the design of the system database is completed.

3 Software Design of Electric Automation Big Data Control System

In this paper, the control algorithm of the system is butterfly operation, the speed of operation directly affects the speed of the whole design, so how to speed up the processing speed of butterfly operation unit is the key to improve the calculation speed of the system operation unit. Butterfly unit is composed of real part and virtual part, and then passes through complex multiplication unit, buffer unit and complex addition unit. Butterfly operation includes complex multiplication and complex addition. Only by increasing the speed of complex multiplication can the processing speed of the butterfly unit be accelerated. The realization of complex multiplication can be realized directly by multiplier operation, but the multiplication operation is difficult to achieve in hardware and the calculation speed is slow [10]. To solve this problem, the CORDIC (Coordinate Rotation Digital Computer) algorithm is used to implement complex multiplication. The CORDIC algorithm can not only convert the complex multiplication into hardware addition and subtraction and shift operations, but according to its iteration principle, the CORDIC unit It can be represented by a pipeline structure, which can make vector rotation parallel processing, which greatly speeds up the speed of butterfly operation [11]. The coordinate rotation digital algorithm includes three rotation systems: circular system, linear system, and hyperbolic system. In this paper, only the circular system is used for optimization, and the plane coordinate rotation is shown in the Fig. 3:

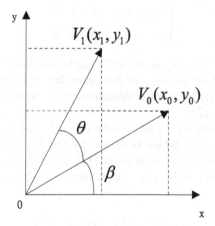

Fig. 3. Plane coordinate rotation

The initial vector V_0 rotates the angle θ to get the vector V_1. The coordinates can be expressed as follows:

$$\begin{cases} x_1 = R\cos(\beta + \theta) = x_0 \cos\theta - y_0 \sin\theta \\ y_1 = R\sin(\beta + \theta) = y_0 \cos\theta + x_0 \sin\theta \end{cases} \tag{1}$$

In the above formula, R is the radius of the circle and θ is the rotation angle. Iterate the above formula and divide the rotation process into multiple steps. Then the rotation formula in step i is:

$$\begin{cases} x_{i+1} = (x_i - y_i \cdot \tan\theta_i) \cdot \cos\theta_i \\ y_{i+1} = (y_i + x_i \cdot \tan\theta_i) \cdot \cos\theta_i \end{cases} \tag{2}$$

In order to achieve convenience in hardware, the following convention is made: each rotation angle θ tangent value is a multiple of 2, namely: $\theta_i = \arctan 2^i$, that is, the sum of all iteration angles is equal to the rotation angle. S_i represents the direction of rotation. Set the angle that differs from θ after n rotations to Z_n, then:

$$Z_n = \theta - \sum_{i=1}^{n} S_i \theta_i \tag{3}$$

Therefore, the i time single-step rotation formula can be changed to:

$$\begin{cases} x_{i+1} = x_i - S_i \cdot y_i \cdot 2^i \\ y_{i+1} = y_i + S_i \cdot x_i \cdot 2^i \end{cases} \tag{4}$$

In this way, CORDIC algorithm can basically realize the function operation needed in the mathematical operation. Because of the generality of CORDIC algorithm in calculation and the simplicity of its implementation, it is suitable to be used as a module unit of data processing [12].

The output error caused by the actual rotation angle and the ideal rotation angle is called the approximate error, which is caused by the limited rotation series and the limited binary digits used to express the angle [13]. If the ideal selection result of vector V_0 is V_1, the ideal rotation angle is Z_0, and the actual angle is Z_0', then the angle error can be calculated:

$$Z_0 - Z_0' = \delta \leq \arctan(2^{n+1}) \tag{5}$$

The approximate error can be expressed as:

$$|V_1 - V_1'| = \sqrt{(\cos\delta - 1)^2 + \sin^2\delta}\,|V_1'| \tag{6}$$

In addition, the limited calculation accuracy in each stage will also cause rounding error, which is caused by the limited operand width. The accuracy of CORDIC algorithm is related to N (rotation Series) and B (operation digit width). The number of iterations is fixed, and the value of approximate error is constant. Therefore, the rounding error can be reduced by increasing the number of processing bits of CORDIC operation unit, so as to reduce the overall error of CORDIC unit. But at the same time, the processing speed will be reduced by increasing the processing bits of CORDIC. The best point must be found between the number of processing bits and the processing speed [14, 15]. By extending the adder and subtracter of CORDIC operation unit to 19 bits, the operation speed of CORDIC operation unit is not greatly affected, but the calculation accuracy is improved. Here, only the adder and subtracter in CORDIC operation unit is extended to 19 bits, ram is still 16 bits wide, 16 bits of data will be sent to the high position of 19 bits adder and subtracter for operation, the calculated 19 bits result will be truncated at the low position, and finally the 16 bits wide result will be output.

4 Experiment

4.1 Experiment Preparation

In order to verify whether the mass data analysis system meets the requirements specification definition, this chapter will conduct system tests to find out areas that are inconsistent or inconsistent with the demand stage, and propose improvements and perfect plans. According to the system function, the system test is divided as shown in Fig. 4:

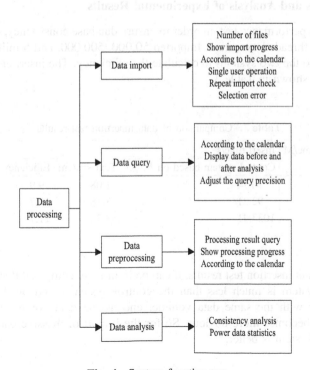

Fig. 4. System function test

In order to verify the effectiveness of the big data control system designed in this paper, the control system based on single-chip microcomputer is used as a comparison method, and the two systems are tested in the same test environment by script. The test environment is shown in Table 2:

Table 2. Test environment

Environmental classification	Project	Parameter
Hardware environment	Blade server	5 units, each with 24 cores
	Memory	65 GB
	Hard disk	2T
Software Environment	Operating system platform	Linux CentOS
	Server platform	Tomcat
	Data base	One MySQL installation, five HBase clusters
	Software platform	Java Web project, Hadoop distributed system

In order to test the performance of the system in this paper in all aspects, we choose to insert performance, query performance and computing performance for testing.

4.2   Statistics and Analysis of Experimental Results

In the insertion performance test, in order to ensure database consistency, data insertion is set to single-threaded execution. Imported 50,000, 500,000, and 5 million pieces of battery data into the two systems to calculate the efficiency. The insertion performance test results are shown in Table 3.

Table 3. Comparison of data insertion test results

Data volume/M	Insert time/s		
	Control system based on MCU	Text system	Efficiency ratio/%
300	10.74	1.08	9.9
600	92.21	8.58	10.7
900	1033.41	88.24	11.7

From the data insertion test results, it can be intuitively compared that the insertion time of this system is much less than the control system based on the single-chip microcomputer with the same data volume, and as the data volume increases, the efficiency gap becomes more obvious. So for the import of massive data, the performance of this system is better.

During the query performance test, the multi-threaded concurrent access to the MySQL database of the simulation system and the distributed HBase database of the system To simulate multi-users reading data at the same time and counting the query time. The query performance test results are shown in Table 4.

Table 4. Comparison of data query

Thread concurrency/one	Query time/s	
	Control system based on MCU	System of this paper
Single-threaded query 100	1.37	0.67
10 threads query 100	2.14	0.68
Single thread query 1000	3.44	0.65
10 threads query 1000	3.76	0.74
100 threads query 1000	24.35	0.79

From the data query results, it can be clearly seen that the query performance of the control system based on single chip microcomputer is still far behind that of the system in this paper, and with the increase of query data and thread concurrency, the gap is more obvious. However, the data volume of tens of millions level is far from the processing limit for the system database of this cluster. For the change of data volume and concurrent thread access volume, the query time tends to be stable.

In the calculation performance test process, the power data variance calculation test is performed on the specified level of data volume by using the microcontroller-based control system and the system of this article, and the data of the two systems are compared calculate ability. Make final analysis and statistics on the test results. The calculation performance test results are shown in Table 5.

Table 5. Comparison of data calculation

Data volume/M	Data calculation time/s		
	Control system based on single chip microcomputer	System of this paper	Efficiency ratio/%
300	873	247	3.5
600	4375	1337	3.3
900	8864	2977	3.0

From the data calculation results, the calculation time of the system in this paper is far less than the calculation time of the control system based on the single chip microcomputer. Although the processing efficiency of the cluster cannot reach 5 times that of a single server due to Hadoop task startup, file reading, and communication delay between clusters, this has greatly improved processing performance. In summary, the performance of the system designed in this paper is significantly better than the traditional data control system.

5 Conclusion

Based on the analysis of some shortcomings in the existing big data control system, this paper optimizes the design from both hardware and software, and designs a big data control system for electrical automation. Starting with data processing, calculation and other methods, the hardware and software configuration has been re-optimized. It is expected that the designed system can have high data processing efficiency. Compared with the existing system, the system in this paper has significant advantages, but there are still some areas that need to be further improved, such as a large number of redundant data in big data, if the redundant data is not processed, the control effect will be affected. The next step will focus on redundant data processing, and further optimize the system to make the system really suitable for the needs of power production.

References

1. Linlin, S.: Design and implementation of big data cloud storage system for monitoring system: an example of monitoring system optimization for port service center in Yangpu Port. Value Eng. **037**(026), 215–217 (2018)
2. Ming, S., Baofeng, W., Yujuan, C., et al.: Design and verification of big data processing system oriented spacecraft control mission. Spacecr. Eng. **27**(6), 69–76 (2018)
3. Xuemei, Z.: Fault diagnosis for electro-hydraulic control system of hydraulic support based on big data. Ind. Mine Autom. **44**(12), 38–42 (2018)
4. Baohua, S., Lei, C., Dong, X., et al.: Design and application of an intelligent operation and maintenance control platform of the power distribution network based on the big data platform. Electr. Autom. **40**(6), 85–88 (2018)
5. Fengjie, S., Chengmin, W., Ning, X.: Frequent pattern network model for association rule mining of big data in smart grid. Electric Power Autom. Equip. **38**(5), 110–116 (2018)
6. Qing, L., Song, L., Xiaojie, L., et al.: Intelligent control system based on big data technology for whole production line of sintering quality. Iron Steel **53**(7), 1–9 (2018)
7. Liu, S., Li, Z., Zhang, Y., Cheng, X.: Introduction of key problems in long-distance learning and training. Mobile Netw. Appl. **24**(1), 1–4 (2018). https://doi.org/10.1007/s11036-018-1136-6
8. Ruiguo, B., Lishan, X., Kuo, B., et al.: Application of big data process quality control system in iron and steel production. China Metal. **28**(8), 76–80 (2018)
9. Haiba, C., Quan, G., Yangcheng, D., et al.: Power grid dispatching intelligent alarm system based on big data. Electron. Des. Eng. **27**(11), 91–95 (2019)
10. Jianguo, X., Haifeng, X.: Realization of a wisdom agricultural internet of things system based on big data. Softw. Guide **190**(8), 133–136+144 (2018)
11. Sifah, E.B., et al.: Chain-based big data access control infrastructure. J. Supercomput. **74**(10), 4945–4964 (2018). https://doi.org/10.1007/s11227-018-2308-7
12. Ndikumana, A., Tran, N.H., Ho, T.M., et al.: Joint communication, computation, caching, and control in big data multi-access edge computing. IEEE Trans. Mob. Comput. **19**(6), 1359–1374 (2019)
13. Gosine, A.: Industrial control systems are converging with Big Data. Control Eng. **65**(3), 40–42 (2018)
14. Shuai, L., Gelan, Y.: Advanced Hybrid Information Processing, pp. 1–594. Springer, Heidelberghttps://doi.org/10.1007/978-3-319-73317-3
15. Fu, W., Liu, S., Srivastava, G.: Optimization of big data scheduling in social networks. Entropy **21**(9), 902 (2019)

Design and Implementation of Walking Control System for Orchard Plant Protection Robot Based on Artificial Intelligence Algorithm

Guang-yong Ji[1]([✉]), Zhen Wang[1], and Rui Zhang[2]

[1] Yantai Vocational College of Culture and Tourism, Yantai 264003, China
suchao20@tom.com
[2] National Energy Penglai Power Generation Co., Ltd., Penglai 265600, China

Abstract. In order to improve the stability of fruit recognition of orchard plant protection robot, the walking control system of orchard plant protection robot was established based on artificial intelligence algorithm. Orchard eppo robot control system design of hardware platform is a rate by machine controller, signal controller, chassis motor drives, cameras and proximity switch of these five parts, is mainly responsible for transferring information to the control system, do matting for software design, on this basis, the set can intelligent power saving communication program and the sensor data acquisition, the recognition data transmission to the robot control system, the implementation is based on artificial intelligence algorithm orchard plant protection design of the control system of walking robot. By combining software and hardware, the research on the walking control system of orchard plant protection robot based on artificial intelligence algorithm is completed. From the results of software and hardware experiments, it can be seen that compared with the traditional robot walking control system, the application of this system for fruit recognition has higher stability and can effectively reduce the workload.

Keywords: Orchard plant protection robot · Walking control system · Sensor data acquisition program · Fruit identification

1 Introduction

At present, intelligent science and nonlinear science have been widely used in various fields. Then some new optimization methods which are different from traditional optimization methods are developed rapidly, namely intelligent optimization algorithm, such as artificial neural network, particle swarm optimization, genetic algorithm, support vector machine, etc. Among them, wavelet neural network and support vector regression machine have the advantages of fast computing speed, strong fitting ability and high precision, and they are widely used methods to solve nonlinear problems with very broad application prospects. Wavelet neural network is a new kind of neural network based on wavelet theory and neural network theory. Support vector machine (SVM) is a new machine learning method proposed by Vapnik and his scientific

S. Liu and L. Xia (Eds.): ADHIP 2020, LNICST 347, pp. 77–87, 2021.
https://doi.org/10.1007/978-3-030-67871-5_8

research team at the end of the 20th century when studying statistical learning [1]. This method can be used to solve practical problems such as small samples, non-linearity and high dimensionality, avoid local minimum points and have high generalization ability. This algorithm is used to solve regression problems and developed into support vector regression machine. The generation of support vector regression machine is of great significance for regression approximation.

It is generally believed that the orchard plant protection robot is a kind of automatic mechanical harvesting system with perceptive ability, which can be programmed to complete the picking, transporting, packing and other related tasks of fruits. Picking robot requires knowledge of mechanical structure, visual image processing, robot walking dynamics, sensor technology, control technology and computational information processing. Orchard plant protection robot can automatically detect the fruit and save the data. When it needs to know the data of a certain aspect of the fruit, it only needs to use the control software to read the required data and carry out certain statistical analysis and chart analysis. The data measured by the lab-type fruit phenotype detection robot system is generally more accurate, but its statistical ability is limited. When too many fruits need to be tested, the lab-type fruit phenotype detection robot system should not be used. Compared with laboratory fruit rapid phenotype detection robot system, the detection speed of greenhouse fruit rapid phenotype detection robot system has been greatly improved, but the cost of greenhouse fruit rapid phenotype detection robot system is generally higher. In order to solve the problem of fruit phenotype detection in fruit breeding, aiming at the problem of rapid detection of fruit phenotype parameters, a walking control system of orchard plant protection robot was established based on artificial intelligence algorithm to improve the reliability and work efficiency of breeding decision.

2   Hardware Design of Orchard Plant Protection Robot Walking Control System Platform

The platform hardware of orchard plant protection robot walking control system mainly includes remote control PC, robot master controller, mechanical arm controller, chassis controller, proximity switch, camera, keyboard, Leap Motion gesture controller, PS2 wireless controller, motor driver, power module and so on.

As shown in Fig. 1, the hardware of orchard plant protection robot control system platform. In order to ensure the reliable performance of orchard plant protection robot control system platform, the choice of robot master controller is particularly important.

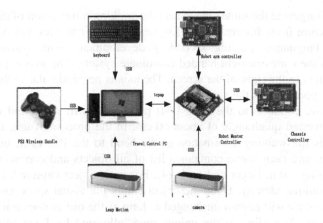

Fig. 1. Hardware structure of orchard plant protection robot control system platform

2.1   Robot Master Controller

The main controller of the robot should be able to send control information to the chassis controller to realize the operation of the chassis, to process the image information collected by the camera, to be able to move, to send control information to the manipulator controller to realize the walking of the manipulator, and to communicate with the remote control PC [2]. This paper selected the Intel Joule platform, which has a wide range of applications in the fields of meter vision, robotics, unmanned aerial vehicles and other highly demanding fields. The platform is based on the module computing system, which can achieve 4K resolution video shooting and display, and support the application of depth camera. The platform is equipped with a 64-bit 1.7 GHz quad-core intel. atomtm processor T5700, equipped with 4 GB LPDDR4 RAM and 16 GBeMMC memory, and provides a wide range of physical interfaces, including USB 3.0, multiple GPIO, I2C and DART interfaces, as well as wi-fi and Bluetooth 4.1. In addition, it supports the application of intel. Real sensetm depth of field sensing camera technology, suitable for computer vision systems with high requirements.

2.2   Gesture Controller

Leap Motion is a new type of intelligent interactive hardware released by Leap, a motion-sensing controller manufacturer. It mainly captures the walking of hands. It tracks and locates hands, fingers and similar tools based on infrared imaging and triangulation ranging principles, and provides these collected real-time information to developers for human-computer interaction. Leap Motion can grasp and wave at will in an effective space to conduct smooth operation of PC space separation, and it can accurately track the movements of hands and fingers no matter how small they are, with an accuracy of 0.01 mm. Leap Motion has two infrared cameras and three LED lights used to illuminate the target. The data is collected in frames, and the maximum frame rate can reach 200 frames/second. It can recognize the continuously changing Motion, track multiple targets at the same time, detect the walking information of the target, and capture the picture from different angles [3]. Leap Motion provides rich API interfaces for different languages, making secondary development more convenient. Leap Motion USES the cartesian right-handed coordinate system. The axis is parallel to the sensor and points to the right of the screen. The axis is perpendicular to the sensor and points to the upper part of the space.

Leap Motion connects to the PC's USB port via a USB cable and recognizes a space of an inverted quadrangle of about 60 cm. In the process of use, Leap Motion regularly sends the walking information of the hand to the PC. Each information is called a frame, and each frame contains a list of all objects and corresponding information, including palm, finger and tools [4]. For each object captured, Leap Motion will assign a unique identity. Once the object enters the visual space and obtains an identity, the identity will remain unchanged as long as the object does not disappear in the visual space. According to the unique mark allocated by Leap Motion for the object, and then according to the axial vector, Angle, translation vector, scale factor and other data generated by the current frame and the data of the previous frame, the information of each walking object can be inquired, and the basic information of hand walking can be obtained.

2.3 Chassis Motor Driver

Arduino is an open source electronic design platform that is fast, flexible and easy to use. The software, namely the hardware schematic diagram of the program development environment ArduinoIDE oArduino in the computer, the IDE software and the core library files are open source, allowing the original design and corresponding codes to be arbitrarily modified within the scope of the open source protocol [5]. The characteristic of Arduino IDE is that it can be used on Windows, Mac and Linux. After simple learning, it can be developed quickly. Based on ATmega2560, the Arduino Mega 2560 adopts USB interface as the core circuit board. It has 54 digital IO ports, 16 analog IO ports, 4 DART ports and one USB port, which is suitable for application scenarios requiring a large number of IO interfaces. It can choose three power supply methods: power supply through the power socket; Power supply through GND and VIN pins; USB interface power supply, this paper adopts the direct power supply method of USB interface [6].

2.4 The Camera

The camera in this paper USES Realsense r200 produced by Intel company, and its target USES mainly include: 3d capture of face, human body and environment, depth enhancement reality, depth enhancement photography and video, measurement, face detection and tracking, etc. Specifications: distance; Depth of field/infrared: 60 frames per second, resolution: RGB: 30 frames per second, 1080; Appearance size:; For start. It should be noted that only the basic function of image acquisition is used in this paper, but considering the needs of follow-up research and application, such a depth camera is specifically selected here [7].

2.5 Proximity Switch

Orchard eppo robot installed a camera and a proximity switch, camera installed at the rear of the car can ensure the PC to be able to see the scene of the homework, in the robot walking forward, can control the robot will not hit the obstacles ahead, from the image and the right and left sides of the car body is the blind area is likely to hit obstacles, in order to prevent the happening of this kind of situation, this article USES proximity switch to realize the obstacle avoidance.

Proximity switch is a position switch that can be operated without direct contact. When the object enters the inductive surface of the proximity switch, the switch can be activated without direct contact and any pressure applied, thus providing control instructions for devices such as computers. Picking robot using two proximity switch, and respectively installed on the both side of the chassis, when on the left side of the barrier into proximity switch movement distance, the left side of the proximity switch to chassis controller switch quantity information, chassis controller receives the information, by controlling the motor realize the obstacle avoidance, the principle of implementation on the right side with the left side is the same. At this point, the hardware design of orchard plant protection robot walking control system was completed [8].

3 Software Design of Orchard Plant Protection Robot Walking Subsystem

For the orchard plant protection robot system, the control system software is the core. This chapter mainly carries on the design to the orchard plant protection robot walking control software, including the intelligent control program, the sensor data acquisition program, each application program and so on.

3.1 Can Intelligent Power Saving Communication Program

The following design is made for the walking control system of orchard plant protection robot: firstly, the can intelligent node communication driver is designed; Secondly, GPS positioning module, INS measurement module and ultrasonic ranging module are designed, which are used to collect data. Thirdly, the main controller, robot controller and related applications of android phones are designed [9]. See Fig. 2 for system composition.

As can be seen from Fig. 2, the main function of can intelligent node communication program is to receive sensor data through relevant interfaces and transmit the data to robot control machine through can wireless network. Therefore, this program mainly transmits some data collected by the sensor to the robot control machine. In the design of can intelligent node communication program of orchard plant protection robot walking control system, it mainly designs can communication program and serial communication program.

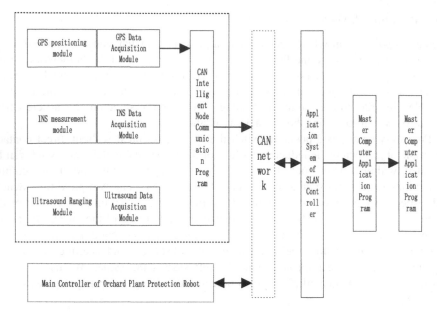

Fig. 2. Software design of robot walking subsystem

3.2 Sensor Data Acquisition Program

Relevant sensors used in this design include sensors of GPS positioning module, sensors of NS measurement module and sensors of ultrasonic ranging module [10]. These sensors can give orchard eppo robot control system provides functions such as coordinate information, including the location information of robot body, the activities of the robot body information, job information of robot body, etc., and the speed of the robot body information, whether there is any obstacle, with obstacle distance how far is it and other related information. In this section, the relevant sensor data acquisition program is designed.

The GPS positioning module used in this research USES the GPS product of ay-gps268 produced by suzhou aiyu technology co., LTD. This receiver has high precision and can realize dynamic reception. After being connected to the chip of can smart node, relevant instruction data can be exchanged. This GPS positioning module follows a certain codestream format and forms a data protocol. According to this protocol, the data format can be analyzed first, and then the data can be collected and processed [11, 12].

Data format analysis: the relevant GPS positioning module adopted in this study can output different information according to the requirements of the system, but the GPS positioning function needs to be simplified. Therefore, the data format should be analyzed to determine the data type and whether the data contains positioning information. Finally, the GPS positioning module will send these information to the serial port, and finally the serial port will receive the information, and then carry out the next step of processing.

According to the relevant code stream protocol, it can be known that the data sent by the GPS positioning module to can smart node is composed of four parts. The first part is the frame header, which is represented by $GPRMC. Second, data valid area; Three is the position inspection; Four is the end of the frame. For the transmission of the data itself, it is need to use a serial port on the can intelligent node, but need to speed up the speed and process, in order to improve the efficiency of data transmission, so the GPS module design, data collection procedures not first to receive all the information, and according to the format of the data itself characteristic, carries on the corresponding data acquisition program design, makes the individual received data more accurate, the process is shown in Fig. 3.

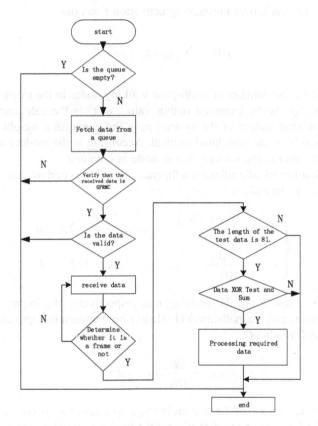

Fig. 3. Flow chart

As can be seen from Fig. 3, in the GPS data frame acquisition program, the first step is to judge the data sent from the serial port. If it is determined that no data is sent, exit the program. If it is confirmed that there is data sent, and the data frame header detected is $GPRMC, it can start the cyclic receiving of characters [13, 14]. Because in the valid area of the data, there is the positioning status identifier of the data, so the validity of the data is judged according to the identifier. If the data is judged to be invalid, the process will suspend the data receiving, and then wait for the new data frame, and send the data again after artificial intelligence calculation. The ideal value is obtained according to the principle of artificial intelligence algorithm, and is defined as formula (1).

$$E = \frac{1}{2} \sum_{S=1}^{S} (y_s - d_s) \tag{1}$$

In formula (1), E is the learning sample, y is the input vector component ordinal number, d is the output vector component ordinal number, s is the learning sample ordinal number.

A linear regression model introducing activation functions:

$$y(t) = \sum_{i=1}^{n} q_i(t)p_i + \omega(t)k \tag{2}$$

In the formula: n is the number of multi-peak walking nodes in the memory file of the control system; $y(t)$ is the expected output value; $q_i(t)$ is the calculated regression factor in the walking section of the system; p_i is the connection weight between the walking nodes; $\omega(t)$ is the calculated residual. According to the model combined with the change parameter k, the walking center node is obtained:

The basic framework of artificial intelligence algorithm is derived from formula (1), which is defined as formula (2).

$$\begin{cases} W_{KI} = \dfrac{W_{KI} + \alpha W_{KI}}{y(t)} \ldots\ldots\ldots\ldots i = 1, 2, \ldots, n \\ v_k = \dfrac{v_k + \beta v_k}{y(t)} \ldots\ldots\ldots\ldots\ldots k = 1, 2, \ldots, l \end{cases} \tag{3}$$

In formula (2), α and β are the learning rate respectively; W_{KI} represents the new network parameter, and the mathematical relationship between its input and output can be expressed as formula (4).

$$v_k = \frac{\partial E}{\partial v_k} = (y - d) \tag{4}$$

Formula (4) ∂E said in one of the first order ∂v_k said assume approximate fitting to get the output of the function and to determine if the final data is valid, you can to continue to receive data, and then to judge the data frame tail, and then if received the character frame is not the end, the need to continue to receive data, and then put into the data buffer. Because for GPS receiver, the received information is not necessarily complete, mainly based on the number of bytes. If the number of bytes is lower than the number of bytes standard, the data is incomplete and can not be processed. This is where the checksum comes in. At this point, the orchard plant protection robot walking control system software design part.

4 Experimental Analysis

After the completion of software and hardware design, the orchard plant protection robot also needs to build an overall software and hardware system for debugging and experiments, in order to test the performance of the system. The system designed in this paper is set as the experimental group, and the traditional robot walking control system is set as the control group.

4.1 Experimental Environment

In the experimental process, the hardware device is firstly built. The ARM controller sends PWM control signal to the driver, which controls the motor rotation. The motor is connected to the reducer, and then to the encoder through the flexible coupling device. Program was written according to the flow chart of the speed ring control software. In the program, the sampling frequency was set to _SOOHz, that is, the sampling time was 2 ms. The sampled data was stored in memory, and then the data in memory was read out to draw a curve with MATLAB for analysis.

First, execute the oscore command on the PC side to start the node manager; Second, Log on to the mobile robot computer through the PC SSH. Last, view the image on the PC.

The comparison between the walking control system of orchard plant protection robot based on artificial intelligence algorithm and the traditional robot walking control system (Fig. 4).

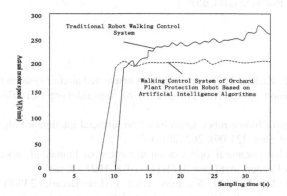

Fig. 4. Comparison of experimental results

Through experimental tests, it is found that the walking control system platform of orchard plant protection robot based on artificial intelligence algorithm achieves the expected effect of Motion control of the mechanical arm when it is controlled by keyboard, Leap Motion controller and PS2 wireless controller. From the perspective of operation effect and convenience, it is similar to the experimental effect of the walking subsystem. The operator is required to be familiar with the key function to achieve accurate operation by operating the mechanical arm movement through the keyboard.

This is the main problem existing in keyboard control, and its main advantage is good versatility. The main problems of Leap Motion control robot arm are poor operation stability, poor movement direction, and tiring operation time. Its advantage is that it can achieve contactless control effect. The PS2 controller is more stable and comfortable than the traditional robot walking control system.

5   Conclusion

Plant protection operation is the most labor-intensive and difficult to realize mechanized operation in the production of shed crops. However, with the development of China's urbanization, the situation of agricultural labor shortage appears, which directly leads to agriculture.

The rising cost of labor. Picking robot mainly solves the problem of labor substitution in the picking process. Its application can effectively reduce labor intensity and production cost, improve product quality and labor productivity. This paper is based on the research of the orchard plant protection robot walking control system based on the artificial intelligence algorithm. The work is mainly carried out in the following aspects: firstly, according to the research object, the targeted research and description of the parts to be used are carried out. Secondly, after the construction of the software and hardware platform, the walking subsystem is tested. The experimental results show that the control system platform has preliminarily achieved the expected control objectives.

Acknowledgements. Scientific and Technological Planning Projects of Colleges and Universities in Shandong Province (J16LB57).

References

1. Yang, S., Yang, X., Mo, J.: The application of unmanned aircraft systems to plant protection in China. Precision Agric. **19**(2), 278–292 (2018). https://doi.org/10.1007/s11119-017-9516-7
2. Dai, L.: Design of bionic robot action based on artificial intelligence algorithms. Electron. Technol. Softw. Eng. **131**(09), 262 (2018)
3. Anonymous: Gait parameter optimization algorithm of humanoid robot based on screw model. Pattern Recogn. Artif. Intell. **31**(11), 1018–1027 (2018)
4. Lu, Z., Lin, G., Lin, B., et al.: Sci. Educ. Educ. J. (Late Decade) **339**(5), 79–80 (2018)
5. Zhou, Z., Li, Y., Yan, R.: Design of orchard picking robot based on OFDM channel and remote monitoring. Res. Agric. Mech. **39**(3), 229–233 (2017)
6. Mao, F., Xu, H., Wang, Y., et al.: Orchard mobile robot path recognition system based on VB/Matlab. Comput. Eng. **43**(12), 315–320 (2017)
7. Leng, W.: Digital video optical fiber transmission system for orchard remote robot based on DWDM. Res. Agric. Mech. **41**(4), 215–218 (2019)
8. Tu, L., Li, L., Lin, G.: Study on global optimal path planning for orchard mobile robots. J. Nanhua Univ. (Nat. Sci. Edn.) **31**(4), 71–74 (2017)
9. Li, B., Li, H., Ge, Z.: Design of autonomous navigation orchard robot based on adaptive genetic algorithm and spline curve. Agric. Mech. Res. **39**(2), 47–51 (2017)

10. Liu, S., Lu, M., Li, H., et al.: Prediction of gene expression patterns with generalized linear regression model. Front. Genet. **10**, 120 (2019)
11. Luo, L., Zou, X., Lu, T., et al.: Virtual simulation and prototype test of picking robot operation behavior. J. Agric. Mach. **49**(5), 34–42 (2018)
12. Liu, S., Bai, W., Srivastava, G., Machado, J.A.T.: Property of self-similarity between baseband and modulated signals. Mobile Netw. Appl. **25**(4), 1537–1547 (2019). https://doi.org/10.1007/s11036-019-01358-9
13. Shuai, L., Gelan, Y.: Advanced Hybrid Information Processing, pp. 1–594. Springer, New York
14. Huang, L., Zhang, K., Hu, W., et al.: Trajectory optimisation design of robot based on artificial intelligence algorithm. Int. J. Wireless Mobile Comput. **16**(1), 35 (2019)

Research on Real-Time Monitoring Method of Communication Network Blocking Based on Cloud Computing

Wei-yan Li[1], Kui Gao[1], Yu Li[1], and Pei-ying Wang[2(✉)]

[1] College of Information Science and Engineering,
Shandong Agricultural University, Taian 271000, China
liweiyan254@sina.com
[2] Tianhe College of Guangdong Polytechnical Normal University,
Guangzhou 510540, China
wangpeiying258@sina.com

Abstract. Aiming at the problems that the traditional method has long response time to communication network congestion monitoring and the detection effect is not ideal, a real-time monitoring method based on cloud computing for communication network blocking is proposed. Firstly, the communication network monitoring point is established, and the communication data collection process is completed by the radio-frequency receiver. On this basis, the real-time traffic calculation of the collected data is performed to determine the existence of abnormal blocking status in the communication network link, and the precise positioning of the blocking point is obtained. The information thus generates an alarm message to obtain a monitoring result. The real-time and accuracy of the monitoring method are analyzed experimentally. It is found that the monitoring method can control the delay time within 0.2 s and the monitoring error rate is low. It can be seen that the monitoring algorithm has high performance.

Keywords: Cloud computing · Telecommunication · Network congestion · Real-time monitoring

1 Introduction

In the process of communication network construction, there are often various interferences that affect the performance of the network; if the interference problem is not cleared, the network optimization work in the network construction is difficult to carry out. Among these interference problems, blocking interference is a systematic, whole network and serious interference problem. If it is not solved, network construction will not be possible. Blocking interference is that when the strong interference signal and useful signal are added to the receiver at the same time, the nonlinear components of the receiver link will be saturated, resulting in nonlinear distortion and blocking the receiver, which is beyond the working range of the amplifier and mixer, making the receiver unable to demodulate normally, interfering with the work of the receiver, resulting in the failure to report the bottom noise level of the communication network normally. When the signal is too strong, the useful signal will also produce amplitude

S. Liu and L. Xia (Eds.): ADHIP 2020, LNICST 347, pp. 88–97, 2021.
https://doi.org/10.1007/978-3-030-67871-5_9

compression, and will block when it is serious. The main reason for blocking is the nonlinearity of the device, especially the multi-step products of intermodulation and intermodulation. At the same time, the dynamic range limitation of the receiver will also cause blocking interference. Blocking will cause the receiver to fail to work properly, and long-term blocking may also cause permanent performance degradation of the receiver.

So, how to confirm that blocking interference does occur? There are the following steps to confirm:

(1) Frequency shift. According to the principle of blocking interference, for the RF terminal whose filter type is if filter, the strong interference signal can be excluded from RF reception by frequency shift (changing the center frequency point of RF reception), so that the signal strength falling into RF reception is less than - 40 dbm. If the RTWP of communication network is reduced, it can be determined that the interference network is blocked.

(2) Att (VGA) attenuation. For the single-mode station suspected to be jammed, if the degree of interference is not very serious, ATT or VGA attenuation can be used to determine whether the communication network is jammed by the interference signal suppression ability. If in the process of att (VGA) attenuation, the RTWP of the communication network has a sudden change, it means that the cell is blocked.

(3) Add wave trap. The notch filter (narrow-band filter, also known as band stop filter) can be customized to attenuate the strong interference signal to a reasonable degree according to the actual situation on site, so that the total received signal in the RF reception is lower than the threshold value of blocking interference, so as to judge whether the disturbed signal network is blocked.

(4) Turn off the interference source. This method is the simplest way to judge. If the network RTWP returns to normal after the suspected interference source (the interference signal does not fall into the wireless network receiving), it can prove that the communication network is blocked. It can be seen from the above content that for the severity of blocking interference, corresponding solutions can be adopted: according to the frequency shift, att (VGA) attenuation, the installation of notch filter, and direct processing of interference source in order. According to the solution of network congestion, this paper introduces cloud computing technology to realize the real-time monitoring of communication network congestion.

In general, cloud computing refers to a business computing model. It distributes computing tasks across resource pools of large numbers of computers, enabling applications to acquire computing power, storage space, and information services as needed. In short, it provides on-demand, scalable, and affordable computing services over the network. Anyone can share and retrieve resources in the world of network communication. The network uses physical links to connect isolated workstations or hosts to form a data link for resource sharing and communication purposes [1–3]. Network communication connects various isolated devices through the network, and realizes communication between people, people and computers, computers and computers through information exchange. However, with the complication of information in the communication network and the large-scale data volume, some network problems occur in the communication network, such as communication network congestion [4]. Communication network congestion is a state of continuous overloaded network. The network transmission performance is degraded due to the limited resources of the storage and forwarding nodes.

As far as the architecture of the Internet is concerned, the occurrence of congestion is an inherent attribute. However, if the blocking condition has a certain persistence, when the cache space is exhausted, the router only discards the packet to ensure that the network avoids the lockup condition. Generally speaking, there are many reasons for the communication network blocking, including the insufficient bandwidth or overload of the server where the target website is located, the network cable problem, the existence of a loop in the network, and the like, and eventually the network speed is slow. In the big environment of cloud computing, in order to maintain the normal communication of the network and avoid the negative impact of congestion on the network, some countermeasures need to be taken to maintain the normal operation order of the communication network.

2   Design of Real-Time Monitoring Method for Communication Network Blocking

2.1   Set Communication Network Monitoring Point

First, a network monitoring device needs to be set up at a certain node in the network to obtain performance parameter data of all links related to this node. Therefore, in order to obtain performance data of all links, it can generally be implemented by setting network monitoring devices on some switching nodes. Therefore, it is necessary to consider which nodes are set up with network monitoring devices, and it is possible to obtain performance data of all links and enable monitoring. The minimum number of devices. Multiple network monitoring devices can be divided into multiple network monitoring areas. One network monitoring device can logically belong to multiple network monitoring areas, that is, multiple network performance monitoring services for multiple users, thereby reducing the number of devices and reducing construction costs [5, 6]. This requires reasonable setting of network monitoring points and rational division and management of network monitoring areas. In order to fully realize the monitoring function of the monitoring point, the structure of the communication network monitoring point is divided into a local area network part, a server system, various workstations, a terminal server, a protocol converter and a remote network device. The structure of the monitoring point is shown in Fig. 1.

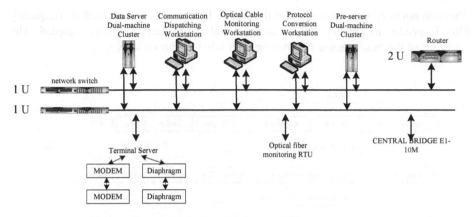

Fig. 1. Schematic diagram of the monitoring point structure

The dual Ethernet LAN device adopts the networking mode of 100M dual Ethernet to ensure the reliability of the system. The monitoring point is configured with dual network switches. Each server and workstation are equipped with dual network interfaces to ensure the failure of any network switch. The following does not affect the function of the system. Set the two network segments, the internal network segment and the external network segment through the bridge function of the network switch to ensure that the data of the internal and external network segments do not interfere with each other. The function of connecting to the remote LAN is realized through the router. The dual-master server cluster adopts the server working mode of the dual-master server cluster, establishes the link between the two servers, and makes the two servers into a cluster working mode. The two servers are hot backups to each other, and the system guarantees that it encounters on any one server. The system function is not affected after the fault.

Add high-performance input and output peripherals such as laser printers to share with online users. The data of each substation is sent to the central station through the bridge. Each bridge is configured with a dual network port to realize the connection with the dual network port [7–9]. At the same time, the dual WAN port is provided to realize the dual E1 channel of the main and standby, and the automatic switching of the dual channel is ensured. In addition, a protocol conversion processor is required to input data from various communication monitoring subsystems of different protocols.

2.2 RF Receiver Collects Communication Network Data

Data collection and storage are the basis for performance monitoring, enabling the acquisition and storage of raw information. With the RF receiver principle, the RF link has an adjustable digital attenuator and VGA to ensure sufficient dynamics to meet the in-band blocking specification. However, if the blocking signal is far from the operating frequency, it may fall within the proximity of the ADC or other Nyquist sampling bandwidth and be sampled by the ADC [10]. If the interference frequency is sampled and falls into the useful signal, causing the in-band signal to alias, the RF and digital filters do not have any suppression of the signal. In this case, an intermediate frequency filter is needed to prevent the unwanted signals from being digitally sampled. The structure of the performance monitoring data table is shown in Fig. 2.

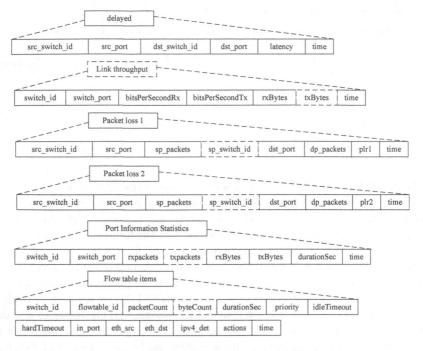

Fig. 2. Communication data acquisition chain diagram

The establishment of the above table and the connection to the database in the controller are implemented by the JDBC interface. Combined with the speed, low cost and open source of the MySQL database, using JDBC as the interface of MySQL has become a common usage. JDBC is a standard API for the Java language to interact with databases. Java applications can access the database directly through this standard interface. JDBC also supports the ability to access different types of databases in a uniform manner. Create a real-time performance data collection thread and store the collected data. The data acquisition and storage implementation relationship module is mainly divided into three implementation packages: the database connection package CLOS.sql, the main function is to realize the connection and management functions of the database; JavaBean package colSto.bean mainly defines the data structure of various performance data tables. The business operation package cloSto.op mainly completes the data receiving and storage functions. When the performance monitoring module obtains the performance parameters, it calls the corresponding class in the business operation package. Data storage operations to enable distributed storage of real-time collected data.

2.3 Communication Network Real-Time Flow Calculation

The data traffic collected in the communication network is calculated in real time, and the real-time computing structure is compared with the maximum storage space and the maximum load of the bandwidth capacity, and it is verified whether the data traffic can smoothly reach the specified storage space through the bandwidth for data storage [11–14]. The calculation formula for the real-time flow calculation of the communication network is as follows:

$$
Q = \min \left\{ \sum_{k \in I_l} \frac{F_l}{\sum_{k \in I_l} P_{lk} y_{lk} - F_l} + G \sum_{\substack{l \in L \\ k \in I_l}} (S_{lk} y_{lk} + d_i m_{lk}) + V \sum_{\substack{l \in L \\ k \in I_l}} (C_{lk} F_l y_{lk}) \right\} \tag{1}
$$

In formula 1, F_l represents the local data traffic on the communication network link; Q is the total data traffic; I_l represents the candidate link model indicator set of the first link; P_{lk} represents the model index of the first link selected as Link capacity at k; S_{lk} represents the candidate route set of the node of the first link; C_{lk} represents the line variable coefficient when the model index of the first link is selected as k; d_i is the length of the first link; m_{lk} represents the communication link I Packet arrival rate; G is a fixed weighting factor; L is a set of all links in the communication network; the constraints in Eq. 1 include:

$$
\sum_{k \in S_p} x_p = 1 (\forall p \in \Pi) \tag{2}
$$

$$
\sum_{k \in I_l} y_{lk} = 1 (\forall l \in L) \tag{3}
$$

Where Π represents the set of all communication node pairs in the network; x_p is the optimization variable and takes a value of 1 when the route is selected as the communication route with other related node pairs, otherwise 0. Where y_{lk} represents the optimization variable. When the model index of the first link is selected as k, the value is 1; otherwise, it is 0; according to the formula, the real-time communication traffic of a certain link can be obtained, and the maximum limit and bandwidth of the storage space can be obtained. The maximum limit of capacity is compared to determine the flow of the flow under normal conditions.

2.4 Abnormal Blocking Recognition and Feature Analysis

In addition to the communication data traffic in the communication network exceeding the maximum bandwidth of the communication network can cause network congestion, data anomalies are also another cause of network congestion. Therefore, it is also very important for the identification of communication network anomalies and system inspections. In the real-time monitoring of communication network blocking, in order to realize real-time monitoring of abnormal data nodes, nodes need to be automatically patrolled. The node monitoring program reads the IP address of the device from the database, and sends a PING command periodically and cyclically. Record the returned result to the working state of the node device. If the returned result times out, the node device is not working properly. Once the network traffic is abnormal, the IP address and port distribution will change. If the network configuration error occurs, the original IP address and the destination IP address will increase, causing the host's packets to increase sharply. According to this feature, the network traffic matrix method is used to analyze the dispersion of traffic distribution characteristics. Suppose the traffic characteristic is A, the total number of samples is B, the number of samples selected is C, and the number of occurrences of a particular traffic characteristic i is n_i. Therefore, the traffic characteristic sample can be selected as:

$$F(x) = -\sum_{i=1}^{C} \left(\frac{n_i}{B}\right) \log_2 \left(\frac{n_i}{B}\right) \qquad (4)$$

If all the selected samples have the same result, then $F(x) = 0$; if all the selected samples have a large degree of dispersion, then $F(x) = \log_2 C$ can describe the abnormal behavior of different flow characteristics, and then perform packet capture processing.

When the traffic monitoring system captures the transmitted Ethernet frame, it needs to parse the data packet first, and then extract the related data, and store the extraction result in the database, which is convenient for real-time analysis of network abnormal traffic. In order to ensure the accuracy of packet capture, the packet capture function interacts with the hardware in real time. In order to ensure the accuracy of the system capture, you need to read the configuration file first; then establish a connection with the database, register the ODBC data source, use the OpenDataBase() function in the Python script file to connect to the database; configure the capture driver to create various types. Timers and threads, real-time monitoring of abnormal traffic using timers; Finally, create a real-time monitoring server, and call the Bind() function to obtain the IP address of the monitoring server from the configuration file, thereby completing the packet capture behavior. The abnormal traffic data packet can be captured in real time, and the real-time monitoring display function can be added, so that the user can view the packet capture result in real time, and then locate the abnormal node.

3   Experiment Analysis

In order to verify the feasibility of the research on the real-time monitoring method of communication network blocking, experiments were conducted. The 50 sets of data in a certain period of time are selected as experimental objects, and the experimental data of the normal state is obtained according to the historical records of previous years, and standardized processing is performed. First of all, set up the experimental environment. Because of the limitations of the lab equipment, it is necessary to use the Mininet to establish a communication network topology in the virtual machine and connect to the remote controller to realize the construction of the communication network environment, and then establish a connection with the remote database for data access. The real-time monitoring error and communication delay time of the communication network are taken as experimental targets. The traditional monitoring method is compared with the real-time traffic monitoring of the communication network. The result is used as the basis for judging the feasibility of the method. The reasons for the communication delay include monitoring the substation measurement and processing delay, data transmission delay, information propagation delay, delay generated by the digital transmission equipment, and delay of the pre-processing of the primary station and the data uploading server. Through the experiment, the traditional monitoring method and the designed real-time monitoring method are compared, and the comparison results are shown in Fig. 3.

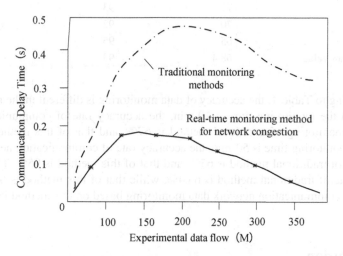

Fig. 3. Communication delay time comparison results

The delay time of the communication network blocking method for real-time monitoring directly reflects the real-time performance of the method. It can be seen from the comparison of the experimental results in the figure that the communication delay time of the traditional monitoring method will be extended with the increase of the data amount. Although there is a clear trend of shortening the delay when the quantity reaches 280M, the delay time is always higher than 0.3 s. In contrast, the communication network blocking real-time monitoring method uses the real-time data acquisition and calculation method, so the experimental results show an ideal delay result, the delay time is always controlled within 0.2 s, and when the data flow After more than 250M, it shows a clear downward trend, which fully reflects the real-time nature of the monitoring method. In addition, the accuracy of the monitoring method monitoring results is also very important. The results of the alarm information obtained by the real-time monitoring method designed by the experiment can be accurately located to the position where the blockage occurs and the blocking value is calculated, which is convenient for timely blocking accidents. Processing, it can be seen that the method has low monitoring error and high accuracy.

In order to further verify the monitoring accuracy of this method, the traditional method and this method are used to verify the monitoring accuracy. The results are shown in Table 1.

Table 1. Monitoring accuracy of communication network data

Monitoring the time/min	Monitoring accuracy of communication network data/%	
	The traditional method	The method of this paper
10	67	89
20	69	96
30	71	93
40	70	92
50	65	95
Mean value	68.4	93

According to Table 1, the accuracy of data monitoring is different in the monitoring time. When the monitoring time is 10 min, the accuracy rate of communication network data monitoring of traditional method is 67%, and that of this method is 89%. When the monitoring time is 50 min, the accuracy rate of communication network data monitoring of traditional method is 65%, and that of this method is 95%. The average accuracy rate of traditional method is 68.4%, while that of this method is 68.4%. The accuracy of communication network data monitoring based on this method is obviously higher.

4 Conclusion

In summary, in the cloud computing environment, in order to ensure the stability of the computer communication network operation, it is necessary to ensure that the network traffic is always in a normal state. Through the real-time monitoring method of a communication network, the real-time running status of the entire communication network is grasped, and the blocking problem of the network is detected in time to facilitate the daily management and maintenance of the network staff. In the design process of the monitoring method, it is found that although the method has high monitoring accuracy, the long-term stable operation of the method has yet to be studied, and a high-performance monitoring method is developed in the future to support real-time monitoring of the communication network.

References

1. Zhao, A.: Research on simulation of blind signal real-time separation in network communication. Comput. Simul. **26**(2), 31–35 (2018)
2. Anonymous: Real-time generation simulation of candidate communication network information in emergency. Comput. Simul. **34**(12), 165–168 (2017)
3. Wang, D., Tian, A.: Real-time monitoring simulation of safety signals in high-speed railway network. Comput. Simul. **38**(4), 25–28 (2018)
4. Lu, N., Lu, K.: Cooperative scheduling method for control and communication in real-time Ethernet system. Comput. Eng. Appl. **53**(7), 15–20 (2017)
5. Liu, H.: Research on control method of blocking jamming in HF communication system. Digit. Technol. Appl. **37**(1), 29–30 (2019)
6. Jiao, Y., Tian, F., Shi, S., Liu, J., Liu, H.: Multipath routing based congestion control strategy for LEO satellite networks. Electron. Des. Eng. **26**(18), 119–123+128 (2018)
7. Liu, S., Lu, M., Li, H., et al.: Prediction of gene expression patterns with generalized linear regression model. Front. Genet. **10**, 120 (2019)
8. Lv, J.: Design and implementation of equalization controller in network communication system. J. Henan Inst. Sci. Technol. (Nat. Sci. Ed.) **47**(2), 60–64 (2019)
9. Huang, Q., et al.: Coordinating SNOP control and AC-line switching for congestion management of hybrid urban grid. Adv. Technol. Electr. Eng. Energy **38**(2), 55–62 (2019)
10. Liu, P., Cai, Y., Lu, G.: Space environment data transfer system based on BBR congestion control algorithm. Chin. J. Space Sci. **39**(1), 117–123 (2019)
11. Hu, L., Yang, J., He, Y., et al.: Urban traffic congestion radiation model and damage caused to service capacity of road network. China J. Highw. Transport **32**(3), 149–158 (2019)
12. Shuai, L., Gelan, Y.: Advanced Hybrid Information Processing, pp. 1–594. Springer, New York (2019)https://doi.org/10.1007/978-3-030-19086-6
13. Zhang, J.: Analysis of TCP packet dropping behaviors in data center. Intell. Comput. Appl. **8**(6), 172–173+176 (2018)
14. Liu, G., Liu, S., Khan, M., et al.: Object tracking in vary lighting conditions for fog based intelligent surveillance of public spaces. IEEE Access **6**, 29283–29296 (2018)

Research on Voluntary Intelligent Reporting System of College Entrance Examination Based on Big Data Technology

Shu-xin Guo[1] and Li Lin[2(✉)]

[1] Jilin University of Finance and Economics, Changchun 130117, China
Cqjtzyxy123@163.com
[2] School of Computer Engineering, Jimei University, Xiamen 361021, China
xd220210@163.com

Abstract. The college entrance examination application is a complex system project, which needs to collect many kinds of information. Aiming at the deficiency of the system research based on the analysis of the domestic mainstream platform, the reference system and Sina simulation system of college entrance examination application, the intelligent application system of college entrance examination application is designed based on big data technology. Considering the scores of examinees, the enrollment plan of colleges and universities, the enthusiasm of application, the prospect of professional development and other factors, the hardware structure of the intelligent filling system for college entrance examination is constructed. Through big data analysis and data mining, a large amount of real and valuable information for college entrance examination filling can be provided for the majority of examinees. It can be seen from the experimental verification results that the system fills in accurate results and has an ideal filling effect, which helps the candidates to apply for the ideal school and improve the admission rate.

Keywords: Big data technology · College entrance examination · Intelligent filling · Data mining

1 Introduction

College entrance examination application is a complex system engineering, which needs to collect many kinds of information, and comprehensively consider such factors as examinee score, college enrollment plan, enrollment enthusiasm, professional development prospect, examinee's personal interest and family situation [1]. Today, when the mobile Internet is highly developed, the Internet is full of various types of information about colleges, universities, majors, admission scores, etc. How to identify true and valuable information in the massive application information to suppress the troubles for the majority of candidates and parents [2]. Due to the lack of information and improper choice of information in college entrance examination, it is common for examinees to fail in high scores and get low marks. It is difficult for examinees and parents to evaluate accurately, and the phenomenon of high score falling out of the list

© ICST Institute for Computer Sciences, Social Informatics and Telecommunications Engineering 2021
Published by Springer Nature Switzerland AG 2021. All Rights Reserved
S. Liu and L. Xia (Eds.): ADHIP 2020, LNICST 347, pp. 98–111, 2021.
https://doi.org/10.1007/978-3-030-67871-5_10

and low score occurs every day [3]. Therefore, examinees and parents are in urgent need of comprehensive guidance on the application of college entrance examination.

The college entrance examination voluntary reference system based on college entrance examination scores or rankings is generally based on the candidates' own scores or rankings, combined with certain directly related information such as the admission scores of colleges and universities in a certain year or years. Candidates recommend some schools or majors [4]. At present, there are many such systems in China, such as the "sunshine college entrance examination" information platform of the Ministry of education, which is developed by the national college student information consultation and employment guidance center of the Ministry of education, the comprehensive reference system of college entrance examination filling and submitting, the Sina college entrance examination simulation filling and submitting system launched by sina.com, the reference system of college entrance examination filling and submitting launched by China Education online, etc. [5]. Some of these systems also provide historical information such as admission scores, enrollment numbers, and employment status in previous years, but they are usually limited to simple data query and statistics. Without in-depth data analysis function, it is difficult to find knowledge, and the laws hidden behind the data can not fundamentally solve the blindness of candidates when filling in the application form.

Based on the analysis of the deficiencies of the mainstream college entrance examination volunteer filling platform in China, this paper designs an intelligent college entrance examination volunteer filling system based on big data technology. The hardware structure of the system includes on-line analysis and processing server, data warehouse server and enrollment information data mining system. On the basis of hardware design, through functional module division, nearest neighbor search based on big data technology, database design Improve the data and recommendation results generation, design system software, so as to complete the system design, solve the problem of unclear candidates' voluntary filling, collect the data involved in the new college entrance examination, and calculate one or more volunteer filling candidate schemes recommended to examinees through big data combining the candidates' filling willingness and College entrance examination scores.

2 System Hardware Structure Design

The hardware structure of the system is as shown in Fig. 1. In terms of program platform, SSH framework of Java EE platform is adopted.

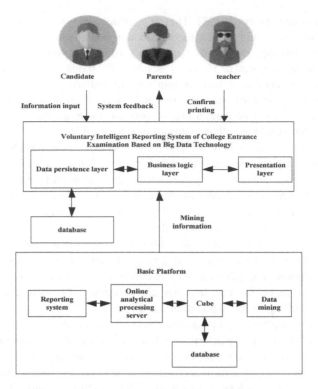

Fig. 1. System hardware structure

In terms of data storage, the SQL Server 2008 platform which has been tested by the enrollment data mining system is adopted. The system adopts a hierarchical structure, from the bottom to the top: basic data platform layer, SSH framework layer and function module layer [6].

2.1 Online Analytical Processing Server

Online analytical processing server is a fast software technology that shares multidimensional information, accesses and analyzes online data for specific problems. The online analytical processing server system is the most important application of the data warehouse system. It is specifically designed to support complex analytical operations, focusing on decision support for decision makers and senior management personnel. It can execute a large amount of data quickly and flexibly according to the requirements of analysts' complex query processing, and provide the query results to decision makers in an intuitive and understandable format [7–9]. The relationship between data warehouse and server is complementary. Modern server system is generally based on data warehouse, that is to extract a subset of detailed data from the data warehouse and store it in the server memory through necessary aggregation for front-end analysis tools to read [10, 11].

The server presents a multi-dimensional view to the user:

Dimension: it is a specific angle for people to observe data, and a kind of attribute when considering problems. Attribute sets form a dimension (time dimension, geographical dimension, etc.);

Dimension level: Observing a specific angle of big data (i.e. a dimension), there can also be various description aspects with different levels of detail (time dimension: date, month, quarter, year);

Member of dimension: a value of dimension, which is the description of the position of data item in a dimension. ("month day of a year" is a description of the position in the time dimension);

Metric: the value of multi-dimensional array;

The basic multidimensional analysis operations of the server include drilling, slicing and slicing, and rotation.

Drilling: It is to change the level of dimension and transform the granularity of analysis. It includes drill down and drill up/roll up. Drilling up is to summarize low-level detail data into high-level summary data on a certain dimension, or reduce the number of dimensions; while drilling up is the opposite, it drills down from the summary data to the detail data to observe or add new dimensions.

Slicing and slicing: it is concerned about the distribution of measurement data in the remaining dimension after selecting values in some dimensions. If there are only two remaining dimensions, they are slices; if there are three or more dimensions, they are slices.

Rotation: is to change the direction of the dimension, that is, rearrange the placement of the dimension in the table.

2.2 Data Warehouse Server

The warehouse control database contains the control tables necessary to store the metadata of the data warehouse center. In the updated version of the warehouse center, the warehouse control database must be a UTF-8 database. This requirement provides extended language support for the storage data warehouse center. If you try to log in to the storage data warehouse center using a database in a non-coding scheme format, the system will receive an error message that you cannot log in. The system can use the warehouse control database management tool to migrate metadata from the specified database to the new coding scheme database [12–14].

2.3 Design of Enrollment Information Data Mining System

The data mining system of enrollment information adopts a distributed system structure, which is based on the database of college entrance examination application, electronic data of general enrollment, application programming interface and large data files provided by some college network applications, using analysis, prediction, association rules, clustering and other mining methods, from a large number of incomplete and fuzzy practical application data, the relationship between colleges and majors hidden in it is found [15]. Analysis, statistics and reasoning of effective information such as school relationship, historical admission score line, enrollment plan, etc. can provide predictive suggestions for candidates to fill in the report. The architecture of enrollment information data mining system is shown in Fig. 2.

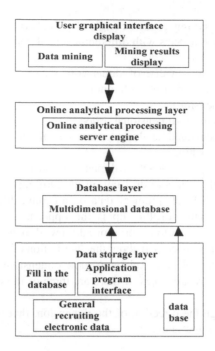

Fig. 2. Enrollment information data mining system

The data mining system of college enrollment information consists of front-end user interface, data preprocessing, data mining algorithm and other modules. The front-end user interface module mainly completes the interaction between the user and the system, and displays the mining results in easy-to-understand forms such as graphics and tables. The data preprocessing module mainly completes the extraction, cleaning, filtering and integration of a large number of complex redundant data in the data storage layer, and provides high-quality rule data for the data mining algorithm module.

3   System Development

3.1   Functional Module Division

The design goal of the college entrance examination voluntary intelligent reporting system is to provide candidates and parents with more accurate guidance for voluntary reporting. In addition, the system also needs to have security, stability and scalability. Therefore, the system is based on big data technology, and realizes six functional modules: human-computer interaction module, user management module, authority control module, voluntary management module, interest preference management module, voluntary recommendation module.

(1) Human-computer interaction module

It mainly provides convenient and beautiful user interface, and has friendly prompt for any operation of the user. At the front desk of the system, JavaScript dynamic scripting technology is adopted to realize the page refresh operation, which can improve the user experience. The purpose of human-computer interaction module is to let users use the filling consultation system at the lowest cost, and obtain good experience effect.

(2) User management module

User management module is the main carrier of user registration and login system, which mainly includes user registration, adding and modifying basic personal information and details, password management, system login, etc. User management module is the main carrier of user registration and login system, which mainly includes user registration, adding and modifying basic personal information and details, password management, system login, etc.

(3) Authority control module:

The permission control module belongs to the background function, which complements the user management. This module is mainly used for the security control of the system. It can realize the system to judge the authority of the currently logged-in user, which can enable legitimate users to browse the appropriate content. The allocation of user rights can be managed in the background.

(4) Volunteer management module

Volunteer management belongs to the background function, mainly the addition, deletion and modification of colleges and majors. This module is used in the voluntary recommendation module, which belongs to the bottom data support. It caches the mining data at the database level and improves the operation efficiency of the system in the voluntary reporting guidance.

(5) Interest preference management module:

This module is a psychological test application module, which mainly includes candidates' professional preference, regional tendency, cost burden and other aspects. Among them, this module is also related to the user's "detailed information" in the user management module, including future, urban or rural, whether ethnic minorities, etc.

3.2 Nearest Neighbor Search Based on Big Data Technology

3.2.1 User Data Standardization
In order to facilitate data processing, according to the candidate's score information and intention information entered by the candidate, after the user attributes are determined, the data is standardized, and min-max standardization is used to standardize the data, as shown in Eq. 1:

$$s' = \frac{s - s_{min}}{s_{max} - s_{min}} \tag{1}$$

In formula (1): s represents the original data of candidates; s_{max}, s_{min} represents the maximum and minimum values of attributes respectively.

3.2.2   Calculate User Proximity

Nearest neighbor search is to calculate the similarity between candidates and their preferences based on their attributes. On the basis of standardized candidate attribute data, the Pearson correlation coefficient is used to perform nearest neighbor search, and the distance between candidates is calculated to express the similar proximity between candidates, as shown in Eq. 2:

$$s = \frac{\sum_{iek} \left(t_{u,i} - \overline{t_u}\right) * \left(t_{v,i} - \overline{t_v}\right)}{\sqrt{\sum_{iek} \left(t_{u,i} - \overline{t_u}\right)^2 * \left(t_{v,i} - \overline{t_v}\right)^2}} \tag{2}$$

In formula (2): $t_{u,i}$ represents the i attribute value of user u; $t_{v,i}$ represents the i attribute value of user v; $\overline{t_u}$ represents the average value of all attributes of user u; $\overline{t_v}$ represents the average value of all attributes of user v.

3.3   Database Design

Database design plays an important role in website development and construction. A good website must be supported by a safe and high-performance database. In the process of candidates applying for a test, the safety, effectiveness, and real-time nature of the data are directly related to whether the candidates can be admitted to their favorite schools. Based on big data, the database of college entrance examination application is constructed by this system, which can obtain the latest enrollment information of colleges and majors through the data mining system of enrollment information. Table 1 and Table 2 are field definitions and descriptions of some data tables.

Table 1. College information table

Field name data type primary key remarks	Field name data type primary key remarks	Field name data type primary key remarks	Field name data type primary key remarks
COLLEGENAME	varchar (50)	Y	University name
PROVINCEID	varchar (50)	Y	Province
BATCHID	tinyint	N	Batch
COLLEGETYPEID	tinyint	N	Category
PLAYTYPEID	tinyint	N	Nature of plan
IS211	tinyint	N	Whether 211 institutions
IS985	tinyint	N	Whether 985 colleges
ISFOREIGN	tinyint	N	Chinese and foreign

Table 2. Batch control score

Field name data type whether primary key remarks	Field name data type whether primary key remarks	Field name data type whether primary key remarks	Field name data type whether primary key remarks
CLASS_ID	varchar (50)	Y	Family ID
BATCH_ID	varchar (50)	Y	Batch ID
YEAR	int	N	Year
CULTURE_SCORES	int	N	Cultural achievements
PROFESSION_SCORE	int	N	Professional performance
PROVINCE_ID	tinyint	N	Province ID
UP_NUM	int	N	Number of people on line

3.4 Complete Material

After entering the system, you need to first improve relevant information, such as personal details, interest preferences, etc. First click on "Personal Center" and enter the "Modify Details" page, you need to follow the prompts to improve all information, the most important of which is the candidate's urban and rural category (rural or urban), the previous category (fresh or past), whether it is a minority This information will be used as a reference when recommending volunteers. Then, click "Report Volunteer" to enter the preference setting mode.

The first step is to select the major categories of interest, which can be multiple choices;

The second step is to choose your own preferred area according to your geographical preference.

The third step is to choose the expense range that you can bear;

The fourth step is to choose your own reporting psychology, which is mainly divided into insurance or sprint types and strong emphasis on professional compliance. The default is to list all possibilities. After preference setting, enter the score setting and select science or liberal arts page. The relevant information set by the user is used as the basis for the voluntary recommendation of the system. If you modify the user profile, you can perform voluntary recommendation analysis in different states.

3.5 Recommendation Result Generation

The recommendation result is based on the admission of universities and professions of neighboring users. First of all, the neighboring users should be determined. Through the calculation and search of similar proximity, the user with the nearest neighbor of 0 is regarded as the nearest neighbor of the target user, and the neighbor is regarded as the neighbor. Corresponding institutions of successful admission are added to the recommendation set. Due to the different difficulty of the test paper each year, the value of the score as the reference value is not stable, so it is more strict and accurate to take the ranking of users as the main reference.

According to the ranking segment of the target user's score, select the x users with the smallest distance as neighbor users, and arrange the universities enrolled by the x users in descending order according to the number of people. And then add the recommendation set in turn, and then screen one by one according to the preferences of candidates. After screening, the recommendation content is put into the new recommendation set until the number of the new recommendation set reaches the target of 50, forming the final recommendation set.

4 Simulation System

4.1 System Login

First, enter the URL to enter the first page of the system, click "register new user" to register, after entering the relevant information, the system will automatically send a verification email to verify activation. Then use the approved account and password to log in, and enter the correct dynamic verification code, you can enter the guidance system.

4.2 System Voluntary Recommendation

At present, the system realizes three batches of voluntary recommendation, one batch of undergraduate, two batches of undergraduate and three batches of undergraduate. Each batch is divided into six volunteer. The function of voluntary recommendation includes shortcut mode and advanced mode. The difference is that the advanced mode can be modified for voluntary selection.

In any batch and any voluntary item, click "Add College Major". In the pop-up page box, you can choose "School Priority" and "Professional Priority". Major priority refers to the arrangement of the most likely major to be admitted, and then to the recommendation of colleges and universities with relevant majors, from high to low; college priority refers to the recommendation of the most likely college, and then to the selection of major. Regardless of whether it is an institution or a major, the interface of selecting an institution will prompt the institution's admission to similar conditions in the past three years.

4.3 System Simulation

Personal details: rural areas are selected for urban and rural areas.

Interest tendentiousness: professional tendentiousness chooses computer, network and technology, regional tendentiousness chooses southwest region, expense interval chooses 4001–10000, mental state system chooses sprint.

Relevant application requirements: The application category is science and the original score is 580 points. After inputting the data, the first volunteer is recommended with the priority of College recommendation. According to the input conditions, the recommended institutions are shown in the Table 3 below.

Table 3. Simulated voluntary recommendation table

School name	2019	2018	2017	Select
List of eligible schools for three consecutive years				
XX University	562.0	544.0	505.0	√
XX University	556.0	533.0	510.0	√
XX University	547.0	525.0	506.5	√
XX University	564.0	520.0	565.0	√
List of eligible schools for two consecutive years				
XX University	565.2	540.0		√
List of schools that meet the standard for one year in a row				
XX University	565.0	598.0	554.0	√

Then, you can choose a college as your first volunteer, and so on, you can choose a second volunteer or other batch of volunteers. Finally, it can print the guidance form for filling in the application form for the convenience of reference.

Design test cases and conduct system tests based on the test cases. First fill in the candidate's ranking, batches, grades of three public courses in Chinese, mathematics and foreign language, and 7 of 3 subjects and corresponding scores, and automatically generate total scores, as shown in Fig. 3.

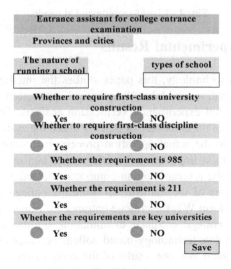

Fig. 3. Candidates' scores

Then, candidates fill in their willingness to report their intentions, including the excluded provinces, municipalities and autonomous regions, the nature and type of school running, whether they require first-class university construction and whether they require first-class discipline construction, and whether they require 985, 211 and key universities.

According to the data in Fig. 3, the intelligent recommendation platform for college entrance examination volunteer filling in is calculated by the server, and finally the recommendation result list is pushed to the Android client, as shown in Fig. 4.

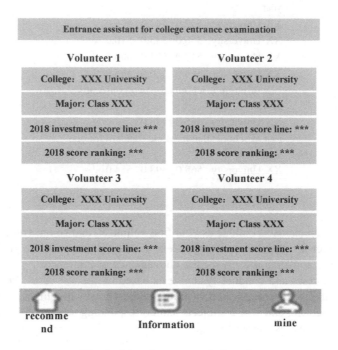

Fig. 4. List of recommended results

5 Analysis of Experimental Results

Based on the big data technology, this paper studies the intelligent filling system of college entrance examination, and takes the practical application of the display system as an example to carry out experimental verification analysis. The specific process of the example is to simulate the examinee or parents to use the system, and input relevant information according to the actual operation process, including account registration, login, personal information improvement, performance and other necessary information input, and display the relevant human-computer interaction screenshots.

Based on the analysis of the mainstream domestic college entrance examination voluntary reporting platform W1, the college entrance examination voluntary reference system W2, the Sina college entrance examination simulated voluntary reporting system W3 and the big data technology-based college entrance examination voluntary intelligent reporting system W4, the results of the comparison and analysis are shown in Fig. 5 As shown.

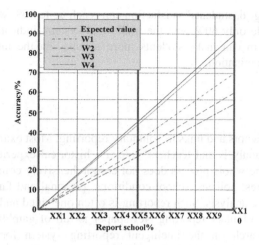

Fig. 5. Comparison and analysis of filling accuracy between the two systems

It can be seen from Fig. 5: The use of the mainstream domestic college entrance examination voluntary reporting platform W1, the college entrance examination reporting voluntary reference system W2, and the Sina college entrance examination simulation voluntary reporting system W3 lack in-depth data analysis functions, unable to find dynamic data, resulting in the system not reaching the actual predicted value Fundamentally solve the blindness of candidates when they fill in the report. Although the intelligent filling system of college entrance examination based on big data technology can't exactly match the actual predicted value straight line, it is also very close. Therefore, the intelligent filling system of college entrance examination based on big data technology has accurate and intelligent filling effect.

In order to further verify the effectiveness of the system, this paper compares and analyzes the student satisfaction of the system and the traditional system, and the results are shown in Fig. 6.

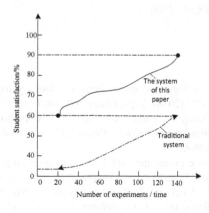

Fig. 6. Comparison of student satisfaction results

According to Fig. 6, students' satisfaction can reach up to 90% with the application of this system, while only 60% with the traditional system, indicating that the application of this system can make students more satisfied in the intelligent filling of college entrance examination.

6 Conclusion

There are unstable factors and links in voluntary reporting. Most examinees and parents rely on subjective analysis and teachers' relevant historical experience to make voluntary reporting. The whole process does not comprehensively combine the objective factors such as scores, policies, school conditions, personal and family conditions to make comprehensive analysis. Such reporting is often one-sided and subjective, which makes examinees' voluntary reporting Newspaper is full of gambling and blindness. Therefore, the research on the intelligent reporting system for college entrance examination based on big data technology is designed.

Through the preliminary trial operation of the project, it shows that the establishment of filling and consulting system based on enrollment data mining is of high feasibility and application value. Summary of the current work, although achieved some results, but there is still a lot of room for improvement, improve or create mining algorithm, in order to get better results of enrollment data mining, it may be necessary to do further research on the algorithm and propose a suitable algorithm.

Acknowledgements. 1. The "Thirteenth Five-Year Plan" Science and Technology Project of Jilin Province Education Department "Research on the Service System of Volunteer Filling for College Entrance Examination Based on Data Mining" (Project Contract Number: JJKH20180466KJ).

2. The 2017 General Planning Project of the "13th Five-Year Plan" of Educational Science in Jilin Province "Research and Practice of Senior High School Students' Career Planning from the Perspective of the New College Entrance Examination Reform" (Project approval number: GH170348).

3. The 2017 Higher Education Scientific Research Key (Self-funded) Project of Jilin Province Higher Education Society "Research on the Decision Model of College Students' Career Planning" (Project Number: JGJX2017C43).

References

1. Ying, Y.: Research on college students' information literacy based on big data. Cluster Comput. **22**(2), 3463–3470 (2018). https://doi.org/10.1007/s10586-018-2193-0
2. Zhang, J., Wang, L., Xing, L.: Large-scale medical examination scheduling technology based on intelligent optimization. J. Comb. Optim. **37**(1), 385–404 (2018). https://doi.org/10.1007/s10878-017-0246-6
3. Bu, F.: An efficient fuzzy c-means approach based on canonical polyadic decomposition for clustering big data in IoT. Future Gener. Comput. Syst. **88**, 675–682 (2018)
4. Jin, S., Peng, J., Xie, D.: A new MapReduce approach with dynamic fuzzy inference for big data classification problems. Int. J. Cogn. Inform. Nat. Intell. **12**(3), 40–54 (2018)

5. Nie, M., et al.: Advanced forecasting of career choices for college students based on campus big data. Front. Comput. Sci. **12**(3), 494–503 (2018). https://doi.org/10.1007/s11704-017-6498-6

6. Chu, S.C., Dao, T.K., Pan, J.S., et al.: Identifying correctness data scheme for aggregating data in cluster heads of wireless sensor network based on naive Bayes classification. EURASIP J. Wirel. Commun. Netw. **2020**(1), 1–15 (2020)

7. Liu, S., Bai, W., Liu, G., et al.: Parallel fractal compression method for big video data. Complexity **2018**, 2016976 (2018). https://doi.org/10.1155/2018/2016976

8. Kumar, P.M., Lokesh, S., Varatharajan, R., et al.: Cloud and IoT based disease prediction and diagnosis system for healthcare using Fuzzy neural classifier. Future Gener. Comput. Syst. **86**, 527–534 (2018)

9. Shuai, L., Weiling, B., Nianyin, Z., et al.: A fast fractal based compression for MRI images. IEEE Access **7**, 62412–62420 (2019)

10. Liu, Z., Hu, L., Wu, C., et al.: A novel process-based association rule approach through maximal frequent itemsets for big data processing. Future Gener. Comput. Syst. **81**, 414–424 (2018)

11. Hu, Z.W., Niu, X.B., Liu, B., et al.: The design and implementation of voluntary filling assistant system under the new college entrance examination reform. Intell. Comput. Appl. **009**(002), 175–179 (2019)

12. Hua, T.: Model for evaluating the classification modes of the China's college entrance examination with hesitant fuzzy information. Int. J. Knowl. Based Intell. Eng. Syst. **21**(4), 265–272 (2017)

13. Lu, M., Liu, S.: Nucleosome positioning based on generalized relative entropy. Soft. Comput. **23**(19), 9175–9188 (2018). https://doi.org/10.1007/s00500-018-3602-2

14. Fu, W., Liu, S., Srivastava, G.: Optimization of big data scheduling in social networks. Entropy **21**(9), 902 (2019)

15. Yao, Y., Zhang, Z., Cui, H., et al.: The influence of student abilities and high school on student growth: a case study of Chinese National College Entrance Exam. IEEE Access **7**, 148254–148264 (2019)

Design of Intelligent Recognition System for Orchard Spraying Robot Path Based on Adaptive Genetic Algorithm

Jie Gao[1(✉)] and Jia Wang[2]

[1] Nantong Polytechnic College, Nantong 226002, China
gaojie3568@sina.com
[2] Mechanical Engineering College, Yunnan Open University,
Kunming 650500, China

Abstract. The problem of excessive pesticide spraying in orchards has caused a great risk of food safety. Because the traditional fueljet robot path recognition system has the problem of low path recognition accuracy, an adaptive genetic algorithm based orchard spray robot path intelligent recognition system is designed to improve the path recognition accuracy. Firstly, the hardware design of the path intelligent identification system is carried out, including the power supply module, the main control board module, the path identification module, the motor drive module and the wireless communication module. Through the division of these modules, the path intelligent identification of the orchard spray robot is realized. Then the design of the path intelligent identification system software system is adopted, and KEIL uVision4 is used as the development environment. The adaptive genetic algorithm is used to accurately identify the path and determine the path control scheme and direction adjustment scheme to improve the path recognition accuracy. Finally, the simulation experiment is compared with the traditional spray robot path recognition system. The intelligent recognition system of orchard spray robot based on adaptive genetic algorithm has higher path recognition accuracy and shorter recognition time.

Keywords: Adaptive genetic algorithm · Orchard spray robot · Intelligent path recognition system

1 Introduction

In the one-year key areas of the national medium- and long-term science and technology development plan and its priority themes, the development of agriculture requires the precise operation and informationization of agriculture [1]. At the "Digital Agriculture" Construction Strategy Seminar, the Ministry of Science and Technology proposed to implement the implementation of digital agricultural science and technology action with "precision agriculture" and "smart agriculture" as the entry point [2]. It can be seen that the implementation of precision agriculture, the widespread application of intelligent agricultural machinery, and the improvement of resource utilization and economic benefits will be the inevitable trend of agricultural development in this century [3]. Due to the long period of biological control and ecological

© ICST Institute for Computer Sciences, Social Informatics and Telecommunications Engineering 2021
Published by Springer Nature Switzerland AG 2021. All Rights Reserved
S. Liu and L. Xia (Eds.): ADHIP 2020, LNICST 347, pp. 112–123, 2021.
https://doi.org/10.1007/978-3-030-67871-5_11

control, weeds and pests and diseases in China's agriculture are still controlled mainly by chemical control [4]. China's pesticide and fertilizer production and application volume rank first in the world. The use per unit area is 2.6 times that of the United States, but the utilization rate of pesticides is only about 30%, the utilization rate of nitrogen fertilizer is only 30% to 35%, the utilization rate of phosphate fertilizer is only 10% to 20%, and the utilization rate of potassium fertilizer is only 35% to 50%. The long-term use and low utilization rate of chemical pesticides have brought serious ecological and agricultural product safety problems such as environmental pollution, chemical pesticide residue exceeding standards and weed resistance. In particular, the problem of excessive pesticide spraying in orchards has caused a great risk of fruit eating safety [5]. Therefore, based on the adaptive genetic algorithm, the path intelligent recognition system is designed for the orchard spraying robot. The orchard spraying robot uses vision sensor instead of human eye to acquire the video information of surrounding environment and convert it into data matrix. The computer replaces human brain to process and analyze the image in software and hardware, so as to realize the task of precise spraying fixed-point variable [6].

2 Hardware Design of Path Intelligent Identification System

2.1 Hardware System Framework Design

The intelligent identification system of the orchard spray robot path uses the idea of modular design in hardware system design. Each part is designed to be controlled by a special structure. The hardware system framework of the orchard spray robot path intelligent identification system is shown in Fig. 1.

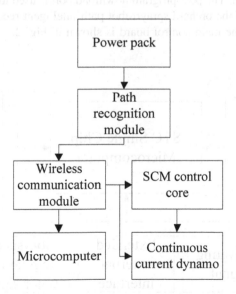

Fig. 1. Orchard spray robot path intelligent identification system hardware system framework

2.2   Power Supply Module

In the power supply module, each hardware component requires a different voltage to drive, and three different supply voltages need to be provided from the power distribution. The power supply is supplied to the main control board and signal processing board respectively, and the motor driving part is used to drive various motors in the intelligent path recognition system of orchard spraying robot. At the same time, for the anti-interference needs, the grounding pole of the battery should be designed separately.

2.3   Main Control Board Module

In the control circuit, the STC12C5A60S2 single-chip microcomputer is selected as the core to form the main control board circuit, which is used to complete the signal processing and control functions of the intelligent identification system of the entire orchard spray robot path. In terms of output, it can complete two types of adjustable speed motor control output and non-speed control, which can control the control of two drive wheel DC geared motors; at the same time, it has an expandable port to facilitate later function expansion. The series of single-chip microcomputers support the program mutual transmission function and provide the basic support for the improvement of the running performance of the Orchard Spraying Robot Path Intelligent Recognition System. The main control board uses a 40-pin MCU with a total of 35 I/O interfaces. Among them, PO, P2 and P4.6 are used for input, a total of 17 ports, P1 and P3 are signals. The output part will be used for device control in the future. The main control circuit board has power input and power output interface, and provides voltage to circuit sensor circuit board of orchard spraying robot path intelligent recognition system through external power part. The start switch is used to control the circuit start of the main control board. The post-program download port is used to download and input the preset program of the orchard spray robot path intelligent recognition system. The signal flow chart of the main control board is shown in Fig. 2.

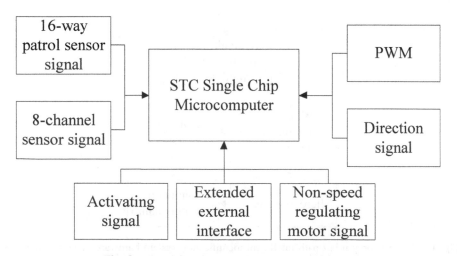

Fig. 2. Signal flow chart of the main control board

The line sensor input interface is mainly used to receive the road line sensor signal. In order to reduce the occupation of the MCU port, two 74HC245 bus driver circuits are used to form the multiplexing of the P2 port of the MCU, and the signal of the P4.4 port of the MCU is used for control. When P4.4 is low, the P2 port receives the low 8-bit signal QQQ-QQ7 of the 16-channel sensor. When P4.4 is high, the P2 port receives the high 8-bit signal of the 16-channel sensor QQ8-QQ15. The 8-channel sensor input interface can be used to connect optoelectronic, proximity or ultrasonic sensors. It can make the Orchard Spray Robot Path Intelligent Recognition System make full use of various sensors to detect external information to determine the distance from the target. The 4-channel sensor input interface uses the P0.0-P0.3 port of the single-chip microcomputer, and the socket J8-J11 is connected to 4 different sensors. The maximum detection distance is 30 cm, 1 m, 2 m depending on the model used. On the intelligent recognition system of orchard spray robot path, proximity switches are mainly used to detect whether some moving parts are in place, and the detection distance is generally less than 5 mm. There are 3 leads on the sensor, which are connected to the 3 pins of J8-J11.

2.4 Path Identification Module

The path identification module is a key part of the orchard spray robot path intelligent identification system control system, which is the "eye" of the orchard spray robot path intelligent recognition system. The quality of the scheme is directly related to the orchard spray robot path intelligent recognition system can deny the road ahead, accurately identify the path information, and provide accurate parameter basis for path tracking to complete the high-quality automatic tracking function of the orchard spray robot path intelligent recognition system. There are several common ways to identify the specific scheme: Scheme 1: Using CCD camera tracking: CCD camera tracking, is to use the CCD camera to collect video image information of the road. It obtains a tracking method by transmitting the image of the front road surface to the microcontroller for processing, obtaining useful path information and then controlling the smart car for path recognition. The advantage is that it has a wide field of vision, can obtain a large amount of road image information, and has complete and comprehensive path information. It is forward-looking and can predict the changes of the road earlier and react quickly and in a timely manner. The disadvantage is that the hardware circuit is complicated, and the video image is separated and binarized. The image processing information is large and the speed is slow, which reduces the real-time control of the orchard spray robot path intelligent recognition system and is expensive. Scheme 2: Photoelectric sensor tracking: Photoelectric sensor tracking, is to use a series of LED diodes and receiver tubes to obtain path information. A light-emitting tube and a receiving tube form a pair, and the light-emitting diode emits light, which is reflected by the ground and received by the receiving tube. In practice, the ability of reflecting light is different because the ground in the working range is composed of green original paint floor and white positioning line. Different voltage values are formed in the receiving tube according to the light reflection condition, so as to determine whether the driving direction of the subsequent path needs to be adjusted. In combination with the actual environment, pre-arranged ground positioning is prone to breakage and discoloration. It is appropriate to consider the first option from the perspective of future practical use. However, considering that the design is currently only applied and practiced, it is not practical as an industrial grade product. Considering the cost and the actual working ability of the chip and platform, the second option is used for design and development [7].

In the design of the line detection circuit, the line sensor needs to be selected [8]. In the path control of the orchard spray robot path intelligent identification system, a photosensitive sensor device arranged in a line is selected. The signal difference caused by the difference in light intensity reflected by the light-emitting diodes on different color grounds is converted into a voltage change value to determine the path offset. In practical applications, the route should be pre-designed, and the vertical horizontal line should be set along the running path at a certain distance to facilitate the calibration of the entire sensor and the distance counting [9]. The sensor uses a 16-way line sensor. The main control board of the sensor signal processing board uses a 12 V power supply circuit for signal processing of the sensor signal processing board, and can be connected to the main control board through the power line. In the sensor signal processing board, the 16-way patrol line sensor sends the collected ground white strip information to the circuit board, and the collected information is first amplified, and the amplified signal is preset with a voltage value for comparison. After filtering, the required effective reflected signal is retained. After filtering, the effective signal is modulated into digital signal for feedback, which is controlled by the main control board. The light and dark changes represented by the signal indicate the positional relationship between the sensing element and the calibration path, thus providing a basis for path adjustment.

Since the voltage value caused by the direct feedback signal is small, a signal amplifying circuit is added to the circuit to amplify the signal for subsequent signal comparison. The ER first signal comparison circuit is connected to the reference voltage by using the LM324 to be connected to the voltage comparator, and the upper comparison result QQQ input is compared [10]. When the upper voltage value meets the predetermined value, the output high level is fed back to the control board, and when the upper voltage input value does not meet the predetermined value, the low level is finally output.

2.5   Motor Drive Module

The basic control principle in this control process is to adjust the average voltage by the duty cycle of the signal, and use the change of the voltage to control the operation of the DC motor. Two MOS tube drive circuits are provided in the circuit to drive the left and right wheel motors respectively. This circuit uses the IR2110 power driver integrated chip, which is a single-chip integrated driver module for dual-channel, gate-driven, high-voltage high-speed power devices with high reliability. The IR2110 input signal is the PWM pulse width signal sent by the main controller, and its output directly controls the on and off of the 75N75MOS tube. In the relay output circuit, L01 and L02 are connected to the DC motor, and the LDIR is the main control board. The left motor direction control signal is sent, and the common contact of the relay is controlled by the triode to operate. When LDIR is low, the relay normally open contact does not operate, the DC motor rotates forward; when LDIR is high, the relay normally open contact closes and the DC motor reverses; The relay circuit is used instead of the general MOS tube bridge circuit as the motor forward and reverse control circuit, which is mainly for the frequent start, stop and forward and reverse operation, the relay circuit is more reliable and safe.

2.6 Wireless Communication Module

Zigbee is a wireless data transmission network platform consisting of up to a wireless data transmission module, which is very similar to the existing mobile communication CDMA network or GSM network. Each Zigbee network data transmission module is similar to a base station of a mobile network, and can communicate with each other throughout the network [11]. The distance between each network node can range from the standard 75 m to hundreds of meters or even several kilometers. The entire network can also be connected to other existing networks. Compared with other solutions, this technology has the advantages of large network capacity, short delay, reliable communication and low power consumption. Therefore, in the design of the orchard spray robot path intelligent identification system, the ZKM101B communication module was selected to realize wireless network transmission. ZKM101B wireless transmission converter can complete two-way wireless data communication between devices and devices based on RS232 and RS485 bus, realize one-to-one communication between one host and multiple slaves. It completely replaces the traditional RS232 and RS485 cables, and the specific functions can be configured according to the user's choice.

3 Design of Path Intelligent Identification System Software System

3.1 Software System Development Environment

In the software development, the KEIL C51 standard C compiler is used to provide the C language environment for the software development of the 8051 microcontroller while retaining the efficient and fast assembly code. The C51 compiler's capabilities continue to grow, allowing you to get closer to the CPU itself and other derivatives. Starting with uVision2, the C51 has been fully integrated into the integrated development environment of uVision2. This integrated development environment includes: compiler, assembler, real-time operating system, project manager, debugger. The uVision2 IDE provides a single and flexible development environment for them. In programming we used KEIL uVision4 as the development environment.

3.2 Accurate Path Recognition Based on Adaptive Genetic Algorithm

Software design and program development for path recognition control scheme using adaptive genetic algorithm. Adaptive genetic algorithm is a computer to complete the control activities described by people in natural language. The adaptive genetic algorithm has many good characteristics. It does not need to know the mathematical model of the genetic object in advance, so it is especially suitable for the control of complex nonlinear systems that are difficult to establish mathematical models. The establishment of the rules is to transform the experience into corresponding variables. It has strong resistance to transformation and performs well in controlling nonlinear hysteresis system. It has the advantages of fast system response, small overshoot and short transition time. Since the processing is classified in the form of establishing rules in processing, the processing speed is faster. In composition, it can be divided into three parts: genetic input, genetic operation, and genetic judgment output. This kind of algorithm is very practical in dynamic and complex environment conditions and practical application environments where obstacles are not fixed. In the adaptive genetic algorithm, it is necessary to convert the exact value that needs to be input into the applicable amount required by the genetic controller rule, the domain of the quantitative value, the lighting level and the quantization factor, and the value of the variable and the inheritance of the membership function is completed. As long as the genetic operation link is based on the established rules combined with the corresponding logical relationship, the logical reasoning process is completed, and the inference result is output. The magnitude of the output is obtained by inference of the genetic relationship equation, not an exact value. Therefore, the genetic output is subjected to de-geneticization, converted into a clear amount within the domain, and then scaled into a precise control. Common methods of de-generization include weighted average method, center of gravity method, and summation method.

In practical applications, the most widely used is the PID control mode. In the process of adjustment, the physical meanings of proportional, integral and differential are mainly applied, and the errors in the actual control are controlled and adjusted to achieve the control objectives. Where r(t) is the actual input value in the entire system; y(t) represents the actual output value; e(t) is the amount of error between the input and output values in the system, which can be expressed by Eq. 1:

$$e(t) = r(t) - y(t) \tag{1}$$

u(t) is the system control signal output by the PID regulator, and is the output value of the controller obtained after the controller performs the PID operation on the error signal. The mathematical formula for its control law is shown in Eq. 2:

$$u(t) = K_p \left[e(t) + \frac{1}{T_1} \varsigma_0 e(t) dt + T_D \frac{de(t)}{dt} \right] \tag{2}$$

Transform it to get the transfer function:

$$G(S) = \frac{U(S)}{E(S)} = K_P \left(1 + \frac{1}{T_{1S}} + T_D s \right) \tag{3}$$

KP is the proportional coefficient in the PID controller, T1 is the integral time constant of the controller, and TD is the differential time constant of the controller. The schematic diagram shows that the system is a closed-loop control system, and the role of the PID controller plays a linear adjustment role. The process of controlling is to input the expected value r(t) into the controller, and detect the actual output y(t), and then compare the two, the error value is obtained by comparison and fed back to the control device with the signal e(t). If the deviation value is 0, the current working status is normal and no adjustment is needed. If the deviation value is not 0, then the PID controller performs adjustment, and the output control signal U(T) adjusts the control object. Through continuous information comparison, the orchard spray robot can be guaranteed to follow the normal route.

3.3 Route Control Scheme

In the orchard spray robot path intelligent recognition system, the tracking function is realized by the detection signal reflected back by the line sensor installed under the trolley to determine the relative position between the current car and the fixed track. Select a word line on the sensor arrangement, which is perpendicular to the direction of advancement in the horizontal plane. In the distribution of the photoelectric sensor probes, according to the principle of equal-angle division, the non-equidistant arrangement is performed according to a certain angle relationship, so that the input and output results are close to a linear relationship. Because the design environment is relatively simple, the offset can be accurately determined for different feedback signals. Here, the adaptive genetic algorithm is selected on the algorithm [12].

3.4 Direction Adjustment Scheme

The two front wheels of the trolley are the driving wheels, each equipped with a DC geared motor to drive the car forward. When the current advancing direction deviates from the fixed direction, the sensors distributed on both sides feedback back different detection signals according to the magnitude of the deflection angle, and have a set of codes formed by different high and low levels. When the signal is fed back to the main control chip, according to the different codes, the angle between the trolley axis and the feedback back to the high-level sensor position is determined, and the adjusted input amount is determined to pass through the two PWM wave output ports of the chip at the same time. By changing the duty cycle and controlling the average voltage per unit time, and changing the motor speed, the rotation speed of the driving wheels on both sides is different, so as to adjust the movement state of the orchard spraying robot [13].

4 Simulation Test

The design of the path intelligent recognition system of the orchard spray robot based on adaptive genetic algorithm is realized by the design of hardware system and software system. In order to verify the path recognition accuracy of the path intelligent recognition system of the orchard spray robot based on the adaptive genetic algorithm, a simulation experiment was designed. In the course of the experiment, an orchard was used as the experimental object, the path intelligent identification system of the orchard spray robot based on the adaptive genetic algorithm was verified by the pesticide spray of the orchard spray robot based on the adaptive genetic algorithm of the orchard spray robot accuracy. In order to ensure the validity of the experiment, the traditional spray robot path recognition system is compared with the path intelligent recognition system of the orchard spray robot based on adaptive genetic algorithm to observe the

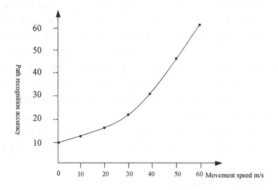

(a)Recognition accuracy of orchard spray robot path intelligent recognition system based on adaptive genetic algorithm

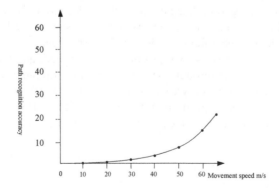

(b)Recognition accuracy of path recognition system for traditional orchard spraying robot

Fig. 3. Comparison of the first path recognition accuracy

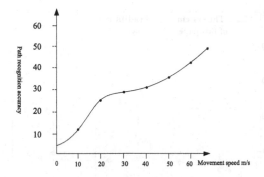

(a)Recognition accuracy of orchard spray robot path intelligent recognition system based on adaptive genetic algorithm

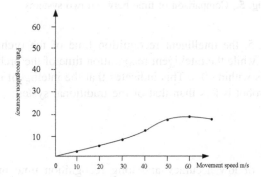

(b)Recognition accuracy of path recognition system for traditional orchard spraying robot

Fig. 4. Second path recognition accuracy comparison

experimental results. The path recognition accuracy of the traditional spray robot path recognition system and the path intelligent recognition system of the orchard spray robot based on the adaptive genetic algorithm is shown in Fig. 3 and Fig. 4.

Experiments show that the path intelligent identification system of the orchard spray robot based on the adaptive genetic algorithm is compared with the traditional spray robot path recognition system, the path recognition accuracy of the orchard spray robot's path intelligent recognition system based on adaptive genetic algorithm is higher.

In order to further verify the effectiveness of the system, the intelligent recognition time of the traditional system and the orchard spray robot in this system is compared and analyzed, and the comparison result is shown in Fig. 5.

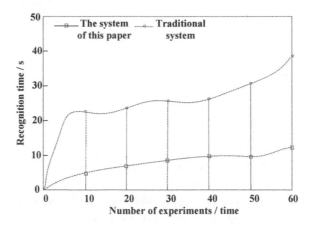

Fig. 5. Comparison of time between two systems

According to Fig. 5, the intelligent recognition time of the orchard spray robot system is within 10 s, while the intelligent recognition time of the orchard spray robot in traditional system is within 40 s. This indicates that the intelligent recognition time of the orchard spray robot is less than that of the traditional system.

5   Conclusion

Due to the problems of low accuracy and long recognition time in the traditional intelligent path recognition system of orchard spraying robot, this paper designs an intelligent path recognition system based on adaptive genetic algorithm for orchard spraying robot, which can broaden the working range of orchard sprayer, improve spraying accuracy and improve work efficiency. It can also improve the level of Agricultural Mechanization in China, which has very important practical significance and broad prospects for the development of precision agriculture in China.

Acknowledgments. 1. Training Program for Young and Middle - aged Scientific Research Backbone in Nantong Institute of Technology (ZQNGG305)

2. College-level Scientific Research Projects 《Design of Intelligent Sorting Equipment for Fruits and Vegetables》 (201820)

3. Design of Intelligent Robot for Fruit and Vegetable Agriculture Management in Nantong City Science and Technology Project JC2018148

References

1. Bin, L., Hui, L., Zhen, G.: Design of autonomous navigation orchard robot based on adaptive genetic algorithm and spline curve. Res. Agric. Mechanization **39**(32), 147–151 (2017)

2. Xin, T., Guangrui, L., Wenbo, Z., et al.: Robot path planning based on improved adaptive genetic algorithm. Mach. Tools Hydraulics **44**(17), 224–228 (2016)
3. Chao, Z., Wang, F., Zhang, C., et al.: Research on the algorithm of measurement path planning for inner wall of air-intake pipe based on spraying robot system. Int. J. Pattern Recogn. Artif. Intell. **31**(9), 1759018.1–1759018.15 (2017)
4. Deng, W., Zhang, H., Li, Y., et al.: Research on target recognition and path planning for EOD robot. Int. J. Comput. Appl. Technol. **57**(4), 325–333 (2018)
5. Liming, Z., Chuan, Y., Yi, Z., et al.: path recognition method of robot vision navigation in unstructured environments. Acta Optica Sinica **38**(8), 0815028 (2018)
6. Shah, H.N.M., Sulaiman, M., Shukor, A.Z., et al.: Butt welding joints recognition and location identification by using local thresholding. Rob. Comput. Integr. Manuf. **51**, 181–188 (2018)
7. Li, Y., Chen, X., Lin, Y., Srivastava, G., Liu, S.: Wireless transmitter identification based on device imperfections. IEEE Access **8**, 59305–59314 (2020)
8. Shuai, L., Weiling, B., Nianyin, Z., et al.: A fast fractal based compression for MRI images. IEEE Access **7**, 62412–62420 (2019)
9. Abolmaali, S., Mansouri-Ghiasi, N., Kamal, M., et al.: Efficient critical path identification based on viability analysis method considering process variations. IEEE Trans. Very Large Scale Integr. (VLSI) Syst. **25**(9), 2668–2672 (2017)
10. Zhengzhi, C., Jiayan, Z.: Robot trajectory planning based on adaptive genetic algorithm. J. Jingchu Insts. Technol. **31**(32), 151–157 (2016)
11. Fu, W., Liu, S., Srivastava, G.: Optimization of big data scheduling in social networks. Entropy **21**(9), 902 (2019)
12. Liu, S., Li, Z., Zhang, Y., et al.: Introduction of key problems in long-distance learning and training. Mob. Netw. Appl. **24**(1), 1–4 (2019)
13. Leitner, T., Sackl, S., V?Lker, B., et al.: Crack path identification in a nanostructured pearlitic steel using atom probe tomography. Scripta Materialia **142**, 66–69 (2018)

Design of Intelligent Lifting System for Real-Time Monitoring Data Expansion in Distribution Station Area

Xin-jia Li[1(✉)], Cheng-liang Wang[1], Yong-biao Yang[2],
and Song Shu[3]

[1] Jiangsu Fangtian Power Technology Co. Ltd., Nanjing 210096, China
lxj987600@126.com
[2] Southeast University, Nanjing 210000, China
[3] Changjiang Polytechnic, Wuhan 430074, China

Abstract. In view of the fact that the monitoring data in large-scale distribution network has the characteristics of quantifiable, real-time, dynamic and so on, and the data storage capacity is insufficient, this paper puts forward the design of the real-time monitoring data expansion intelligent upgrading system in the distribution station area. Realizes the monitoring data expansion capacity intelligence enhancement. Through designing the hardware module of expanding capacity and installing the data acquisition interface, the hardware design of intelligent upgrading of monitoring data expansion capacity is realized. On this basis, the hierarchical extended storage mechanism is used to store the data node information. The real-time reading and querying function of the data is realized, and the capacity ratio of the monitoring data is calculated. Finally, the intelligent upgrading system of the real-time monitoring data expansion capacity is realized.

Keywords: Distribution network · Data storage · Data acquisition interface · Capacity expansion

1 Introduction

The demand for electric energy is increasing day by day in modern society, at the same time, the demand for power quality is higher and higher. Electricity is a special commodity, can not be stored on a large scale, flexible, can only use as much electricity as possible. However, electricity is very important in modern human civilization. If a long period of blackout occurs, it will bring serious damage and influence to human life and production [1]. At present, the information construction level of the distribution network is relatively high. A large number of liquidity data generated during the operation of the power network are recorded in the dispatching, operation and maintenance, marketing system, the storage amount is in the TB level [2], the data types are diverse, and the variables are various. The storage and monitoring data capacity of distribution network is not enough. Based on the existing problems, the design of real-time monitoring data expansion and capacity-expanding intelligent upgrade system is proposed. With the increase of the distribution network data, it is more and more

© ICST Institute for Computer Sciences, Social Informatics and Telecommunications Engineering 2021
Published by Springer Nature Switzerland AG 2021. All Rights Reserved
S. Liu and L. Xia (Eds.): ADHIP 2020, LNICST 347, pp. 124–133, 2021.
https://doi.org/10.1007/978-3-030-67871-5_12

important to design the real-time monitoring data expansion and capacity enhancement system in the distribution station area. There are a lot of complex and abnormal power network data in the distribution network. If there are loopholes or errors in the transmission process, the big data structure of the power network will tend to become infinitely complicated and reduce the generation efficiency of the distribution network. Therefore, it is extremely necessary to enhance the capacity of the distribution network.

Distribution network real-time monitoring data expansion intelligent promotion, with the intelligent lifting system as the core, through the use of a variety of communication methods to complete the distribution system capacity expansion intelligent upgrade, and through the integration of relevant application system information, To realize the scientific management of power distribution system capacity expansion. So as to improve the data storage space and power supply quality, strengthen the reliability of power supply data storage, and increase the amount of monitoring data storage [3], increase the economic benefits of power supply enterprises and strengthen the level of enterprise management, and optimize the operation of the power grid. Through the design of the expanded capacity intelligent upgrading system, the data storage capacity [4] has been improved, the work difficulty of the overhauling personnel has also been reduced, and big data's storage and demand information extraction has been realized. Therefore, the intelligent upgrade system for data expansion in the distribution station area is designed, and the hardware design for the intelligent upgrade of monitoring data expansion is realized by designing the expansion hardware module and installing the data acquisition interface. And use a layered extended storage mechanism to store data node information. Real-time data reading and query functions are realized, and the capacity ratio of monitoring data is calculated. Realize the intelligent upgrade system of real-time monitoring data expansion capability, and conduct experimental verification.

2 Hardware Design of Intelligent Lifting System for Real-Time Monitoring Data Expansion

The real-time monitoring data expansion intelligent upgrading system should have a certain good monitoring performance and excellent stability. The hardware configuration of the monitoring data expansion intelligent upgrading system is as follows: 2 data acquisition ports, 7 analog signal isolators, master plate, power board each, computer host computer, upper computer, scanner, serial port, keyboard, AD conversion interface, printer, etc., mainly complete the adjustment and operation of each signal. Among them, the power supply module adopts input 220 V AC, output 12 V, 5 V and 3.3 V DC, to provide the other modules with the power supply required for the stable operation of the system. The main control module communicates with the switch data module and the man-machine interaction module through the 422, 485 interface, receives the data of the switch quantity module, at the same time, judges the data and stores the data. If the fault is found, it will be recorded in the memory. If the request command of the human-machine interaction module is received, the data is sent to the human-machine module for display. The main control module provides fieldbus communication interface, communicates with other devices through CAN, MODBUS or Ethernet [5], and provides remote operation and management functions. The electric life can be calculated by using the circuit breaker breaking current, and the double functions of collecting and fast calculation can be realized, so as to ensure the speed and accuracy of the signal processing. The system module composition is as follows (Fig. 1):

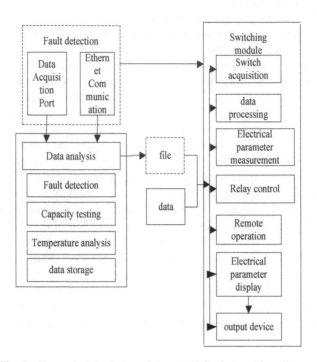

Fig. 1. Expansion hardware system module composition diagram

The hardware module of the expansion capacity is shown in the diagram. In order to realize the data expansion capacity safely, stably and quickly, the data acquisition interface is installed, through the data collection interface, the user electric meter and the data collection work in the intelligent station area are realized. The platform provides all interfaces for current mainstream production control systems, consumer meters, and power-related systems [6]. Considering the possible network failure of WAN, the data caching mechanism is established at the sending end of the interface. When the transmission of the network is interrupted, the field data can be stored online for more than 1 month. After the hardware system is powered on, the data is read from the sensor through the serial port, the corresponding frequency signal is collected, and the digital input is used to process the data. Adjust the voltage to the specified range for data input, carry out frequency conversion in the hardware system [7], analyze the data, and finally send the results of the analysis and diagnosis through the serial port.

3   Materials and Methods

The software design should take into account the real-time, reliability and maintainability of the control system. In addition, as a powerful and complete system, the software design should take into account the real-time performance, reliability and maintainability of the control system. Real-time monitoring system also needs to exchange information with remote equipment for remote control. The information between the system software is transmitted and exchanged, and the data size of the power network is judged, and the communication between the system software and the remote equipment is discussed.

3.1   Store Data Node Information

The hierarchical extended storage mechanism is used to realize the real-time reading and querying of data so as to improve the availability and storage ability of the system and make the intelligent control of the power grid more rapid and accurate. In a grid, when the sensor node perceives the data, the Data message is generated and sent to the computing IED node in the grid. An intelligent electronic device is placed at the center of each grid, which is defined as computing the IED node, receiving the sensing data sent by the sensor node in the grid, calculating the purpose of storing the grid and forwarding the data. The data matching the storage type of the grid is received and forwarded to the appropriate storage node, the storage IED node in the grid is managed, and the dynamic expansion of the storage node is realized.

3.2   Calculating the Capacity Ratio of Monitoring Data

On the basis of storing IED node information, data integration or access to the same grid of computing IED nodes through sensors, all-round, multi-angle data collection, so as to improve the accuracy of data acquisition. Start the judgment, carry on the sampling data in the sampling interval to carry on the calculation, judge whether the detection data exceeds the bearing range, and carry on the inquiry to the status of the main control board, when there is no data transmission, Continuously carry on the data query [8], when the capacity is insufficient, will monitor the data to carry on the data debugging. Through the establishment of relational database HBNDU for data programming, using database modeling, circuit breaker and motor operation data acquisition, real-time transmission and timely recording of the capacity ratio of data [9].

The power monitoring data and queue output rate are denoted by $y(t)$ and $r(t)$. The power monitoring data flow arrival rate and interference flow arrival rate are $Z_y(t)$ and $Z_d(t)$, respectively. The power monitoring data and interference data entering the buffer queue form the data flow arrival rate expressed by $Z(t)$, And $Z(t) = Z_y(t) + Z_d(t)$.

Let A denote the output link bandwidth when the time is t, and the output rate meets the link bandwidth constraint. D_w discrete random dynamic process is used to represent the change in the length of the output queue. At this time, the change in the queue can be expressed as an expected value. According to the balance of the queue, the change formula of the length of the information data queue $p(t)$ is as follows:

$$\frac{dp(t)}{dt} = Z(t) - r(t) - G(t) \tag{1}$$

$$r(t) = \begin{cases} D_w, p(t) + Z(t) > D_w \\ \min\{D_w, (t) + Z(t)\}, p(t) + Z(t) \le D_w \end{cases} \tag{2}$$

Let the large amount of power information data be transmitted in the form of a single packet. When the maximum buffer of the queue cannot accept the information packet length, the new information data is lost. Select the Droptail data packet loss strategy; set the new data packet to be completely discarded when the queue length is maximum, and the time is t. When using this power information data transmission technology to transmit the power system packet loss rate formula is as follows:

$$G(t) = \begin{cases} 0, \ p(t) + Z(t) - r(t) \le p_{max} \\ p(t) + S(t) - r(t) - p_{max}, \ p(t) + Z(t) - r(t) > p_{max} \end{cases} \tag{3}$$

The online real-time transmission of massive power information data is described by formula (1)-formula (3).

The minimum support F_{min} and the maximum confidence F_{max} of data association rules are set as constant functions, and the maximum capacity of distribution network is taken as standard to record the data capacity presented by F_{min} and F_{max}, as shown in Table 1.

Table 1. Distribution network monitoring terminal data

Thread	Transformer substation	Transformer substation	Interactive terminal	Monitor
$F_{min}(1)$	1	1	1	0
$F_{min}(2)$	0	1	1	0
$F_{min}(3)$	0	0	1	0
$F_{max}(1)$	0	0	1	1
$F_{max}(2)$	0	0	0	1
$F_{max}(3)$	1	1	1	1

For expansion intelligence enhancement, first calculate the capacity ratio of the received data [10], retain part of the aggregation attributes of the data during the calculation, and then use the line evaluation protocol to calculate its support. Furthermore, the accuracy of real-time monitoring data of distribution network is improved. The data processing calculation is as follows:

$$w(j) = \sum AjeG(t) \tag{4}$$

In the formula, $w(j)$ represents the query target parameters of monitoring data and A represents the original data set. $w(j)$ is formally defined. If and only if $w(j)$ contains transaction item x_i, the form of its capacity ratio can be recorded as $w(j) \Rightarrow x_i$.

According to the relevant requirements of calculation, without considering data missing and quantity value, $w(j) \Rightarrow x_i$ is taken as the minimum support rule coefficient of monitoring data, and then the capacity ratio of monitoring data of distribution and substation is obtained. The formula is as follows:

$$F = \frac{B_0/f(t)}{w(j)^2 \Rightarrow x_i} \tag{5}$$

In the formula, point F is the association rule of large data in distribution network; point B_0 is the constant of data capacity; and point $f(t)$ is the total amount of data mined in t time. This calculation does not do directional analysis.

On the basis of calculating the capacity ratio of monitoring data, the capacity expansion of intelligent lifting monitoring data is realized.

3.3 Realizing the Intelligent Improvement of Monitoring data expansion capacity

In view of the phenomenon of large monitoring data in distribution network, combined with the characteristics of real-time monitoring data, the transmitted data are sorted out, and the POL technology [11, 12] is put forward. The principle is that when the received data exceeds a certain fixed value, The capacity of the distribution network can be increased by automatically starting the capacity expansion function and storing the excess data in another database. The distribution network terminal is used to monitor the distribution station area in real time, collect its operating parameters, and take these data as the basis for expanding the capacity. The specific capacity expansion process is shown in Fig. 2.

In Fig. 2, x_0, x_1 and x_2 represent the process of data expansion, V_0, V_1 and V_2 represent the process of data transmission and exchange, and F_1 and F_2 represent the process of data clustering.

Using a simple expansion capacity calculation function, the expansion capacity enhancement calculation formula is as follows:

$$Enc(D_i) = \bmod m + c_i \tag{6}$$

$$Dec(D_i) = c_i - k \bmod m \tag{7}$$

In the formula, $Enc(D_i)$ represents the normal data transmission parameters; $Dec(D_i)$ represents the expansion capacity parameters; this calculation does not do directional analysis.

Through the software calculation, the real-time monitoring data expansion capacity is improved by a deduction calculation [13, 14], and the excess data is inversed on the computer to solve the capacity problem.

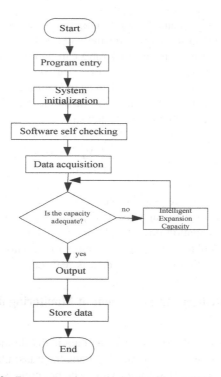

Fig. 2. Data expansion capacity Intelligent lifting proc

4 Results

In order to test whether the real-time monitoring data expansion intelligent lifting system in distribution network distribution station meets the requirement of capacity expansion intelligent upgrading, and to verify the feasibility and effectiveness of the system, an experiment is carried out on a certain platform. The traditional intelligent capacity expansion system is compared with the hardware system and software system designed in this paper.

4.1 Parameter Setting

Sequence $f(u)$ is used as the intelligent lifting function for capacity expansion. The higher the value, the stronger the expansibility of the representative system. The calculation formula is as follows:

$$f(u) = \frac{1}{c}\left(\sum_{s=0}^{k} \gamma^k u_s\right) \tag{8}$$

In the formula, c represents the data parameters of the basic capacity, γ represents the parameters to be estimated, and u_s represents the expanded capacity parameters.

4.2   Result Analysis

Before the contrast experiment of the proposed system, due to the weak stability of the collected data, it is necessary to normalize the collected data, and the processing process is shown in Fig. 3.

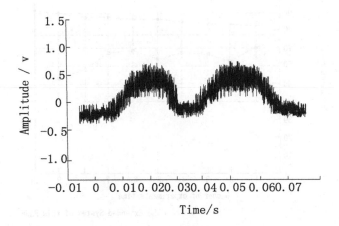

(a) Pre-processing

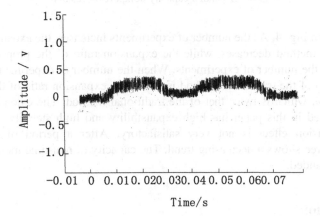

(b) After processing

Fig. 3. Data normalization process

As shown in Fig. 3, by normalizing the data and removing some incomparable data, the amplitude of the data tends to be zero, the activity is reduced, and the stability is higher, laying a good foundation for experimental verification.

On this basis, taking the data expansion as the test index, comparing with the traditional method, the data expansion ratio of different methods is tested. The test results are shown in Fig. 4.

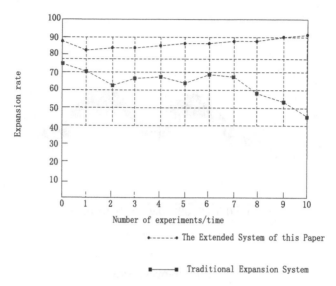

Fig. 4. Expanded capacity comparison results

As shown in Fig. 4, As the number of experiments increases, the expansion ratio of the traditional method decreases, while the expansion ratio of the proposed method increases with the number of experiments. When the number of experiments is 10, the expansion ratio of the traditional method is 45%; The expansion ratio of the proposed method is 91%, which is twice that of the traditional method. The extended capacity system proposed in this paper has high expansibility and high stability, and the traditional expansion effect is not very satisfactory. After a period of testing, the expansion degree shows a decreasing trend. The capacity of real-time monitoring data cannot be expanded.

5   Conclusion

In order to effectively improve the real-time monitoring data expansion capacity of distribution network distribution station, an intelligent lifting system based on real-time monitoring data expansion is designed. The technology realizes the intelligent enhancement of real-time monitoring data expansion capacity from both hardware and software, and reduces the amount of data loss in the process of data upgrading. Based on the intelligent enhancement of real-time monitoring data expansion capacity, the capacity expansion efficiency of distribution network is improved effectively. The experimental results show that this technique can improve the capacity expansion of real-time monitoring data and ensure the stability of the expansion. Therefore, it can be said that in the real-time monitoring data expansion process of distribution network distribution station area, the real-time monitoring data expansion process can be achieved. The design of the expansion system proposed in this paper is suitable for the data expansion of the distribution network.

References

1. Tang, Y., Cheng, L.F., Li, Z.J., et al.: Design of advanced analysis system for economic operation and optimization of distribution network based on intelligent station area. Power Syst. Protection Control **44**(15), 150–158 (2016)
2. Chen, P.: Visualization of real-time monitoring datagraphic of urban environmental quality. EURASIP J. Image Video Process. **2019**(1), 42 (2019)
3. Duan, X.J., Wang, J.L., Feng, D.Z., et al.: Design and implementation of intelligent decision-making system for construction and renovation of distribution stations. Power Grid Technol. **41**(8), 2709–2715 (2017)
4. Li, M., Kong, R., Han, S., et al.: Novel method of construction-efficiency evaluation of cutter suction dredger based on real-time monitoring data. J. Waterw. Port Coast. Ocean Eng. **144**(6), 710–714 (2018)
5. Zhang, T., Li, X., Qi, W., et al.: Real-time reversible data hiding based on multiple histogram modification. J. Real-Time Image Proc. **16**(3), 661–671 (2019)
6. Sanchari, D., Kari, T., Karuna, K., et al.: Impact of electric vehicle charging station load on distribution network. Energies **11**(1), 178 (2018)
7. Sun, B.H., Chen, L., Xia, D.: Design and application of intelligent operation and maintenance management and control platform for distribution network based on big data platform. Electr. Autom. **40**(6), 85–88 (2018)
8. Zheng, P., Shuai, L., Arun, S., Khan, M.: Visual attention feature (VAF): a novel strategy for visual tracking based on cloud platform in intelligent surveillance systems. J. Parallel Distrib. Comput. **120**, 182–194 (2018)
9. Shuai, L., Weiling, B., Nianyin, Z., et al.: A fast fractal based compression for MRI images. IEEE Access **7**, 62412–62420 (2019)
10. Feng, S.Y., Fu, J.M., Guanli, et al.: Deepening application research of operation and distribution data based on intelligent distribution network construction. Electr. Appl. **S2**, 322–326 10.Feng, S.Y., Fu, J.M., Guanli, et al.: Deepening application research of operation and distribution data based on intelligent distribution network construction. Electr. Appl. **S2**, 322–326 (2015)
11. Zhang, J., Wang, L., Wang, N.: Studies on explosion flow distribution and personnel protection in the subway station. J. Chem. Eng. Jpn. **52**(3), 293–299 (2019)
12. Song, J., Xie, H.N., Yang, Z.H., et al.: Statistical analysis of distribution network fault information based on multi-source heterogeneous data mining. Power Syst. Prot. Control **44**(3), 141–147 (2016)
13. Liu, S., Li, Z., Zhang, Y., et al.: Introduction of key problems in long-distance learning and training. Mob. Netw. Appl. **24**(1), 1–4 (2019)
14. Gui, G., Yun, L. (eds.): ADHIP 2019. LNICST, vol. 301. Springer, Cham (2019). https://doi.org/10.1007/978-3-030-36402-1

Dynamic Monitoring System of Big Data Leakage in Mobile Network Based on Internet of Things

Yan-ning Zhang and Ying-jian Kang[✉]

Beijing Polytechnic, Beijing 100016, China
witgirl_ninger@126.com, kangyingjian343@163.com

Abstract. Aiming at the problem of large acquisition time synchronization error caused by data explosion in traditional monitoring systems, a dynamic monitoring system for mobile network big data leakage based on Internet of Things is designed. The wireless sensor network is arranged in the system, the multi-channel base station node is designed, and the sensors are arranged in different channels to achieve the purpose of data diversion. Ep3c16q240 chip is selected as the core control chip of the multi-channel base station node, Based on the above hardware design, cluster monitoring, node performance monitoring and job operation monitoring functions are designed to upload the big data status information of the mobile network for job operation monitoring step by step to meet the needs of dynamic monitoring of data leakage. So far the overall design of the system is completed. The experimental results show that: compared with the traditional monitoring system, the designed monitoring system based on the Internet of things has smaller acquisition time synchronization error, better data acquisition synchronization performance, and improves the dynamic monitoring accuracy of mobile network big data leakage.

Keywords: Internet of Things · Mobile network · Big data leakage · Dynamic monitoring

1 Introduction

As an important milestone in the modern history of human development, the Internet is a symbol of human innovation and wisdom, and an important symbol of the rapid development of science and technology [1]. From its emergence to the present, the Internet has brought great changes to the industry and life of the whole world. It not only realizes many people's entrepreneurial dream, but also makes people's life inseparable and mutual influence. It also makes great changes in the organizational structure of the society [2]. The Internet has a wide and far-reaching impact, people's lives have become more convenient, and the ways of disseminating information have also become diverse. With the rapid development of science and technology, the development of the Internet has always maintained a relatively fast level. With the rapid development of the scale of the network, the services provided by the network are also diversified, which greatly facilitates people's lives [3]. The development of the Internet has become more in-depth, especially the development of mobile clients has

© ICST Institute for Computer Sciences, Social Informatics and Telecommunications Engineering 2021
Published by Springer Nature Switzerland AG 2021. All Rights Reserved
S. Liu and L. Xia (Eds.): ADHIP 2020, LNICST 347, pp. 134–146, 2021.
https://doi.org/10.1007/978-3-030-67871-5_13

profoundly changed the lives of Internet users. With more and more mobile applications such as mobile banking, train ticket ordering and takeaway ordering entering mobile clients, the Internet has begun to fully meet the needs of users, making people's life increasingly "networked" [4].

In the first decades after the emergence of computer network, it was mainly used by University researchers to send e-mails and by company employees to share printers. In these cases, security will not be noticed [5]. Nowadays, millions of ordinary people use the Internet. What we want to do in the real world is to be done on the Internet; make private calls, save personal documents, sign letters and contracts, vote online, electronic publishing, and handle Banking and shopping require security protection [6]. Network security is the basic condition for the existence of the Internet, which makes the computer network from an important business tool of academic concern. Security limitations have also become the limitations of the Internet [7]. Security vulnerabilities have been discovered one after another, and network security has become a hot topic that people pay close attention to. At this time, facing the problem of big data leakage in mobile networks, it is particularly important to introduce a monitoring mechanism into the network security management summary and establish a powerful dynamic monitoring system for big data leakage in mobile networks [8]. Using this monitoring mechanism, the data security of mobile network can be monitored in real time, and the abnormal situation such as data leakage can be warned. In case of failure, the management personnel should be informed in time after the problem occurs, so as to ensure the data security of mobile network.

In previous research, there are many mature monitoring technologies and open source monitoring systems in foreign countries. The monitoring system can provide information data about the network and system operating status, and also provide abnormal notification functions. Both local and remote servers can be monitored. Only need to modify the configuration file [9, 10]. At present, facing the problem of data leakage in mobile network, many domestic enterprises begin to study cloud computing and study different big data monitoring solutions. Obvious results are ZigBee-based monitoring system and web-based monitoring system. However, in the face of the current state of data explosion, the above two systems are difficult to ensure the synchronization of data collection, and there is a problem of large synchronization error in collection time. In view of this phenomenon, this paper proposes and designs a mobile network big data leakage dynamic monitoring system based on the Internet of things. The hardware of the monitoring system is designed. Ep3c16q240 chip is selected as the core control chip of multi-channel base station node. On the basis of hardware design, the system software is designed, including cluster monitoring function, node performance monitoring function and operation monitoring function Through the system hardware design and software design, the mobile network big data leakage dynamic monitoring system design based on the Internet of things is completed, which reduces the acquisition time synchronization error, improves the mobile network big data leakage dynamic monitoring accuracy, and improves the data acquisition synchronization performance.

2   Hardware Design of Dynamic Monitoring System for Mobile Network Big Data Leakage Based on Internet of Things

Internet of things technology is to connect any object with the network through information sensing equipment, according to the agreed protocol, and to exchange and communicate information through information media, so as to realize intelligent supervision and other functions [11, 12]. The wireless sensor network is adopted in the monitoring system, multi-channel base station nodes are designed, and the sensor nodes are arranged in different channels to realize data exchange monitoring. The node arrangement in the monitoring system using the Internet of Things technology is shown below (Fig. 1).

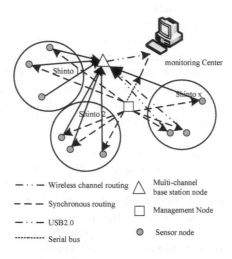

Fig. 1. Node layout of monitoring system

The multi-channel base station node and the monitoring center are connected through the USB2.0 bus, which can meet the requirement of simultaneously uploading 8 wireless channels of data to the monitoring center at a high speed and complete [13, 14]. The network data transmission rate is doubled from the original 250 Kbps of a single channel. Can reach 2 Mbps. The speed of USB2.0 bus can reach more than 12 Mbps, which can solve the problem of throughput limitation caused by using serial port.

The hardware structure of the designed multi-channel base station node is shown in Fig. 2.

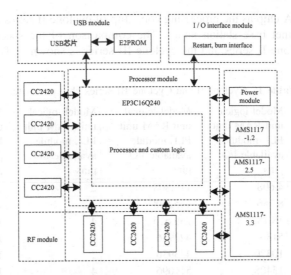

Fig. 2. Hardware structure of multi-channel base station node

The hardware of the multi-channel base station node includes FPGA chip, USB chip, 8 independent CC2420 radio frequency modules, electrically erasable programmable read-only memory chip, power management module, clock module and peripheral circuits of each chip. Among them, 8 RF modules, each pin of USB chip and part of FPGA chip can be connected with I/O pin, and functional configuration of available I/O pin of FPGA chip can be realized through program, so as to realize the operation of PLD module on RF module and communication module. The CC2420 radio frequency module is responsible for receiving and sending wireless data, and its initial configuration and work are controlled by the FPGA chip. Multiple CC2420 RF modules are used to receive data in different channels in parallel, and there is no competitive interference between them. After the USB chip is powered on and initialized, the data stream is uploaded under the control of FPGA chip.

FPGA is the product of further development based on programmable array logic, general array logic GAL, complex programmable logic devices and other programmable devices. It appears as a semi-custom circuit in the field of special integrated circuits, which not only solves the shortcomings of custom circuits, but also overcomes the shortcomings of the limited number of gates of original programmable devices. Considering the universality of multi-channel base station node design, FPGA chip selection mainly investigates Altera's products, which produce general-purpose FPGA chips. The research contents include FPGA chip selection manual, chip manual and online quotation provided by Altera. Research shows that Altera's main products include low-end chip series (cyclone IV Series), middle end chip series (ARIA Series), etc. With the further introduction of the series products, the Cyclone series can provide more available pins and memory cells, and further reduce the static power consumption, which can achieve the function of a multi-channel base station node, and the cost is significantly lower than the Arria series chips. Through the analysis, the multi-channel base station node needs more available I/O pins and logic units, so choose the appropriate FPGA chip as the processing core chip of cyclone II series and cyclone III series multi-channel base station nodes. The comparison of the specific parameters between the cyclone II series and the cyclone III series is shown in Table 1.

Table 1. Cyclone II and Cyclone III series chip parameters

Model logical unit RAM unit PLL module available I/O pins	Model logical unit RAM unit PLL module available I/O pins	Model logical unit RAM unit PLL module available I/O pins	Model logical unit RAM unit PLL module available I/O pins	Model logical unit RAM unit PLL module available I/O pins
EP2C5	4.608	119.808	2	142
EP2C8	8.256	165.888	2	138
EP2C20	18.752	239.616	4	142
EP3C5	5136	423936	2	106
EP3C10	10320	423936	2	106
EP3C16	15408	516096	4	160
EP3C25	24624	608256	4	148

Considering the actual application of monitoring, we must complete the configuration of the USB chip function, control the USB chip to realize the function of uploading data at high speed; obtain the status of the RF module in real time, and control the 8 RF modules to complete the data receiving operation in parallel at high speed according to the reception status of the data packet; Encapsulate and verify the received data packets, and use the HDLC mechanism to parse the data packets to ensure that the host computer program can be completed at high speed; CRC verification mechanism is adopted to ensure the correctness of data packets uploaded to the monitoring center. In conclusion, the ep3c16q240 chip of the cyclone III series is selected as the core chip of the multi-channel base station node.

3 Software Design of Dynamic Monitoring System for Big Data Leakage in Mobile Network

Based on the above hardware design, the software part of the monitoring system is designed. The dynamic monitoring of big data leakage in mobile network can be divided into three levels: cluster monitoring, node performance monitoring and job operation monitoring.

Managing and scheduling nodes is an important part of big data clusters. Whether the cluster is operating well or not directly determines the efficiency of the entire cluster, so the system needs to monitor and manage the various information resources of each node in real time. The administrator can adjust the cluster in real time according to the cluster performance, improve the cluster bottleneck, and ensure the good operation of the job.

The cluster monitoring agent is mainly responsible for the collection of monitoring index data of the cluster. For cluster monitoring, you can view the cluster status information, the total number of machines in the cluster, cluster users, host groups and other information. The acquisition of cluster performance data is mainly achieved through the command line. Such as uptime: view the load of the machine, output the number of processes waiting for CPU resources and the number of processes blocked in uninterruptible IO; mpstat-pall: display the occupancy of each CPU; pidtat1: CPU output of continuous output process will not cover before the data.

The cluster monitoring plug-in is mainly responsible for receiving the monitoring indicator data sent by the agent process. When monitoring the cluster, the implementation process of the plug-in process is to use the Icings expansion mechanism to encapsulate the command to receive cluster performance data, and follow Icings' custom plug-in expansion mechanism to form a plug-in, change the Icings configuration file, and plug the plug-in Deployed on Icings server. Restart the icings service to view the cluster performance indicators.

Node performance monitoring is mainly to monitor the underlying nodes and collect various performance indicators, so that managers can timely and comprehensively grasp the performance information of cluster nodes. The node performance monitoring module is mainly completed by Icinga. Node performance monitoring agent is mainly responsible for collecting node performance data and data processing, including performance monitoring data collection module and performance monitoring data processing module.

The monitoring data collection module includes monitoring of public services and monitoring of private services. For public service monitoring, you can access by using some standard protocols of the public network, such as HTTP, POP3, IMAP, FTP, and SSH. Some Icinga's own plug-ins can directly collect performance data. As a result, agents do not need to be installed on nodes to monitor these services. Private service is the opposite of public service. Private service cannot be accessed through the public network. Information cannot be accessed directly through the network. Therefore, when monitoring private services, the corresponding agent must be installed on the monitored object.

In the performance monitoring data receiving module, after the Icinga agent returns the performance index data, Icinga supports two methods to process the performance data, that is, use the command line to process or write directly to a specific file, which can be configured through the configuration file. If the first method is used, define the processor of plug-in performance data in the icinga configuration file. After executing the plug-in, execute the performance data processor. At this time, the performance processor obtains the data from the environment variables. Icinga starts the plug-in through the command line to perform the corresponding status check, and then captures the standard output stream of the plug-in to obtain the results after execution. The plug-in The execution result of icinga includes at least one line of readable text to represent the current state, and the performance data related to the plug-in is also included in the result. If the second method is used, icinga can directly put the obtained performance data into the performance data file, use the component to process the file, or configure a command in icinga to periodically process the performance data file.

Job operation monitoring is an important part of the monitoring system, and the monitoring object is the mobile network data of operation status. Job monitoring includes data acquisition module, data processing module and data sending module. The object of data collection is the logs generated during the operation of the mobile network. IDEA packages the executable shell code and the dependent jar packages through submit, generates the corresponding jar files, and submits them to the cluster through the command line. For tasks, save the running logs in the work directory, and create a new folder for each task to save its log files and dependent jar packages. The status information of mobile network big data can be obtained from the log.

After the data collection is completed, the data is filtered and extracted, stored in a certain data structure, and completed by the agent resident on the cluster node. In the data processing module, the user calls the data processing function module after the performance data collection. First, the mobile network operation task generates logs, which are stored in the work folder. Using the log collection method based on text analysis, readfile analyzes and processes the collected logs, and obtains the task Ido of the operation task Readformfile processes the collected logs, extracts the running indicators of each task, including monitoring indicators such as user, start time, duration, running status information, and stores them in a predetermined data structure. In SNMPUtilSend, all monitoring indicators are aggregated to form an SNMP data packet, ready to be sent.

The data sending module uses the event driven mechanism to send monitoring information through SNMP. It does not need the monitoring server to poll, saves bandwidth and server utilization, and provides system performance. So far, the design of a dynamic monitoring system for mobile network big data leakage based on the Internet of Things is completed.

4   Experimental Research on Dynamic Monitoring System of Big Data Leakage in Mobile Network

In order to verify the time synchronization effect of the mobile network big data leakage dynamic monitoring system, a time synchronization effect function verification experiment platform was built. In the experiment, the monitoring system is used to collect the fixed frequency waveform generated by the signal generator in real time. After the experiment, the acquisition time synchronization error of the sensor node is obtained by analyzing the collected data, and then the acquisition synchronization performance of the traditional monitoring system and the designed mobile network big data leakage dynamic monitoring system is compared.

4.1   Experimental Scheme Design

It is found that most of the experiments are based on computer simulation. In addition, based on the experimental platform of wireless sensor network for structural health monitoring, foreign researchers have verified the synchronization effect of network acquisition. The experimental method they used was to use a signal generator to generate a fixed-frequency waveform for the sensor network to collect, and then analyze the data collected by all nodes. Based on this method, an experimental method based on standard waveform acquisition is designed to verify the synchronization performance of the monitoring system.

The equipment used in the experiment included 16 Telosb sensor nodes, a dual RF relay node, management node, AFG3021 arbitrary waveform generator and laptop. Afg3021 arbitrary waveform generator has 14 bit output accuracy. Among them, 16 nodes are divided into 8 different channels. In the experiment, the AFG3021 arbitrary waveform generator was set to generate a standard signal under test for 16 nodes to collect. The waveform generated by the measured signal is set as a triangular wave signal, and the output end is connected to the ADCO and GND pins of 16 nodes with multiple leads. The voltage range of the output waveform is 0-two point five 5. To meet the allowable range of voltage input of AD sampling channel of telosb node and avoid damaging telosb node. The slope of the rising edge of the output triangle wave is set to 300 V/s. Since the AD sampling accuracy of the Telosb node is 12 bits, the range of the AD sampling value can be calculated from 0 to 4960. The reference voltage selected by the Telosb node is 2.5 V, and the voltage resolution calculation formula is as follows:

$$P = \frac{1}{\mu} \times 2.5\,V \tag{1}$$

In the formula, μ represents the AD sampling value. According to the above formula, when the AD sampling value changes by 1, the measured voltage value correspondingly changes by P. Therefore, when the AD sample value changes by 1, the corresponding time change on the rising edge of the output waveform is calculated as follows:

$$t = \frac{P}{\kappa} \tag{2}$$

Where κ is the slope of the rising edge of the output waveform. When multiple sensor nodes collect data at the same time, if the collection time is fully synchronized, the data collected from the same serial number data should be consistent. When there is a sequential error at the node's collection time, you can use the AD sampling values collected by different nodes in the same collection cycle to calculate the voltage value collected at this time and the average collection voltage value of all sensor nodes. Then, the voltage value collected by each node is subtracted from the average voltage value, and the time synchronization error of the node relative to the average sampling time is calculated using the formula. The formula is as follows:

$$\Delta t = \frac{\Delta V}{\kappa} \tag{3}$$

In the formula, ΔV represents the difference between the collected voltage value and the average voltage value.

The entire experiment lasted 2 h. Before the experiment, turn on the arbitrary waveform generator, adjust the output waveform accurately according to the set value, connect the positive voltage output terminal to the ADCO pins of all sensor nodes, and connect the negative voltage output terminal to the GND pins of all sensor nodes. After the experiment starts, the host computer control management node sends a start command to all sensor nodes to start monitoring the network. During the working process, the management node synchronizes the collection time of all sensor nodes in the network according to its own program. The base station node in the monitoring system receives the data packet sent by the sensor node in real time and uploads it to the upper computer for storage, waiting for the end of the experiment for analysis.

4.2 Experimental Results and Analysis

After the experiment, according to the package number of the data package stored in the upper computer, select three time points randomly, record the AD sampling values collected by all sensor nodes at the three time points, and convert them into voltage values, as shown in the table below.

Table 2 shows the experimental measurement data of the proposed physical network-based mobile network big data leakage dynamic monitoring system, and the synchronization effect cannot be intuitively seen. For this reason, the measurement data of the other two monitoring systems are not listed one by one, directly Compare the time error jitter and synchronization effect measurement results. The experimental results of the traditional ZigBee Based monitoring system are as follows (Table 3).

Table 2. Synchronous effect experimental measurement data

Node number	Measuring voltage (V)		
	1	2	3
3	1.1684	1.2019	1.3412
6	1.1721	1.2341	1.3681
9	1.1745	1.2471	1.3124
11	1.1617	1.2036	1.3369
13	1.1702	1.2571	1.3154
15	1.1629	1.2347	1.3274
17	1.1794	1.2367	1.3096
19	1.1624	1.2903	1.3264
21	1.1454	1.2036	1.3746
22	1.1564	1.2461	1.3091
23	1.1754	1.2412	1.3325

Table 3. Synchronization effect analysis of monitoring system based on ZigBee

Index	Measuring voltage		
	1	2	3
Maximum	1.2936	1.4216	1.6047
Minimum value	1.1021	1.3011	1.4214
Maximum difference	0.1915	0.1205	0.1833
Average value	1.1974	1.3824	1.5691
Synchronization error	232 μs	304 μs	357 μs

The jitter of time error is as follows (Fig. 3):

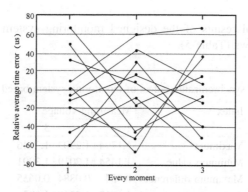

Fig. 3. ZigBee-based system time error jitter

The experimental results of the web-based monitoring system are shown below (Table 4).

Table 4. synchronization effect analysis of web-based monitoring system

Index	Measuring voltage		
	1	2	3
Maximum	1.3147	1.5007	1.6214
Minimum value	1.1724	1.2374	1.4025
Maximum difference	0.1423	0.2633	0.2189
Average value	1.2371	1.3725	1.506
Synchronization error	201 μs	262 μs	304 μs

The time error jitter is as follows (Fig. 4):

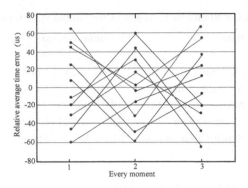

Fig. 4. Web-based system time error jitter

The experimental results of the designed monitoring system based on physical network are as follows (Table 5).

Table 5. Analysis of synchronization effect of monitoring system based on physical network

Index	Measuring voltage		
	1	2	3
Maximum	1.1794	1.2903	1.3746
Minimum value	1.1454	1.2019	1.3091
Maximum difference	0.034	0.0884	0.0655
Average value	1.1663	1.2360	1.332
Synchronization error	29 μs	38 μs	49 μs

The jitter of time error is as follows:

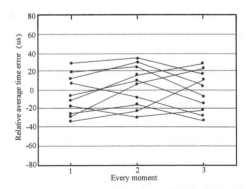

Fig. 5. System time error jitter based on the Internet of Things

The points in the figure represent the nodes. Observing the above results, it can be seen from the time error jitter results of the relative average in Fig. 5 that the time error jitter range of the relative average in Fig. 5 is narrower than that of the other two results, and the synchronization error shown in the synchronization effect analysis table of the measured voltage is always within 90 μs, while the synchronization error of the monitoring system based on ZigBee and web is above 200 μs, which is far away Out of standard time error range. In conclusion, the designed mobile network big data leakage dynamic monitoring system based on the Internet of things has a smaller time synchronization error. Within the allowable range of normal application of the system, the system has a good collection synchronization function.

In order to further verify the effectiveness of the system in this paper, the traditional ZigBee Based System and the Internet of things system proposed in this paper are used to compare and analyze the dynamic monitoring accuracy of mobile network big data leakage. The comparison results are shown in Fig. 6.

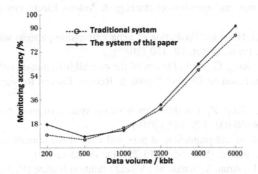

Fig. 6. Monitoring accuracy comparison results

According to Fig. 6, the maximum dynamic monitoring accuracy of mobile network big data leakage in this system can reach 98%, while the highest dynamic monitoring accuracy of mobile network big data leakage based on ZigBee system is only 82%, which shows that the dynamic monitoring accuracy of mobile network big data leakage of this system is higher than that of traditional mobile network big data leakage monitoring based on ZigBee system.

5 Conclusion

Mobile network big data as the trend of social development, its challenges in network security need more attention. Using the characteristics of the Internet of things and the characteristics of big data, this paper designs a mobile network big data leakage dynamic monitoring system based on the Internet of things. Taking ep3c16q240 chip as the core control chip of multi-channel base station node, the system hardware is designed, and the system software is designed through the functions of cluster monitoring, node performance monitoring and job operation monitoring Complete the design of mobile network big data leakage dynamic monitoring system based on Internet of things, solve the problems existing in the traditional monitoring system, and lay a good foundation for the future development of monitoring system [15].

References

1. Zhang, L., Shao, F.: Design of abnormal risk monitoring system for network big data platform. Mod. Electron. Tech. **41**(22), 143–146 (2018)
2. Du, S.: Application of online monitoring system in urban rail transit based on the big data. Urban Mass Transit **21**(S2), 30–33 (2018)
3. Chen, Z., Sun, J.: Simulation of mobile network information transmission security defense under big data. Comput. Simul. **35**(05), 207–210 (2018)
4. Zhang, Y., Liu, K., Yang, L., et al.: Platform construction and data processing application technology in coal industry monitoring big data. Coal Sci. Technol. **47**(03), 75–80 (2019)
5. Leng, X., Chen, G., Jiang, Y., et al.: Data specification and processing in big-data analysis system for monitoring and operation of smart grid. Autom. Electr. Power Syst. **42**(19), 169–178 (2018)
6. Liu, S., Lu, M., Li, H., et al.: Prediction of gene expression patterns with generalized linear regression model. Front. Genet. **10**, 120 (2019)
7. Liu, W., Zong, L., Xing, C., et al.: Design of the overall information collection of wind farm monitoring system based on EDPF-CP system. Renew. Energy Resour. **36**(08), 1204–1208 (2018)
8. Deng, Z., Cui, J., Liang, Z.: Condition monitoring system of mine hoist based on storage test. Coal Technol. **38**(05), 179–181 (2019)
9. Deng, M.: Regulation and protection of personal data in the context of big data. J. Beijing Univ. Posts Telecommun. (Soc. Sci. Ed.) **21**(01), 19–25 (2019)
10. Zheng, P., Shuai, L., Arun, S., Khan, M.: Visual attention feature (VAF): a novel strategy for visual tracking based on cloud platform in intelligent surveillance systems. J. Parallel Distrib. Comput. **120**, 182–194 (2018)
11. Fu, X., Gao, Y., Luo, B., et al.: Security threats to hadoop: data leakage attacks and investigation. IEEE Netw. **PP**(2), 12–16 (2017)
12. Buesing, H., Vogt, C., Ebigbo, A., et al.: Numerical study on CO_2 leakage detection using electrical streaming potential (SP) data. Water Resour. Res. **53**(1), 455–469 (2017)
13. Liu, S., Liu, D., Srivastava, G., et al.: Overview and methods of correlation filter algorithms in object tracking. Complex Intell. Syst. (2020). https://doi.org/10.1007/s40747-020-00161-4
14. Blackford, J., Artioli, Y., Clark, J., et al.: Monitoring of offshore geological carbon storage integrity: implications of natural variability in the marine system and the assessment of anomaly detection criteria. Int. J. Greenhouse Gas Control **64**, 99–112 (2017)
15. Lu, M., Liu, S.: Nucleosome positioning based on generalized relative entropy. Soft. Comput. **23**(19), 9175–9188 (2018). https://doi.org/10.1007/s00500-018-3602-2

The Design of Philosophy and Social Sciences Terms Dictionary System Based on Big Data Mining

Han-yang Li[1,2(✉)]

[1] School of Humanities, Tongji University, Shanghai 200092, China
lihanyang1685@163.com
[2] School of Journalism and Communication, Minnan Normal University,
Zhangzhou 363000, China

Abstract. Aiming at the problem that the traditional dictionary system cannot use the big data mining technology for term calculation, which leads to the long response time of the dictionary system retrieval, a philosophy and social science term dictionary system based on big data mining is designed. The hardware part designs the system controller and connects the single-chip microcomputer connection circuit. The software part first divides the term dimensions according to the characteristics of philosophy and social science terms, mines the corpus according to different dimensions, completes the calculation of philosophy and social science terms, sets up the term database structure, and finally completes the software design of the dictionary system. The experimental results show that: Compared with the traditional dictionary system, the search time of philosophy and Social Sciences terminology dictionary system based on big data mining is the shortest, and the detection accuracy of philosophy and social science terms is higher, which is suitable for all applications.

Keywords: Big data mining · Philosophy and social sciences · Term dictionary · Reaction time

1 Introduction

Glossary is an important tool for providing knowledge services in professional fields. However, there are still some problems in the compilation of existing term dictionary. For example, the knowledge content of terminology dictionaries is mostly simple, mainly providing explanations, English translation and other content, and the organization and description of deep knowledge needs to be improved. The degree of automation of term dictionary compilation is relatively low. Many term dictionary compilations still follow the traditional manual method. The process of term collection, collation, classification, typesetting, and proofreading is mainly done manually, and lacks the necessary automated auxiliary tools. These simple and repetitive manual labor are extremely error-prone and inefficient, leading to the compilation of term dictionary

S. Liu and L. Xia (Eds.): ADHIP 2020, LNICST 347, pp. 147–158, 2021.
https://doi.org/10.1007/978-3-030-67871-5_14

lagging behind the development of science and technology and the change of language facts, and it is difficult to achieve resource sharing [1]. How to deeply describe terminology knowledge from the perspective of knowledge organization, and then design a semi-automatic terminology dictionary compilation system, is an important topic in the current terminology dictionary research field and undoubtedly has very important significance.

Essentially, the compilation of terminology dictionary is an important part of knowledge production, and it is a frontier cross-cutting field of multiple disciplines such as lexicology, terminology, library and information science, computational linguistics [2]. The design of the terminology dictionary compilation system must first be based on knowledge organization, accurately reveal all kinds of knowledge behind the terminology, and form a unified and standardized knowledge representation framework. This requires the relevant achievements of lexicology, terminology and knowledge organization theory. Second, to achieve semi-automatic compilation of terminology dictionaries and improve the efficiency of knowledge production, it is necessary to actively absorb the achievements of computational linguistics in corpus construction, new word discovery, and terminology calculation. Finally, the terminology dictionary compilation has strong knowledge engineering features, and needs to realize the co-construction and sharing of knowledge, interactive collaboration and dynamic update from the perspective of project management.

The design of the term dictionary compilation system should rely on knowledge organization to form a more standardized and semi-automatic knowledge production process. The term dictionary is a tool that provides professional knowledge services, and needs to reveal in depth the objective things or knowledge content referred to by the term. Therefore, the compilation of term dictionary requires editors to have not only language knowledge, but more importantly, professional knowledge [3]. The term dictionary focuses on the concept of terms, and expresses these concepts in terms of words, which are generally sorted in order of topic. The conceptual category of terms and the relationship between category members are an important part of the term dictionary research. The term dictionary compilation is based on terminology and lexicography, applying basic methods and techniques of knowledge organization and computational linguistics. Norms, knowledge descriptions, knowledge links, etc., form a human-machine knowledge resource.

2 Hardware Design of Term Dictionary System

2.1 Design System Controller

The internal chip of the controller selects STC89C52 single-chip microcomputer as the processing core, and receives and processes the system-related index data [4]. Allocate MCU port to realize the control function of MCU. According to the I/O drive of MCU, the STC89C52 MCU has four groups of I/O ports P0, P1, P2 and P3. An external pull-up resistor is connected to the foot to ensure that the transistors in the internal output circuit of all ports are in the off state and the lower transistors are in the open state. Control the P0 port in parallel with two pull-up resistors to ensure the output of "0" and "1" processing instructions. The connection diagram of the single-chip microcomputer as the output port is shown in Fig. 1:

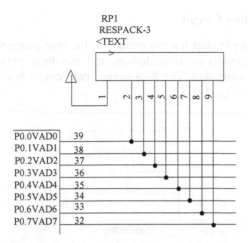

Fig. 1. Wiring diagram of the single-chip P0 group port used as output

As shown in Fig. 1 above, the P0 group port is connected to the circuit that drives the LCD display, the P1 group port is used to store philosophy and social science terminology data, the P2 group port controls the LCD display drive signal of the dictionary system, and the P3 group port is used The change of terminology and the detailed allocation of I/O ports are shown in Table 1:

Table 1. MCU I/O port allocation table

Serial number	I/O port	Allocation function
1	Group P0	Provide data signal for LCD display
2	P1.0 pin	Receive display output
3	P1.1 pin	Receive data signal output
4	P 1.2–p 1.4 pin	Provide driving signal for LCD display
5	P1.5 pin	Driver chip controller
6	Group P2	Drive control chip
7	P3.2–p3.4 pin	Receive key input
8	P3.5 pin	Receive controller output

According to the port function shown in Table 1, connect each port of the chip, design the control circuit of the controller, and complete the design of the hardware part [5].

2.2 SCM Connection Circuit

The system uses a single-chip microcomputer as the core processing unit, combined with resistors and capacitors and other devices, and uses the single-chip microcomputer as the smallest processing unit. The final control core circuit is composed of a block diagram, as shown in Fig. 2:

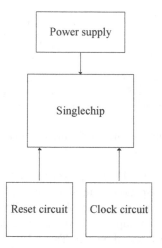

Fig. 2. The minimum circuit composition of the STC89C52 microcontroller

As shown in Fig. 2, the left side of the single-chip microcomputer is mainly connected to the power circuit, and the right side is connected to the clock circuit and the reset circuit. When the single-chip microcomputer is actually working, the time for the single-chip microcomputer to access the storage from the ROM is defined as a machine cycle. Store access data for one machine cycle [6]. Use the XTAL1 and XTAL2 ports of the oscillator as the oscillator input/output ports. The XTAL1 port of the oscillator uses an internal and external clock to connect a quartz crystal, and then an external capacitor is mounted to form a parallel resonant circuit to allow the internal oscillation circuit to generate self-oscillation (Fig. 3).

In order to prevent the SCM from being disturbed by the environment and causing the dictionary system to malfunction, the reset circuit of the SCM system is adjusted to a level switch reset method, so that the capacitor charge is in a short circuit state when

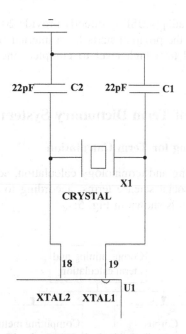

Fig. 3. MCU clock circuit diagram

the dictionary system is turned on, and the reset pin is adjusted Connect to high level [7]. After the power supply is stable, the reset pin is grounded through the resistor, so that the capacitor plays the role of isolating the DC level. Design the reset circuit of the single chip microcomputer as shown in Fig. 4:

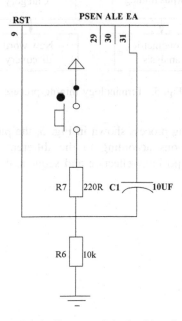

Fig. 4. Reset circuit diagram of single chip microcomputer

The power circuit part adopts USB to directly provide 20 V DC power. In order to ensure that the reading of the program starts from internal storage, the EA pin in the above figure is connected to a high level to complete the hardware design of the dictionary system.

3   Software Design of Term Dictionary System

3.1   Use Big Data Mining for Term Calculation

When using big data mining and terminology calculation, according to the characteristics of philosophy and social science terms, according to the following process of mining, the mining process is shown in Fig. 5:

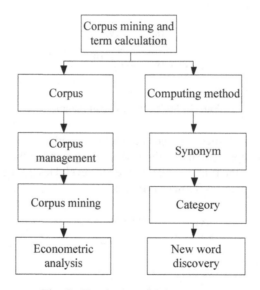

Fig. 5. Terminology mining process

According to the mining process shown in Fig. 5, the philosophical corpus is first divided into four dimensions according to the different learning styles, sensory/intuitive, visual/linguistic, positive/reflective and sequential, as shown in Table 2:

Table 2. The four dimensions of philosophy and social science terms

Dimension	Classification of learning styles	Source of corpus
Dimension 1	Active type	Philosophy innovation theory
	Reflective type	Problem
Dimension 2	Language type	Literal interpretation
	Visual type	Media publicity
Dimension 3	Sensory type	Examples, facts
	Intuition type	Overview
Dimension 4	Global type	Academic research
	Sequential type	Material science

According to the four dimensions shown in Table 2, each dimension corresponds to the source of philosophical and social science corpus of different dimensions, mining the corpus according to different dimensions, summarizing all the above corpus data, and describing the characteristics of social science terms using Jones matrix Words:

$$M_j(w) = e^{\beta(w)} U(w) \tag{1}$$

In the above formula, β indicates the amount of terms that do not affect the description mining, w indicates the frequency of each material component relative to the term, U indicates the frequency function, and j indicates the number of types of terms. To process the philosophical corpus data summarized above, assuming that at this time R_{in} represents all philosophical material data, and R_{out} represents output philosophical social science terms, then the two processing processes can be expressed as:

$$\begin{cases} R_{in} = \begin{pmatrix} e^{-\varphi} \cos \xi & -e \sin \xi \\ e^{-\varphi} \sin \xi & e \cos \xi \end{pmatrix} \\ R_{out} = \begin{pmatrix} e^{-\psi} \cos \xi & -e \sin \xi \\ e^{-\psi} \sin \xi & e \cos \xi \end{pmatrix} \end{cases} \tag{2}$$

In the above formula, (φ, ξ) and (ψ, ξ) are used to describe the input and output of the philosophy and social science PSP, and the final output of the above formula is R_{out} philosophy and social science terms. Due to the repetitiveness of the mining objects, the excavated terminology is synonymous. Use the obtained PSP to determine synonyms. The calculation formula is:

$$M_d(\Delta t) = \begin{pmatrix} e^{jw} & 0 \\ 0 & e^{-jw} \end{pmatrix} \tag{3}$$

In the above formula, $M_j(w)$ represents the scope of philosophy and social science terms, so for any philosophical literature, the discovery process of philosophy and social science terms can be expressed as:

$$\overline{J}_{out} = M_j(w)R_{in} \tag{4}$$

In the above formula, $\overline{J}_{out}$ represents the output of philosophical and social science terms. After using big data mining and terminology calculation, the calculated philosophy and social science terms are aggregated into a dictionary, and then the dictionary is converted into a philosophy and social science terminology database, and the software design of the philosophy and social science term dictionary is finally completed [8].

3.2 Design Terminology Database

Before designing the terminology database, set up the structure of the terminology database according to the philosophical terminology data dictionary obtained from the above calculation. The designed database structure is shown in Table 3:

Table 3. Designed database structure

Terminology name	Identifier	Type	Null value
Dialectics	id	varchar	no
Whiteboard	openid	varchar	can
Superman	opendate	varchar	no
Transcendental	applicationid	varchar	can
Contemplation	applicationdate	varchar	can
The world of existence	maingenusid	varchar	can
List	genusid	char	no
Dionysian	applicationcontryid	varchar	can
Public will	contryid	char	can
Leap of Faith	patenttype	varchar	no
Heisenberg Uncertainty Principle	name	char	no
Generated world	claim	char	no
Pascal's bet	interapplication	varchar	can
True self	interopen	char	can
Thinking self	incomedate	varchar	no

According to the database structure shown in Table 3, the philosophy and social science terminology is converted into different social science categories, and this database structure is managed using the existing management form of the computer in order to realize the retrieval function of the philosophy and social science terminology system [9, 10]. The database index is not designed for full-text indexing. When using keywords to query, the database search process becomes a traversal process similar to page-by-page flipping. For an efficient retrieval system, the key is to establish a reverse indexing mechanism similar to the scientific and technological index. When storing the data source, such as a series of articles, in sorted order, there is another sorted keyword list.Used to store the mapping relationship between keywords and articles, the retrieval process is the process of turning fuzzy queries into a logical combination of multiple precise queries that can use the index [11, 12]. As a result, the efficiency of multi-keyword query is greatly improved. Therefore, the term retrieval problem is ultimately a structural ranking problem. Therefore, Lucene is used to redefine the terminology database composition structure during retrieval. The defined composition structure is shown in Table 4:

Table 4. Database search composition structure

Serial number	Composition structure	Search function
1	Org. apache. Lucene. search/	Search entrance
2	Org. apache. Lucene. index/	Search entrance
3	Org. apache. Lucene. analysis/	Search entrance
4	Org. apache. Lucene. queryParser/	Search entrance
5	Org. apache. Lucene. document/	Search entrance
6	Org. apache. Lucene. store/	Underlying IO storage structure
7	Org. apache. Lucene. util/	Some common data structures

According to the database search composition structure as defined in Table 4, the retrieval of the philosophy and social science terminology database is finally realized, and the software design of the philosophy and social science terminology dictionary system is completed.

4 Simulation Experiment

4.1 Build System Experiment Framework

When constructing the experimental system framework, enter keywords in the system to search and query according to the requirements of the dictionary system, give relevant explanations, the server performs the search operation after receiving the searched query, and returns the search results and displays to Queryer, so the experimental framework of the system can be built as:

According to the system experiment framework shown in Fig. 6, two traditional philosophical and social science term dictionary systems and a big data mining-based philosophical and social science term dictionary system are used to conduct experiments, and three systems are set to query 100 to 500 philosophies respectively. Social science terms, comparing the reaction time of three design systems to different amounts of philosophical terms.

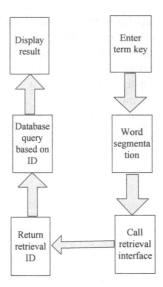

Fig. 6. The system experiment framework built

4.2   Analysis of Experimental Results

Based on the above experimental preparation, the system's detection program is called, and the received data packets of the three systems are detected to be in the normal receiving state. The definition begins to be retrieved. The interface displays the term interpretation as a reaction time. The query time fed back by the dictionary system under the number, and the final experimental results of the five systems, are shown in Table 5:

Table 5. Experimental results of three dictionary systems

Number of terms	Average time of system response (ms)		
	Traditional dictionary system 1	Traditional dictionary system 2	Designed dictionary system
100	10.54	8.64	3.58
200	9.65	6.57	3.62
300	10.20	7.23	2.68
400	10.68	8.32	3.57
500	9.88	6.48	3.84

During the experiment, the debugging program is called to detect the reaction time of the system to a philosophical and social science term. From the experimental results, it can be seen that the traditional dictionary system 1 has the longest response time to a term retrieval among the three dictionary systems, and the system's timely response is weak Compared with traditional dictionary system 1, the traditional dictionary system 2 has a shorter retrieval response time than the traditional dictionary system 1, but there is still a delay. The average retrieval response time of the philosophy and social science term dictionary system based on big data mining is faster than the two traditional ones. System, the timeliness of the system is strong, and it is more suitable for practical application.

On this basis, the accuracy of the three systems for the detection of philosophy and Social Sciences terms is analyzed, and the comparison results are shown in Fig. 7.

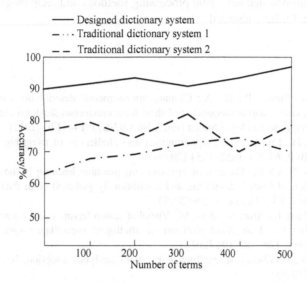

Fig. 7. Accuracy test in terms of philosophy and Social Sciences

It can be seen from Fig. 7 that the average retrieval accuracy of traditional dictionary system 1 for 500 words is 68%, that of traditional dictionary system 2 for 500 words is 74%, while that of philosophy and Social Sciences terminology dictionary system based on big data mining is 94% on average. Thus, it can be seen that the philosophical and social science term dictionary system based on big data mining has a positive impact on philosophy The accuracy of social science terminology detection is high.

5 Conclusion

The term dictionary compilation requires a more general knowledge organization model to provide a framework for the design of dictionary compilation systems. Furthermore, the concepts of user interaction, dynamic update, and term calculation in knowledge organization research are introduced into the dictionary compilation process, and a semi-automatic term dictionary auxiliary compilation system is designed.

This design properly integrates process management, term calculation, user interaction, etc., and helps to improve the quality and efficiency of term dictionary compilation. Using the existing professional literature database as a rough corpus, it is convenient for compilers to choose vocabulary, quantitative analysis and knowledge extraction, and improve work efficiency. The dictionary data is multi-dimensionally linked according to the semantic structure of knowledge organization to form a multimedia representation, which helps users understand the relationship between different concepts and improve the efficiency of knowledge learning. Strengthening the research on terminology calculation, scientific and technological corpus construction, and the formation of term-oriented automatic processing methods and technologies is a subject that needs to be further enhanced.

References

1. Wagreich, M., Sames, B., Hu, X.: Climate-environmental deteriorations in a greenhouse earth system: causes and consequences of short-term cretaceous sea-level changes (a Report on IGCP 609). Acta GeologicaSinica (Engl. Ed.) **93**(S1), 144–146 (2019)
2. Jia, P., Xue, H., Liu, S., et al.: Opportunities and challenges of using big data for global health. Sci. Bull. **64**(22), 1652–1654 (2019)
3. Susan Smith, N.A.S.H.: The role of big data and machine learning in the integration and implementation of historical, current, and continuously gathered earth data. Acta GeologicaSinica (Engl. Ed.) **93**(S1), 56–58 (2019)
4. Zheng, P., Shuai, L., Arun, S., Khan, M.: Visual attention feature (VAF): a novel strategy for visual tracking based on cloud platform in intelligent surveillance systems. J. Parallel Distrib. Comput. **120**, 182–194 (2018)
5. Boldosova, V.: Deliberate storytelling in big data analytics adoption. Inf. Syst. J. **29**(6), 1126–1152 (2019)
6. Zhao, Y., Zhong, H., Xu, S., et al.: Quantitative expression of paleogeographic information based on big data. Acta GeologicaSinica (English Edition) **93**(S1), 83–85 (2019)
7. Li, M., Fu, J., Chen, A., et al.: The connation of geological big data mining based on cloud computing. China Min. Mag. **28**(S1), 343–346+348 (2019)
8. Zhao, Y., Zhang, Y.: Design of text information mining system in big data environment. Mod. Electron. Tech. **41**(1), 125–128 (2018)
9. Chen, B., Sun, L., Han, X., et al.: A bridge-based lexicon learning method for semantic parsing. J. Chin. Inf. Process. **33**(5), 24–30 (2019)
10. Bao, H., Li, L., Xu, W., et al.: Data mining of psoriasis prescription for external use: based on dictionary of traditional Chinese medicine prescriptions. J. Tradit. Chin. Med. **60**(11), 974–978 (2019)
11. Shuai, L., Gelan, Y.: Advanced Hybrid Information Processing, pp. 1–594. Springer, Heidelberg
12. Liu, S., Lu, M., Li, H., et al.: Prediction of gene expression patterns with generalized linear regression model. Front. Genet. **10**, 120 (2019)

Design of Urban Air Quality Monitoring System Based on Big Data and UAV

Ying Zhao[1], Peng-yao Shi[1,2], and Wen-hao Guo[3(✉)]

[1] Nanchang Institute of Science and Technology, Nanchang 330108, China
Zhaoying85236@163.com
[2] Nanchang Key Laboratory of VR Innovation Development and Application,
Nanchang 330108, China
[3] Business School, Suzhou University of Science and Technology,
Suzhou 215009, China
liwen33218@163.com

Abstract. In order to effectively evaluate China's air quality and provide technical support for maintaining a good atmospheric environment, the design scheme of urban air quality monitoring system based on big data and drone is studied. In the research process, a system hardware environment including a drone platform, an air quality sensor and anti-jamming equipment was constructed, which provided the basis and support for the development of system software. In the software design, the monitoring terminal program and the air quality information acquisition module are designed according to the system requirements, and the data received by the drone is restored, analyzed and stored and managed by the data multi-thread receiving module. The experimental results prove the effectiveness of the urban air quality monitoring system based on big data and drone. Applying the system to actual monitoring is beneficial to better analysis of the atmospheric environment and better maintenance of the atmospheric environment.

Keywords: Big data · Drone · Air quality · Monitoring

1 Introduction

At present, monitoring equipment for atmospheric environment in most cities relies on air quality monitoring stations and portable air quality detectors [1]. The air quality monitoring station is the basic platform for air quality monitoring and assessment of air quality, but it can only monitor the average air quality in the local area, and the cost and maintenance costs are high. In addition, the conventional air quality monitoring methods on the ground are also difficult to monitor the pollution sources in the vertical direction.

Uav air monitoring and real-time data processing can solve this problem. As the third generation of rocker technology, uav telemetry and induction technology is the key to uav air quality monitoring, which can collect, store and transmit urban atmospheric data. It has been paid more and more attention by scholars and experts at home and abroad, and is a hot topic in the academic field and enterprise research and

S. Liu and L. Xia (Eds.): ADHIP 2020, LNICST 347, pp. 159–169, 2021.
https://doi.org/10.1007/978-3-030-67871-5_15

development field. Telemetry and sensing of uav is a comprehensive system, in which the core technologies include telemetry sensor, data storage and real-time transmission technology. Airborne air quality monitoring sensor is the main equipment of uav to monitor air quality. Its work types mainly include sample collection, particle detection, infrared scanning, microwave radiation, etc. The data storage and real-time transmission system of uav is an important part of remote sensing system, which directly determines the scale and quality of the whole uav used for air quality monitoring. The information transmission between uav platform and ground console is realized through data link [2].

This study proposes a city air quality monitoring system based on big data and unmanned aerial vehicles to make up for the shortage of existing equipment. The drone has good stability and can fly at a long distance. The flight control is simple and reliable, and the cost is low. The hovering height is suitable for air quality monitoring. The airborne air quality sensor samples and processes the monitored area, avoiding the adverse effects of a few human errors on air quality monitoring results due to limited or unreasonable monitoring stations. The data storage and real-time transmission system of the drone is an important part of the remote sensing system, which directly determines the scale and quality of the entire UAV for air quality monitoring [3]. The information of the drone platform and the ground control station (including control commands, location information, task data) is transmitted through the data link. At present, the domestic research and development of air quality monitoring for drones has started shortly. The main indicators of monitoring are only particle concentration, NO2 content and O3 content. The encrypted big data is stored in the database; when the data is extracted, the data is decrypted by the chaotic encryption method to complete the safe storage of the urban air quality monitoring big data.

2 System Hardware Design

This UAV air monitoring system consists of two systems, the UAV platform and the ground station, which can carry out more sophisticated air quality monitoring for industrial areas, large construction projects and other areas [4]. When the drone performs the monitoring task: first place the drone platform in the open space, connect the power of the drone and the ground station, and initialize the hardware and software; then load the task data; the drone platform receives the ground. The flight instruction of the station begins to lift off; when the scheduled altitude is reached, the drone platform begins to level off to the mission point; after reaching the mission point, the drone begins to collect the air quality data at this point and simultaneously transmits it to the ground station; After the air monitoring task of this point, the drone platform flies to the next task point; after the two steps of the cycle, until the monitoring of all the mission points is completed, the drone returns spontaneously; if the drone platform experiences mechanical failure or power during the execution of the mission If the alarm is too low, the ground station should send the return flight instruction in time. Otherwise, the ground station will not receive the return instruction within 3 min, and the drone platform will return to the self-reliant emergency.

2.1 Hardware Component Framework

According to the design ideas and application analysis, the system hardware platform is built. The system uses the coaxial anti-slurry unmanned helicopter to carry the design hardware air quality monitoring system as the monitoring terminal. The ground-connected PC acts as the ground terminal and passes the collected air quality data information. The GPRS network is transmitted back to the ground terminal display, and finally the air quality monitoring function is realized. According to the design monitoring system requirements, the air quality data acquisition and processing part includes the acquisition module (sensor), the control processing chip, the storage module, the positioning module, the transmission module, etc. [5]. The data receiving part of the ground includes a data storage module, a data processing module, a display module, and the like. The use of various modules to coordinate and cooperate with the GPRS network to complete the collection, reception, processing and display of air quality data. The block diagram of each system module is shown in Fig. 1.

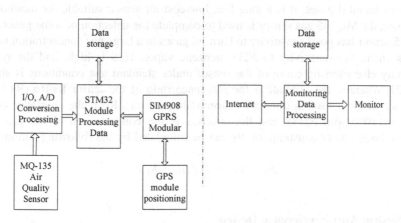

Fig. 1. System composition framework

2.2 UAV Platform Design

This section focuses on a relatively clear introduction to the hardware modules of the system data link, and integrates the modules to complete the hardware design of the air quality monitoring system. The air quality monitoring system device designed by the system is designed and built on the self-developed coaxial anti-slurry unmanned helicopter. The global coverage of the GPRS network is used to monitor the air quality of the area in real time, and the hovering characteristics of the helicopter are also It is easy to apply air quality to the vertical space of a specific location [6–8].

ARM (Advanced RISC Machines), a name derived from Reduced Instruction Computer (RISC) technology, is a popular name for a class of microprocessors. The main system of the ARM core consists of the core ICode bus, system bus, DCode bus and GP. - DMA four drive units constitute [9]. The STM32 chip selected by the system is an ARM microprocessor based on Cortex-M3 core. The architecture also includes three passive units in addition to four active units: internal SRAM, internal flash memory, AHB to APB bridge (AHB2APBx).), and adopted the ARMv7-M architecture.

2.3 Air Quality Sensor

The attitude measurement sensor module is mainly used to measure the three-dimensional posture information of the aircraft when flying in the air, that is, three attitude angles, angular velocities and the like. Accurate real-time acquisition of the attitude of the aircraft is the basis for the stability control of the four-rotor attitude, which directly determines the stability of the aircraft control. The entire attitude measurement module consists of a three-axis gyroscope, a three-axis accelerometer, and a three-axis magnetometer. In this paper, the MPU6050 micro inertial device and the AK8975 magnetic sensor are used to design the attitude measurement module [10].

The MQ135 gas sensor is highly sensitive to sulfides, NH3, aromatics, and benzene vapors and can also monitor smoke or other harmful gases. Because it can detect a variety of harmful gases, it is a long-life, low-cost air sensor suitable for monitoring. Therefore, the MQ135 gas sensor is used to complete the detection of some gases. The MQ135 sensor has good sensitivity to harmful gases in a large gas concentration range. Among them, the sensitivity to NH3, benzene vapor, H2S is high, and the typical sensitivity characteristic curve of the sensor under standard test conditions is shown [11, 12]. Wherein, the ordinate is the resistance ratio of the sensor Rs/Ro (Rs is the resistance of the sensor in different concentrations of gas, Ro is the resistance of the sensor in 1000 ppm NH3), and the abscissa is the concentration of the gas. The sensitive body power consumption Ps can be calculated by the following formula:

$$Ps = Vc^2 \times R_s / (R_s + R_L)^2 \tag{1}$$

2.4 System Anti-interference Design

Whether a system has anti-interference ability or not depends on whether the system is capable of fulfilling its functional mission is also the main indicator to prove the reliability of the system. In the hardware design system, its anti-interference ability must be fully considered [13]. In order to reduce interference during hardware design, it is necessary to take anti-interference measures and implement the following measures:

(1) The independent power supply line is used, and the battery on the unmanned helicopter is not used, and a power supply that can ensure the power consumption of the system is specially designed, so that the power supply line of the air quality monitoring system and the power supply part of the drone that is prone to interference are separately supplied with power. They are unaffected by each other, reducing the mutual coupling of the public power sources and improving the reliability of the system function circuits.

(2) Add a decoupling capacitor between the system power and ground. The decoupling capacitor can bypass the high-frequency noise of the device and can also be used as

the storage capacitor of the integrated circuit. The charge and discharge energy generated by the circuit switch gate can be provided or absorbed by it.

(3) Widening the broadband of the system power supply and ground wire, and selecting a thicker copper wire as the bottom line. The width is in order: ground wire > power wire > signal wire. Wiring avoids small angles and minimizes high frequency noise.

(4) A more reasonable layout. The system that has been miniaturized is mounted on the head of the unmanned helicopter, far from the motor part of the drone, and will not cause interference due to the close distance of the components. The internal power supply and high-frequency circuit parts of the system should be kept as far as possible, and the GPS antenna should be pulled to the tail rod for fixing, which is conducive to signal reception.

(5) Use a larger amount of magnetic beads in circuits such as power supply circuits and SIM908 modules. The magnetic beads have the effect of suppressing RF noise and peak interference on the transmission line, eliminating electrostatic pulses, etc., and largely attenuating the high-frequency current when passing through the wire.

(6) The clock signal is very susceptible to noise interference, and is also the source of some noise. When designing, try to make the clock circuit should be placed in the memory as much as possible. The crystal oscillator does not take the signal line between the two pins. The crystal oscillator case is connected to the capacitor and grounded.

3 System Software Design

3.1 System Software Development Tools and Environment

The system applies RealView MDK, an RM embedded development tool, which contains C language program, assembly language compiler, real-time kernel, debugger and other components. It has powerful functions and can help users complete the corresponding engineering tasks. Terminal application STM32 series microprocessors, the processor chip for Cortex-M3 kernel. The system microprocessor supports a variety of program download modes, such as ULink, J-Link and other online simulation debugging programmers, serial port download, Flash download and so on. J-link simulation debugger is selected in this study. The system can be connected to PC through USB port, which is convenient for online debugging when the hardware is running and for direct application of zero-cost serial port download.

3.2 Monitor Terminal Programming

The monitoring data terminal of the air quality monitoring system mainly completes the data collection, processing, positioning and other work, and finally transmits the data to the ground display terminal. On this basis, combined with the hardware design requirements to complete the corresponding software program design. After the system is started, the initialization of each module is completed first, and the network and information positioning are searched. The air quality sensor is a heat sensitive module, which needs to be warmed up for a certain period of time. After the process is finished, the module will collect. The air quality information of the location is connected to the system control unit through the AO port to process the AD conversion data, and the internal register storage related information is backed up by DAM. After the GPRS network connection is successful, the data information is sent to the server through the established IP protocol, and the remote display air quality of the ground display terminal is monitored.

3.3 Acquisition Process Module Design

The main task of the air quality data acquisition system is to complete the collection, processing and analysis of atmospheric dust particle concentration, carbon monoxide, temperature and humidity data, and transmit the collected data to the ground monitoring terminal. System software is designed in a modular fashion. After the system is powered on, firstly initialize the communication interface and related function modules used by the system, and wait for the acquisition of the acquisition signal. Once enabled, each detection sensor is driven to start collecting atmospheric data. After the processing is completed, the data is saved and saved to the SD card. Finally, it is sent to the flight control system through the serial port, and then the flight control passes the wireless data transmission module to forward the data to the ground data monitoring terminal.

(1) Dust particle concentration collection: The dust particle detection sensor PMS5003 uses the serial communication protocol to transmit data to the MCU through the serial port. To collect dust particle concentration data, it is necessary to initialize the serial port first, and then design the serial port driver. The carbon monoxide sensor and data transmission module also communicate through the serial port and are connected by different serial ports.

(2) Carbon monoxide concentration collection: The ZE07–CO type carbon monoxide detection sensor used in the system provides a UART output mode, and the data from the sensor can be read only through the serial port of the STM32F103ZET6 processor. When designing the carbon monoxide concentration acquisition program, first initialize the UART3 serial port. After the configuration is complete, open the serial port to start receiving interrupts. Then send a request to read the data frame, wait for the ZE07-CO sensor to respond to the output data, and then check the received data frame. If the verification is correct, save and parse the data frame to obtain carbon monoxide concentration data; If it fails, it will return to re-read the serial port data.

(3) Temperature and Humidity Data Acquisition: The system uses the temperature and humidity sensor SHT21 to communicate with the MCU through the standard I2C protocol. I2C is a two-wire serial bus consisting of data lines and clock lines. It is commonly used for communication between microcontrollers and peripherals. There are three types of signals, start, acknowledge, and end, during data transmission over the I2C bus. The data is transmitted and received through the cooperation of the clock SCL and the data line SDA.

The temperature and humidity acquisition process is as follows: first, the sensor is powered on and then enters the idle state; after being stabilized, the sensor is ready to receive commands from the MCU. MCU first sends start transmission signal, that is, when SCL is high, SDA is converted from high level to low level, then sends acquisition command (00000011 represents acquisition temperature, 00000101 represents acquisition relative humidity), waits for the end of acquisition, reads temperature and humidity measurement data in turn, and checks the data through CRC check code. Finally, the measurement signal is converted into actual temperature and humidity data, and the relative humidity signal and temperature signal output by the sensor are:

$$RH = -6 + 125 \times \frac{S_{RH}}{2^{16}} \tag{2}$$

$$T = -46.85 + 175.72 \times \frac{S_T}{2^{16}} \tag{3}$$

3.4   Data Multithreaded Receiving Module

In order to make the functions of the data receiving program not conflict, the receiving program of the design display terminal adopts multi-threading technology. First, a main thread is created, and then a sub-service thread is created by the main thread. When the display terminal program starts, the sub-service thread is used to wait for the connection request of the client, that is, the drone air quality monitoring system, to collect the terminal data. After the request is received, the main thread creates another data receiving sub-thread to process the request, and then returns a connection request waiting for another client. The three threads do not affect each other, operate independently, and send data to transfer data between threads.

4   Simulation Experiment and Result Analysis

In order to verify the practical application performance of the urban air quality monitoring system based on big data and UAV designed in this study, the following simulation experiment is designed.

After completing the hardware and software design of the entire UAV atmospheric environment monitoring system, it is necessary to debug the functions of each module of the system to ensure stable operation of the system. The UAV and air quality acquisition system were debugged separately, the feasibility of the system was tested, and the test results were analyzed. Finally, the whole system function was verified.

After completing PID parameter setting and basic debugging, flight test is carried out. Test system performance based on actual flight data. During the test, the flight parameters were sent to the ground terminal in real time through the wireless data transmission module, and the data was saved and analyzed by MATLAB. The experimental workflow is shown in Fig. 2.

The experimental environment is as follows: ambient temperature: 20–24 °C; wind speed: level 2; uav flying altitude: 100 m.

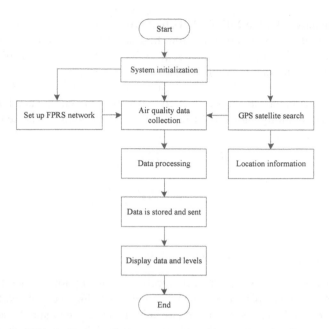

Fig. 2. Schematic diagram of the experimental process

The working environment of uav is as follows: the dual-frequency board card has good shielding function and anti-interference ability, the 80-pin pin provides a rich communication interface, it has very low power consumption of BD920 and the standard data update rate of 20 Hz. The dual-frequency board card is responsible for the output of high-precision base station coordinate data; The u-blox positioning module USES the ds-26-u8l GPS/GLONASS/Beidou module, which is a super-high power consumption and ultra-high sensitivity, ultra-small appearance GNSS receiving module, built-in SAW+LNA, can support multi-system positioning.

4.1 Anti-interference Performance Test

When the drone is hovering, the aircraft is disturbed by artificially displacing the frame to test the anti-jamming performance of the system. In the case of interference, the attitude response is shown in Fig. 3.

Analysis of Fig. 3 shows that at 86.5 s, the UAV's pitch and roll angles are disturbed, the pitch angle has a disturbance amplitude of up to 10°, and the roll angle interference is less than about 3.5°. The aircraft then responds quickly to adjust the flight attitude. After about 1.5 s, it gradually returns to the hovering state, and the adjustment speed is faster. It can be seen that the roll angle and the pitch angle have strong anti-interference performance. In about 116 s, the yaw angle of the drone is disturbed, the interference amplitude reaches 26°, and the interference is removed at about 116.8 s. Then the aircraft adjusts the yaw angle posture by itself, and returns to the initial state after about 4 s. In the yaw angle adjustment process, a 12° overshoot is

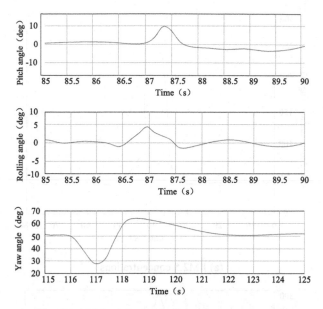

Fig. 3. Anti-jamming attitude angle response curve

generated, and the overshoot is too large. After the analysis, the integral action is too large, and the integral parameter can be further optimized and controlled.

4.2   Air Quality Data Acquisition System Debugging

The air quality data acquisition system is designed based on the STM32F103 minimum system module. After completing the hardware platform construction and software design, the function of each sensor module is debugged. First, the dust particle detection sensor, carbon monoxide sensor, temperature and humidity sensor are tested through the serial port debugging assistant. It can work normally, and the standard monitoring instrument is used to compare and calibrate the collected data to ensure the accuracy of the collected data. After the data of each sensor is tested normally, the air quality data acquisition module is mounted on the drone platform and jointly debugged with the ground monitoring terminal.

Firstly, in the indoor environment, when the drone is in a static state, the indoor air quality is detected, and the sensitivity of the system is tested by means of artificially manufacturing pollution. The ground monitoring terminal displays and records each concentration data, and sets the data to be updated once in 1 s. The air pollution was simulated by igniting the wooden block. After the system was powered on, it was detected without any interference. The burning wood block was approached to the four-rotor atmospheric environment monitoring system, and then the wooden block was immediately removed. The test results are shown in Fig. 4.

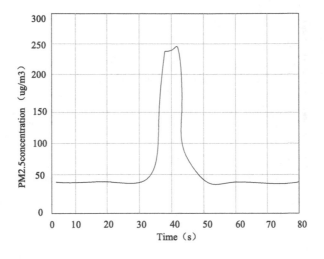

(a) PM2.5 concentrations

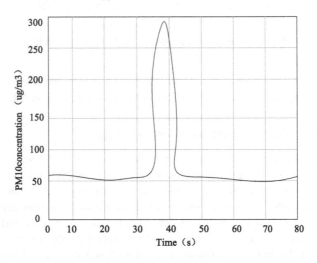

(b) PM10 concentrations

Fig. 4. Concentration variation curves of PM2.5 and PM10

According to Fig. 4, at the beginning of detection, the concentration of PM2.5 and PM10 was kept at about 46 ug/m and 53 ug/m respectively. When the burning block was brought close to the aircraft for about 32 s, the concentration of PM2.5 detection instantly reached 232 ug/m, while the concentration of PM10 detection rose to 289 ug/m. When the block was removed and the fire source was extinguished, the monitoring data gradually returned to normal. It can be seen from the curve in Fig. 4 that the air quality data acquisition system designed has a high sensitivity.

5 Conclusion

This study designed an urban air quality monitoring system based on big data and unmanned aerial vehicles, and proved its effectiveness through experimental results. However, due to the limitations of research time and experimental resources, the system still has some deficiencies, such as high energy consumption and slow start-up speed. Therefore, in the following research, the focus will be on improving the start-up speed of the system and reducing the system energy consumption.

Acknowledgements. Jiangxi Provincial Department of Education Science and Technology Research Project (GJJ171108).

References

1. Xu, Z., Puqiang, Z.: A geographic information service system for air quality in networked cities. Surv. Mapp. Sci. **42**(2), 47–52 (2017)
2. Yilin, Z., Xiaoyan, W., Wei, W.: Analysis of environmental air quality characteristics and predictive influencing factors in major cities in Northwest China. Environ. Monit. China **33**(5), 17–22 (2017)
3. Cong, D., Jian, W., Xiangfeng, et al.: Design of an early warning platform for environmental air quality prediction in typical plateau mountainous cities. China Environ. Monit. **33**(5), 57–62 (2017)
4. Wang, Y., Wang, Z., Jiping, et al.: Spatial distribution types and weather background of air quality in the Pearl River Delta region. Environ. Eng. (12), 77–81 (2017)
5. Jinjia, W., Menglan, S., Zhi, H.: Sparse group lasso penalty vector autoregressive model for atmospheric pollutant prediction: Beijing-Tianjin-Hebei case study. High-Tech Newslett. **27**(6), 567 (2017)
6. Zhou, Z., Hu, D., Liu, Y., et al.: Visual analysis method for spatial and temporal multidimensional attributes of air quality monitoring data. J. Comput. Aided Des. Graph. **29**(8), 1477–1487 (2017)
7. Liu, S., Liu, D., Srivastava, G., et al.: Overview and methods of correlation filter algorithms in object tracking. Complex Intell. Syst. (2020). https://doi.org/10.1007/s40747-020-00161-4
8. Anonymous. Dynamic change characteristics of urban air quality index (AQI) in China. Econ. Geogr. **38**(9), 264–284 (2018)
9. Zheng, P., Shuai, L., Arun, S., Khan, M.: Visual attention feature (VAF): a novel strategy for visual tracking based on cloud platform in intelligent surveillance systems. J. Parallel Distrib. Comput. **120**, 182–194 (2018)
10. Liu, S., Bai, W., Zeng, N., Wang, S.: A fast fractal based compression for MRI images. IEEE Access **7**, 62412–62420 (2019)
11. Chen, Y., Lin, Q., Xia, J., et al.: Drone-borne mobile monitoring system for urban air quality. J. Electron. Meas. Instrum. **33**(10), 1–6 (2019)
12. Liu, S., Lu, M., Li, H., et al.: Prediction of gene expression patterns with generalized linear regression model. Front. Genet. **10**, 120 (2019)
13. Lv, N., Zhao, J.: Evaluation method for urban air quality based on gis technique. J. Donghua Univ. (Nat. Sci. Ed.) **45**(05), 730–734 (2019)

Design of Intelligent Monitoring System for Air Visibility Data Based on UAV

Jun Zhang[1(✉)] and Jun-jun Liu[2]

[1] College of Electrical Engineering, Zhengzhou University of Science
and Technology, Zhengzhou 450064, China
zhaoxia1812@163.com
[2] Department of Information Engineering Information Engineering,
Zhengzhou University of Science and Technology, Zhengzhou 450064, China

Abstract. UAV has the advantages of small size, low cost and strong performance. How to use UAV to realize the intelligent monitoring of atmospheric visibility data has become one of the research hotspots. Therefore, an intelligent monitoring system of visibility data based on UAV is designed. Through the overall design of the system, the requirements for the collection, processing, transmission, display, monitoring and positioning of atmospheric visibility data are realized. The airborne air visibility sensor module is used to collect gas data, the microprocessor is used to process data, the global positioning system is used to complete monitoring and positioning, and the general packet wireless service is used to complete data transmission and data display of the ground terminal. The experimental results show that the air visibility data intelligent monitoring system based on the UAV has good accuracy and high monitoring efficiency It can meet the requirements of data monitoring system.

Keywords: Drone · Air visibility · Data intelligent monitoring · System design

1 Introduction

With the development of science and technology, the application of UAV is more and more extensive. Its excellent performance and wide application are the reasons. The research of UAV has been carried out in China and remarkable results have been achieved. UAV, especially unmanned helicopter, has the characteristics of high reliability, low noise, small vibration, simple structure, low cost, no need for landing field and runway, vertical landing, free hover, high mobility, and many kinds of remote control or automatic control modes. It is an important rescue equipment for building ground air rescue network [1]. At present, it is competent in earthquake investigation and rescue, medical rescue, mountain search and rescue, water risk rescue, high-rise fire rescue and other difficult rescue work, anti terrain racism rescue, etc. Based on the concept of rescue and detection, we are committed to the research of UAV monitoring capability. Air visibility is also a major issue related to the national economy and people's livelihood. Air pollution is the top priority of air pollution [2]. The monitoring of air visibility has become an important issue of air pollution. Therefore, relevant researchers have conducted a lot of research.

S. Liu and L. Xia (Eds.): ADHIP 2020, LNICST 347, pp. 170–181, 2021.
https://doi.org/10.1007/978-3-030-67871-5_16

In reference [3], a data link of UAV air quality monitoring system based on single-chip microcomputer is designed for air environment protection, so as to realize the functions of air quality data collection, analysis, display and output, and pollution source location. Through the UAV equipped with air quality monitoring module, the collected data are analyzed by the main control chip; through the GPS positioning system to accurately locate the monitoring area, the GPRS communication protocol network is used to quickly and accurately send the monitored air data and positioning data back to the ground terminal, track the pollution source and timely warn and process the regional air quality. According to the route planning of the positioning system, the system can accurately monitor and collect data, but the design of the system is complex and the efficiency of monitoring is poor. Literature [4] is mainly aimed at the improvement of today's environmental monitoring technology. Through the air data acquisition system carried by UAV, the air environment quality of various areas (such as farms, heavy industry factory gathering areas, traffic network intensive areas, etc.) can be comprehensively assessed. The detectable air data include PM2.5, sulfur, carbon, nitrogen and other air pollution parameters. The system can effectively monitor the quality of atmospheric environment, but the visibility monitoring of the system is rarely considered, resulting in the low accuracy of monitoring. Literature [5] In order to improve the detection speed of UAV air data (altitude, airspeed) and optimize the flight control performance, a fast data detection system based on GM 8123 chip is designed. The parallel communication of three asynchronous serial port data is completed by using serial port expansion technology, and then the synchronous acquisition of dynamic and static pressure data and real-time response to the correction command of flight control computer are realized. The performance index of the system is better than that of the original detection system without serial port expansion technology, which is of high practical value for improving the dynamic performance of UAV air data detection. But the system is easy to be affected by the external environment and has poor anti-interference ability. The design of mobile multi baseline visibility test system is proposed in reference [6]. The system consists of four modules: transmitting, receiving, moving and signal processing. The 532 nm laser is used as the transmitting light source. The mobile car on the sliding track receives the transmitted signal at fixed point to realize multi baseline measurement. Then, the initial signal is processed by time averaging and least square method to obtain the final atmospheric visibility, The measurement variance and baseline sampling point model of the equipment are ana-lyzed theoretically. The noise simulation experiment is carried out to prove its anti noise ability. This method can accurately transmit and receive the visibility signal, and the monitoring accuracy is high. However, the cost of this method is high and it is not suitable for general use. Reference [7] proposed the design of automatic monitoring data acquisition system based on intelligent wireless nodes of the Internet of things. In order to improve the ability of data detection, diagnosis and analysis, the design method of data acquisition system based on dynamic gain control and DSP high-speed signal processing is proposed. Based on microcomputer bus technology, the overall design of wireless node automatic monitoring data acquisition system is carried out. The system mainly includes DSP processor and PCI bus. The sensor array of data acquisition is composed of wireless monitoring nodes with multi-sensor information fusion. The transceiver conversion circuit and power amplifier circuit are designed to realize the signal amplification and digital to analog conversion of collected data. The dynamic gain control method is used to amplify and filter the collected data. The integrated design of automatic monitoring data acquisition system is based on DSP signal processor. The system can achieve continuous real-time data acquisition and recording of more than 15 MB/s, but the process of data acquisition is complex and has certain limitations.

In view of the above problems, this paper designs an intelligent atmospheric visibility monitoring system based on UAV. It takes the electric coaxial twin propeller UAV as the carrier. It has the advantages of low noise, no launch interference, hover detection, low altitude flight and so on, so it is suitable for environmental monitoring platform of urban atmospheric visibility or emergencies. Whether based on rescue or monitoring, UAV can be used to monitor the air visibility in a specific area more easily and achieve vertical monitoring. In addition, the UAV is easy to operate and driverless, which reduces the cost and improves the safety performance. Therefore, UAV is applied to air visibility. Degree monitoring is a very convenient method. The data link of UAV air visibility data monitoring system is designed, including data display platform, data acquisition terminal (coaxial twin propeller UAV helicopter), GPRS wireless transmission network. GPS positioning system is used to accurately locate the monitoring area, and GPRS communication protocol network is used to determine the air visibility data the bit data is quickly and accurately transmitted back to the ground terminal to complete the intelligent monitoring of regional atmospheric visibility data.

2 Design of Air Visibility Data Intelligent Monitoring System

2.1 System Overall Structure

Fire control system, command control system and flight control system of UAV often need to perform flight tasks based on real-time atmospheric data, so an indispensable function of modern UAV includes real-time measurement of various atmospheric parameters. This function is mainly accomplished by using special air visibility data measurement equipment. The performance and quality of air visibility data measurement equipment are directly related to the flight quality and even flight safety of UAV. Nowadays, with the deepening of research, air visibility data measurement equipment has gradually developed into a special direction. A special air visibility data measurement system has been developed on UAV, including fixed-speed flight, trajectory estimation, mission estimation and decision, etc. A special control or interface is generally designed on ground station to display and monitor the air visibility information of UAV. The equipment on the UAV has the requirements of safety, reliability and maintenance. The air visibility data measurement system must be designed according to these requirements. At the same time, different UAV air visibility data measurement systems may have different functional requirements. But in these requirements, many things are universal. In the design of the air visibility data measurement system, many things are common. It can be embodied.

In order to meet the requirements of UAV, the air visibility data measurement system must have good real-time performance and be able to carry out data correction and real-time adjustment online. Its measurement of air visibility data source is mainly realized by static pressure and dynamic pressure sensor components. Its solution depends on the different requirements of air visibility data parameters, such as the number of requirements, errors, accuracy, number of channels, real-time and interface requirements, and the system scale will also be different. Its data processing and data transmission requirements meet the requirements of real-time, reliability and accuracy, in order to meet the requirements of rapid dynamic response and control of UAV. The UAV system of this subject puts forward almost strict requirements on the weight, size and power consumption of the airborne electronic equipment, and at the same time, it puts forward high accuracy and real-time requirements for the air pressure height and airspeed in flight. According to the requirements of the system, we design and implement an UAV air visibility data measurement system. The block diagram of its working principle is shown in Fig. 1.

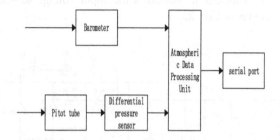

Fig. 1. The working principle block diagram of the system

It can be seen from the figure that the air visibility data measurement system includes three parts: measurement, data processing and output. Pitot tube, differential pressure sensor (space speed sensor) and AD acquisition interface complete differential pressure data acquisition. The barometer completes the collection of static pressure data. The operation processing unit based on DSC completes the output of fish data. The static pressure sensor is a sensor with digital SPI interface based on MEMS technology. The differential pressure sensor includes a pressure gauge and a simple airspeed tube suite. The differential pressure signal is output of analog signal, which needs to be converted to the processor by AD. Considering the low AD accuracy of F28335, this topic chooses to use external AD to complete the differential pressure analog signal. Collection. The interface between external AD and DSC processor is a parallel interface. As an external extended memory of the DSC, the external AD interacts with the DSC through address and data bus. The atmospheric data measurement computer based on DSC receives pressure data, temperature data of SPI interface and differential pressure data of external parallel interface respectively, and obtains air pressure height Hp and airspeed V through certain function operation. The results are transmitted to the UAV flight control system through serial interface in real time for its control, operation or transmission to other airborne equipment.

2.2   System Hardware Design

A system that can achieve complete functions needs the perfect combination of hardware and software, and the hardware part is the backbone of the system, which supports the connection of various f realization systems. This chapter focuses on the hardware module of the system data link to make a relatively clear introduction, and integrate the various modules together to complete the hardware design of the air quality monitoring system.

(1) Composition of Circuit Module

The operating voltage of UAV ranges from 3.2 V to 4.8 V, and burst data transmission may lead to low voltage. Therefore, it is necessary to provide a peak current up to 2A. The battery supply voltage we use is 5 V. In order to ensure the stability of the input voltage, a regulator MIC29302 is added, and a bypass capacitor of 100 mF is connected beside VBAT. This circuit stabilizes the input voltage to 4.2 V. The specific circuit diagram is shown in Fig. 2.

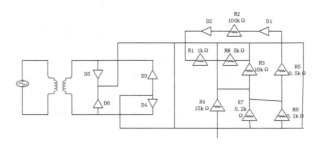

Fig. 2. Power supply circuit diagram

The start-up of the circuit is accomplished by pressing the key PWRKEY for at least one second during the opening process. At this time, the interface voltage is also pulled up to 3 V inside the module, so there is no need to pull up any more. The schematic diagram of the power start circuit is shown in Fig. 3, 4 and 5. A GPIO interface of STM32 chip connects the PWRKEY pin through a NPN triode. A high level switch pulse (>1 s) is used to start the module. The output voltage of test pin status (if greater than 2.7 V) can be used to test whether the startup is successful. When the startup is completed, URC is sent to indicate that the module is ready to operate at a fixed baud rate.

(2) Modular Serial Communication

Module is the DCE (Date Communication.) and the data, it provides two unbalanced asynchronous serial interfaces. One serves as a serial port for communication and the other as a debug port. Its main pin definitions are shown in Table 1 below.

Table 1. The table of definition at foot

Category	Pin name	Pin number	Function
Power interface	VBAT	62.63	Voltage supply
Switch interface	PWRKEY	3	Used to turn on/off modules
serial interface	RXD	68	Receive data
RF interface	GSM-ANT	79	Receiving antenna

Data debugging and firmware upgrade can be accomplished by debugging interface and error tracking can be achieved. It receives AT instructions through serial port, which is connected with STM32RBT6 chip to complete the function realization. Connections between TXD and RXD pins are interactive. The remaining serial ports of the module are connected to the GPIO port of the processor. Because the working voltage of the processor is 3.3 V, the two can be connected directly [8].

Air Visibility Sensor

There are many kinds of air components, and there are many factors affecting visibility. According to scientific research conditions, we can not monitor all related gases. Mq135 gas sensor can also monitor smoke or other harmful gases, such as sulfur compounds, NH3, aromatic compounds, benzene vapor sensitivity. Because it can detect various harmful gases, it is a low-cost air sensor suitable for long-life monitoring. Under the condition of only experimental stage, this paper selects a relatively wide gas sensor on the market to complete partial gas visibility detection, in order to achieve the effect of system function [9]. MQ135 will change in conductivity for output signal echoes with the gas concentration in the circuit is very simple. The basic test circuit as shown in Fig. 3.

Two voltage is applied to the sensor, the heater voltage VH and test voltage VC. The temperature sensor provided by VH. Load voltage VRL resistance RL on VC was measured with the sensor in series, because the MQ135 sensor has weak polarity, so the power to use dc [10].

The power consumption and resistance of the sensor can be calculated by the following formula:

Sensor power consumption P_S:

$$P_S = V_C^2 \times R_S / (R_S + R_L)^2 \tag{1}$$

Sensor resistance R_S:

$$R_S = \left(V_C / V_{RL}^{-1}\right) \times R_L \tag{2}$$

It should be pointed out in particular that the heating is normal after the sensor is electrified, and the data measured by preheating about 20S is needed to be stable, which is related to the material of the sensor itself.

2.3   System Software Design

As we have known before, the system is a combination of hardware and software. When the hardware part of a system is available, the software system is needed. If the hardware part is the flesh and blood of the system, then the software part is the brain and soul that dominates the operation of the body. Therefore, the design of the software is the core part of the realization of the system function, and the design part of the system software of the subject needs to be introduced emphatically. The software adopts modular design, which mainly includes height processing and calculation module, airspeed acquisition and calculation module and serial port transmission module [11].

Height Processing and Solution Module

Because the altimeter used in this subject is less than 5000 m, and considering the influence of air pressure by environmental factors, the altitude range of the design is less than 6000 m, so the air pressure and altitude are suitable for the formula:

$$H_P = 4430.76 \left[1 - \left(\frac{P_S}{101.325} \right)^{0.190263} \right] \tag{3}$$

The air pressure corresponding to the 0-bit height (sea level) is 1013.25 mbar. Because the calculation of air pressure and height conversion is relatively complex, it needs relatively large floating-point resources to complete the calculation. In order to avoid the complex calculation, the piecewise approximation method is generally used. The formula of piecewise approximation is as follows:

$$h = j_n - (p - p_{lower}) \times i_n / 2^{11} \tag{4}$$

In the formula h is the pressure height (m), p is the pressure (0.1 mbar), i_n and j_n are the sectional coefficients. In theory, by this piecewise linear calculation, the maximum error of measurement in the height range of -700 m to 9000 m will be less than 5 m. The height error is relatively small at low altitude (high pressure). With the decrease of the pressure at the measuring point, the height error increases gradually. The absolute error is less than 5 m in the measurement range of the project's target demand.

Design of Airspeed Solution Module

The main measurement methods of airspeed meter are differential pressure type, runner type and hot wire type. This project plans to adopt differential pressure airspeed measurement method. Although the measurement accuracy of differential pressure airspeed meter is relatively low, it is reliable, simple and easy to maintain. At present, differential pressure airspeed meter is mainly used in aircraft [12].Pitot tube is the earliest and most widely used pressure flow sensor. It is cheap, simple in structure and easy to manufacture and use, so it is still widely used in airspeed measurement and other fields. A small hole B in a swimming place is called a static pressure hole. When the pitot tube is placed in the flow field, the velocity at point A is zero due to the obstruction of the pitot tube. Therefore, besides the static pressure of the flow field, there is also a pressure (dynamic pressure) converted from kinetic energy, that is, the pressure at this point is the sum of static pressure and dynamic pressure (total pressure). Assuming that the flow of fluid is an ideal steady flow of incompressible fluid, according to Bernoulli equation of ideal incompressible fluid, the following relations can be listed for point A and point B:

$$P_0 = \frac{1}{2}\rho V^2 + P \tag{5}$$

In the formula: P is the total pressure; V is the fluid velocity; the density of the measured fluid; P is the static pressure.

According to the characteristics of UAV, the flight speed range of UAV is 38–81 km/h (10–23 m/s), and the corresponding differential pressure measurement range is 60–400pa (air density is 1.225 kg/m). Because the flight speed of UAV is not high, and the working relative height span is less than 150 m, the influence of air pressure change caused by flight height change on air speed can be neglected. The working time is short, the height span is small, and the temperature span is small, so the influence of temperature can be neglected.

Design of Serial Port Output Module

Serial port is sent by interrupt, and the processing flow is as follows:

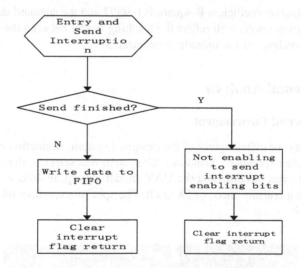

Fig. 3. Serial Port Send Interrupt Processing Flow

The relationship between EMF and air visibility satisfies a certain curve, and the corresponding parameters are shown in Table 2.

Table 2. Corresponding parameters

Serial number	X_1 ($CO_{2concentration}$)/ ppm	Y_1(EMF)/ mV	X_2 ($CO_{2concentration}$)/ ppm	Y2(EMF)/ mV
1	100	440	1500	298
2	200	375	2000	294
3	300	350	2500	289
4	400	325	3000	286
5	500	320	4000	281
6	600	315	6000	275
7	800	308	8000	268
8	1000	301	10000	263

The relationship between EMF and carbon dioxide concentration of the sensor satisfies the exponential mathematical model. The data in the table are fitted exponentially. The number of terms is chosen as 2 (two peaks of data). The fitting formula is as follows:

$$y = 3735x^{-0.6643} + 263.5 \tag{6}$$

The determination coefficient R-square is 0.9897 and the standard deviation RMSE is 4.911. The fitting results well reflect the teaching model between the sensor element and the corresponding carbon dioxide concentration.

3   Experimental Analysis

3.1   Experimental Environment

In order to verify the effectiveness of the designed system, simulation experiments are carried out. In the experiment, Windows XP system was selected, the system memory was 8 GB, CPU was 3.6 ghz, and the UAV model was sprite 4rtk, which was a four rotor UAV with a cruising speed of 58 km/h. The specific experimental environment is shown in Fig. 4:

Fig. 4. Experimental UAV

3.2 Experimental Parameter Design

The specific experimental parameter design is shown in Table 3:

Table 3. Experimental parameter design

Parameter	Value
UAV body length/mm	650
UAV fuselage width/mm	650
Duration of endurance/min	60
Maximum working height/m	6000

Based on the above parameters and experimental environment settings, simulation experiments are carried out.

3.3 Analysis of Experimental Results

The test validation of the whole air visibility data measurement system mainly includes two parts: one is function validation, that is, checking whether the hardware platform has the function of pressure and airspeed measurement, and can correctly measure and transform the required atmospheric information; the other is flight validation, that is, through actual flight test, combined with other sensor data information of UAV platform, through flight. Analysis and processing software, check or proofread the function of air visibility data measurement system. After several flight calibrations, the air visibility data computer achieves the measurement requirements of air pressure height and airspeed for the UAV.

The other part is to test the validity of the intelligent monitoring system based on UAV air visibility data, and compare the data accuracy with the traditional monitoring system. The experimental results are shown in Table 4.

Table 4. Experimental comparison

Number of experiments	Accuracy of intelligent monitoring system for air visibility data based on UAV (%)	Accuracy of traditional monitoring system (%)
1	89.19	61.09
2	90.12	63.13
3	88.15	62.54

According to the data in the table, it can be seen that the intelligent monitoring system based on UAV's air visibility data has good accuracy and can accurately monitor the air visibility data. The whole design achieves the expected purpose and requirements.

In order to further verify the feasibility of the proposed system, this paper compares the energy consumption of the system and the traditional system in operation. Among them, the lower the running time energy consumption, the more advantages the system has. The experimental results are shown in Fig. 5.

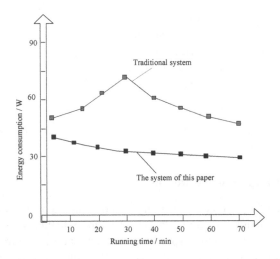

Fig. 5. Energy consumption comparison of different systems

From the analysis of Fig. 5, it can be seen that there is a certain difference between the operation energy consumption of this system and the traditional system for visibility monitoring under the same experimental environment. Among them, the energy consumption of the system in this paper is lower, while the energy consumption of the traditional system is higher. In contrast, this system has more advantages in the implementation.

4   Conclusion

With the rapid development of China's economy and aviation industry, the number of flight missions has increased rapidly. This puts forward higher requirements for flight, which not only ensures flight safety, but also makes effective use of space and time, speeds up air traffic flow and enhances the limited area capacity per unit time. Air visibility data computer is an indispensable airborne equipment for aircraft. Its altitude, airspeed and other related information are one of the key parameters of the aircraft measurement and control system. The accuracy of information plays a vital role in the safe flight of aircraft. In this paper, an intelligent monitoring system based on UAV air visibility data is designed and implemented. The overall structure and working process of the system are described in detail. The software design of the display terminal of the system is introduced. The functions of the display terminal, the conversion principle of the collected air data, the software design and interface design of the display terminal are introduced. The intelligent monitoring system of UAV air visibility data uses GPS positioning system to locate accurately, and transmits the monitored air data and positioning data to the ground terminal quickly and accurately through GPRS communication protocol network. It completes the regional air quality monitoring and achieves the overall operation function of the system.

Acknowledgments. The Henan Science and Technology Department Foundation of China under Grant No. 172102210114.

References

1. Zhuang, L., HeRui, J., Yonghong, D.: Overall design of an UAV air quality monitoring system. Ind. Des. **15**(2), 130–131 (2018)
2. Qingzhan, Z., Tianyi, Z., Chen, H.: Design of air quality monitor for small four-rotor UAV. Mod. Electron. Technol. **40**(24), 94–99 (2017)
3. Liang Dong, T., Jinpeng, Y.: Design and implementation of UAV rudder system controller based on TMS320F28069. Aviat. Precis. Manuf. Technol. **53**(1), 16–20 (2017)
4. Chao, W., Zhigang, J., Hui, C., et al.: Design and implementation of UAV ground station resource management system based on Visual Net. Electron. Technol. Softw. Eng. **19**(5), 76–78 (2017)
5. Shuai, L., Gelan, Y.: Advanced Hybrid Information Processing, pp. 1–594. Springer, Heidelberg
6. Liping, X., Lazhen, Q., Jiang, K.: Design of UAV precision variable spraying system based on PWM. South. Agric. **11**(27), 108–109 (2017)
7. Liu, S., Glowatz, M., Zappatore, M., Gao, H., Gao, B., Bucciero, A.: E-Learning, E-Education, and Online Training, pp. 1–374. Springer, Heidelberg
8. Zhuoyi, D., We, W., Jianzhong, G., et al.: New challenges in high altitude long endurance solar UAV high efficiency aero dynamic design. J. Aerodyn. **35**(2), 156–171 (2017)
9. Jinpeng, Y., Dong, L., Tao, Y., et al.: Appl. Electron. Technol. **21**(5), 86–89 (2017)
10. Rui, H., Weiwei, L., Danyang, S.: Research on UAV logistics distribution selection based on different external environment and load weight. Air Transp. Bus. **19**(12), 51–55 (2017)
11. Zheng, P., Shuai, L., Arun, S., Khan, M.: Visual attention feature (VAF): a novel strategy for visual tracking based on cloud platform in intelligent surveillance systems. J. Parallel Distrib. Comput. **120**, 182–194 (2018)
12. Rong, W., Ming, B.Y., Peng, et al.: Aerodynamics health optimization design for plane configuration off lying-wing UAV. J. Aeronaut. **38**(1), 77–84 (2017)

Design of Short-Term Network Congestion Active Control System Based on Artificial Intelligence

Shuang-cheng Jia[(⊠)] and Feng-ping Yang

Alibaba Network Technology Co., Ltd., Beijing 100102, China
xindine30@163.com

Abstract. Traditional network congestion active control system has the problems of large amount of network congestion data and uneven distribution of main control nodes. Therefore, this paper proposes a short-term network congestion active control system based on artificial intelligence. In the congestion control framework, the network motor and congestion control nodes are connected to complete the construction of the system hardware operating environment. The control strategy of artificial intelligence node is used to determine the congestion location, improve the logic control standard, optimize the system software running environment, and complete the design of short-term network congestion active control system based on artificial intelligence. The results show that, compared with the traditional random detection control technology, the short-term network based on artificial intelligence is feasible. After the design of congestion active control system, the total amount of congestion data is significantly reduced, and the master node presents an ideal uniform distribution state.

Keywords: Artificial intelligence · Network congestion · Active control · Main control frame · A congestion node · Sudden transmission mechanism

1 Introduction

Artificial intelligence is a branch of computer science that attempts to understand the essence of intelligence and to produce a new intelligent machine that responds in a similar way to human intelligence. Research in this field includes robots, language recognition, Image recognition, natural language processing and expert systems. Since the birth of artificial intelligence, the theory and technology have matured day by day, and the application field has also been expanded. It can be assumed that the scientific and technological products brought by artificial intelligence will be the "container" of human intelligence in the future. Artificial intelligence can simulate the information process of people's consciousness and thinking. Artificial intelligence is not human intelligence, but it can be thought like a man Examination, may also exceed human intelligence [1, 2].

In reference [3], a network congestion control system based on random detection algorithm is proposed. Using LAIDS/lids host random detection architecture, the congestion controller and network data filter are adjusted to complete the hardware

© ICST Institute for Computer Sciences, Social Informatics and Telecommunications Engineering 2021
Published by Springer Nature Switzerland AG 2021. All Rights Reserved
S. Liu and L. Xia (Eds.): ADHIP 2020, LNICST 347, pp. 182–192, 2021.
https://doi.org/10.1007/978-3-030-67871-5_17

operation environment of the new system. Through the definition of transmission data congestion, the possible congestion in the network is classified, and then the appropriate network congestion control mode is selected according to the specific judgment results, so as to realize the software operation environment of the new system. Combined with the structure of software and hardware, the design of network congestion control system based on random detection algorithm is completed. This method can reduce the network congestion delay, but can not solve the problem of uneven distribution of master nodes. In reference [4], a congestion control algorithm for computer networks with multiple delays is proposed, and a mathematical model of the congestion control system for computer networks is established. Based on the optimal control theory, an optimal congestion control algorithm for multi delay computer networks based on active queue management is proposed. By solving the linear matrix Riccati equation, the state feedback control law is obtained to achieve the optimal performance index. This method can improve the network congestion tracking effect, but the total amount of congestion data is large.

With the progress of science and technology, how to better control the congestion of short-term networks has become the main research direction. The existing control technology makes use of the idea of cloud platform to build an embedded real-time micro-framework, and realizes the timely elimination of congestion network data through the active response of the database. However, this method can solve the congestion data relatively limited, and can not make the host node maintain a uniform distribution state for a long time. In order to solve the above problems, a new type of short-term network congestion active control system based on artificial intelligence is designed. From the aspects of software and hardware, the position information of the control node is defined, and a number of hardware operation modules such as network motor are added. And in the follow-up application process, the effectiveness of the new system is verified.

2 Hardware Design of Short-Term Network Congestion Active Control System

On the basis of the congestion main control framework, the network control motor module and the short-term congestion node are designed to complete the hardware operation environment of the active control system, and the specific operation method thereof can be carried out according to the following steps.

2.1 Network Congestion

Network congestion is a kind of network state. When the load in the network is too low, the data packet transmitted by the network increases with the increase of the network load. When the network load increases to a certain stage, the network reaches the maximum transmission capacity and reaches the optimal state. Along with the rising of the network load is the ideal state of network, network transmission of data is also increasing, but this is clearly impossible, when the network load increasing, the network will enter a state of congestion, the network load transfer, the greater the network congestion state, the corresponding data transmission group showing a negative correlation, began to decrease. The state of the network is shown in Fig. 1.

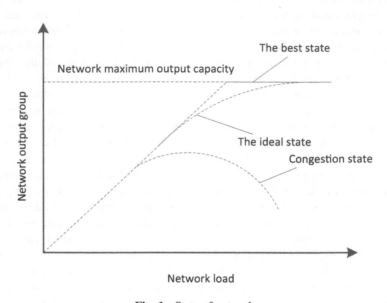

Fig. 1. State of network

Network is a multi - to - one transmission network. In the process of wireless network operation, data fusion, classification and other computing processes are needed, but the processing speed of sensor nodes is very low, so there is a certain possibility of cache overflow. When a sudden event occurs in the WSN, a large number of data streams are generated, and the cache space and available bandwidth of the output stream of these sensor nodes are limited. Meanwhile, the environment also interferes with the network, so the network cannot timely forward the groups in the cache and thus cause congestion. Unbalanced distribution of network resources and network traffic will also increase the probability of WSN congestion.

In summary, the root cause of WSN congestion is the insufficient maximum transmission capacity of the network. Shared bandwidth resources, limited node cache space and multi-hop communication mode all limit the network transmission capacity. The excellent control algorithm can detect the congestion in time and control the congestion, which can prolong the life cycle of the network and take into account the fairness of the network. Due to the particularity of network application, the network congestion algorithm needs to be specially optimized according to the needs of special occasions, and in-depth research needs to be carried out around the two cores of WSN congestion control algorithm, congestion detection and congestion processing.

2.2 Design of Congestion Control Framework

The congestion control framework of the new network system consists of two main parts: microcontroller and core application platform. Among them, the microcontroller can capture the basic connection of the network during the short-term execution time, and establish the mapping space of the network congestion data with the support of the specific hardware IP. According to the existing form of the data, the function model suitable for the data is determined, and the data acquisition operation before the operation of network control processing is completed. The core application platform contains a large number of system connection protocols, and each protocol maintains a one-to-one correspondence with each congestion state data [5, 6]. In order to ensure the short-term network can have a strong ability of timely feedback in artificial intelligence environment, DSP detection equipment is added between the microcontroller and the core application platform in the main control framework of the system. Normally, the detection results are always verified, and the active transmission of the data is always maintained at a higher level. When the detection results are negative, the DSP device takes advantage of its self-regulating function to speed up the operation of the feedback operation. To achieve the purpose of maintaining high efficiency of data transmission. The network congestion control framework is shown in Fig. 2.

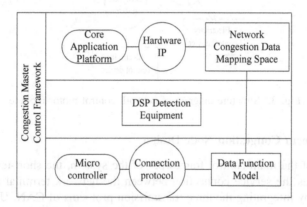

Fig. 2. Congestion control framework architecture

2.3 Network Control Motor Module Design

The network control motor module is connected with a system microcontroller by a remote terminal, and the module is used as a core device by an external motor with a plurality of rated voltages of 380 V and a rated current of 8.4 A. In order to avoid the phenomenon of insufficient power supply of the system, each external motor is connected with a remote electric quantity control device of the DSP. When the short-term network begins to execute the operation instructions, the system enters the active control state, at which time all hardware universal processing layer devices are in standby state. With the increasing running time of the system, the remote terminal processor begins to sense the power request of the system and sends the request to the multi-DC control motor through the incoming device [7–9]. And when the real-time control chip of the motor module senses the request, the real-time control chip of the motor module issues an opening instruction to the connection general switch to enable the network to control the motor module to enter the continuous power supply state. The specific module structure of which is shown in Fig. 3.

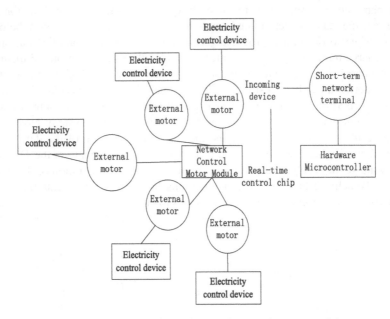

Fig. 3. Structure diagram of network control motor module

2.4 Short-Term Congestion Node Design

On the basis of the main control framework of the system, the short-term congestion node establishes the service connection between the network terminal and the active control client by integrating the three transmission protocols of CAN, UART, SRAM. When the system input module collects the execution state data of the short-term network congestion, the related operation module will change the number of real-time control nodes of the connected system according to the real-time feedback information of the input system. With the delay of the running time of the system, the congestion control data collected by the system input module also increases. In order to ensure that each data can be effectively recorded, the number of short-term control nodes of the system must also increase. All three transport protocols of CAN, UART, SRAM have high data dependency. A large number of short-term network congestion control state data are connected to real-time control nodes under the three protocols mentioned above, which improves the feedback efficiency of the system to a certain extent [10, 11]. When the network congestion control state changes, the feedback data received by the relevant operation module also changes. In order to ensure the stable running state of the system, the control motor in the congestion active control framework structure is not affected. Under the regulation of real-time control node, it is always in the state of variable frequency power supply, and the output current and voltage amplitude are always proportional to the number of nodes.

3 Software Design of Active Control System for Short-Term Network Congestion

According to the definition of artificial intelligence control node, congestion packet burst mechanism is determined, logical control standard is perfect operation flow, software running environment of the system is built, and the hardware operation conditions are combined and stable. To realize the short-term network congestion active control system based on artificial intelligence running smoothly.

3.1 Definition of Artificial Intelligence Control Node

Artificial intelligence control node (AI) means that when congestion occurs in short-term network, the congestion can be alleviated by increasing the supply of resources in the congestion area, that is, forming an additional path to share the load on the original path. After congestion occurs, multiple paths are established for data transmission, and the new paths are removed after congestion relief. This node control method costs a large amount of storage and communication, and the start-up time is long. When routing is established, two forwarding nodes are configured for each node, the default parent node and the alternate parent node [12]. When a node sends data, the alternate parent node listens for and caches packets. Once the default parent node becomes congested, the alternate parent node forwards the received packet. This method is simple, but the spare parent node consumes more energy. Priority mechanism is a method to alleviate congestion by first transmitting, delaying or dropping packets according to the importance of packets after congestion is detected. According to the idea of priority scheduling, the data is divided into three kinds of priorities. When congestion occurs, data with the highest priority is sent first. The nodes with high priority can get more speed and bandwidth by assigning different priority to the data stream. The purpose of the protocol is to ensure the transmission effect of high-priority data on the main path. After the congestion is detected, the low-priority data is forwarded through other paths to alleviate the congestion on the main path. Let U representing the artificial intelligence identification condition of the control node, w represents the maximum order of magnitude parameter of the control node, simultaneous U, w results of the definition of the artificial intelligence control node can be represented as:

$$\lambda = U \cdot (\frac{qt}{w} + p) \tag{1}$$

Where q area range coefficient representing congestion for short-term networks, t represents the basic congestion vector, p represents congestion control conditions.

3.2 Determination of the Transmission Mechanism of the Congestion Package

When the data congestion occurs in the short-term network, the control node listens to the busy situation of the surrounding media through the carrier sensing mechanism. If listening to the channel state is busy, the sending node delays sending data until the channel state is idle. If the monitored channel state is idle, in order to avoid conflicts and collisions with other nodes sharing the channel, the node enters a random Backoff state before sending network congestion data. The competitive window value has an important effect on the Backoff time, and thus realizes the active control and adjustment of congestion behavior [13–15]. When congestion occurs, multiple packets are cached in the cache queue of the node that occurs congestion. These congestion data nodes use the congestion control mechanism of packet burst to make the detected congestion node compete to the channel. In the current cycle, multiple packets can be sent continuously. According to the congestion condition, the node determines the number of continuous control instructions sent by the node in a working cycle. Let y control parameters representing packet burst mechanism, e on behalf of congestion control coefficient, the simultaneous formula (1) can express the result of congestion packet burst mechanism as follows:

$$R = \frac{\lambda \sqrt{\frac{yd \cdot |f|}{se}}}{g^2} \tag{2}$$

Where g represents a shared condition for short-term network congestion data, d representing the underlying control of the packet burst mechanism, $|f|$ represents the maximum capacity of the congestion control channel, s represents the maximum active control vector of a short-term network.

3.3 Perfect Logical Control Standard

Short-term networks do not have a separate transport layer. The functions of current transport protocols, such as reliable transmission and congestion control, are transferred to applications, supporting libraries and forwarding policy modules. When the application needs reliable transmission, the application itself or its support library monitors the status of the sent interest packets and retransmits them if necessary. Logical control standards can be used for congestion control. This control standard means that a network congestion packet forwarded across a link causes the link (in the opposite direction) to forward a packet that matches it, because the PIT records the incoming port of the interest packet. When a router receives a matching packet, it forwards the packet to the incoming port of the interest packet based on the information recorded in the PIT. Unlike TCP/IP, where the terminal is responsible for security, the publisher of logical control standard content encrypts and signs each packet to directly protect the data itself. The publisher's signature ensures the integrity of the data, which enables the user to determine the origin of the data and separate the user's trust in the data from the manner in which the data is obtained and the location of the data. Let β representing the logical coefficient condition of a short-term network, the simultaneous formula (2) defines the logical control criteria as:

$$M = R \sum_{x=1}^{c} [\beta(Lk) + G] \qquad (3)$$

Where c, x represents the top of the system control domain, L representing data congestion molecules in short-term networks, k representing the initiative control factor, G represents the fault-tolerant vector of the system channel. The design of short-term network congestion active control system based on artificial intelligence is completed by integrating all the above-mentioned numerical and hardware execution conditions.

4 Experiment and Discussion

In order to verify the practicability of short-term network congestion control system based on artificial intelligence (AI), a comparative experiment is designed as follows. Supported by the basic platform of Linux, two virtual computers with the same configuration are used as the experimental objects, and recorded in the same experimental environment, after applying the new active control system and the common control technology, respectively. The change of data parameters in short-term network environment, the former as the experimental group and the latter as the control group.

4.1 Pre-experimental Preparation

For the sake of authenticity, the configuration of relevant experimental equipment and preparation of experimental parameters can be completed according to the following Table 1.

Table 1. Lab preparation table

Experimental parameters	Numerical condition	Experimental parameters	Numerical condition
Classification of simulation platform	Linux Foundation Platform	Experimental time	100 min
Control Node Distribution Coefficient	0.69	Basic congestion coefficient	0.43
Maximum Distribution Uniformity	About 70%	Limit congestion data volume	$7.59 \times 10^9 T$

In order to ensure the authenticity of the experimental results, the experimental group and the control group always keep the same experimental parameters.

4.2   Total Congestion Data Comparison

Under the condition that the basic congestion coefficient is equal to 0.43, 100 min is used as the experimental time to record the change of the short-term network congestion data after the application of the experimental group and the control system. The experimental comparison details are shown in Table 2.

Table 2. Total congestion data comparison Table

Experimental time/(min))	Control group total congestion data/($\times 10^9$T)	Total congestion data volume of experimental group/($\times 10^9$T)
10	7.12	3.42
20	7.15	3.21
30	7.19	3.18
40	7.23	3.07
50	7.28	3.03
60	7.36	3.35
70	7.47	3.36
80	7.55	3.29
90	7.64	3.24
100	7.79	3.24
Average value	7.38	3.24

It can be seen from Table 2 that when the experimental time is 10 min, the total congestion data of the experimental group is 3.42×10^9T, and that of the control group is 7.12×10^9T。 When the experiment time is 50 min, the total congestion data of the experimental group is 3.03×10^9T, and the total congestion data of the control group is 7.28×10^9T. When the experimental time is 80 min, the total congestion data of the experimental group is 3.24×109t, and the total congestion data of the control group is 7.55×10^9T. The total congestion data of the designed system is significantly lower than that of the control group. With the increase of experimental time, the total amount of short-term network congestion data in the experimental group shows a trend of decline, rise, decrease and stability, and the average level in the whole experimental process is far below the ideal maximum value. The total amount of short-term network congestion data in the control group showed an upward trend, and the maximum value of the whole experiment was 7.79×10^9T, which was much higher than that of the experimental group.

4.3   Comparison of Distribution Uniformity of Main Control Nodes

Under the condition that the distribution coefficient of the control node is equal to 0.69, the change of the distribution uniformity of the main control node is recorded in the experiment group and the control group by using 100 min as the experimental time and the control system respectively. The details of the experiment are shown in Fig. 4.

Distribution Uniformity of Main Control Nodes/（%）

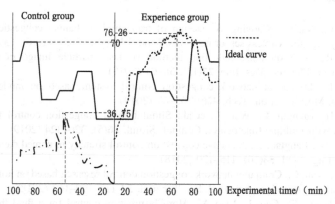

Fig. 4. Comparison of master node distribution uniformity

Figure 4 shows that the distribution uniformity of the main control nodes in the experimental group keeps a high level all the time, and the maximum value of 76.26% exceeds the ideal extreme value of 70%. The distribution uniformity of the main control nodes in the control group was relatively low, and the maximum value was only 36.75%, which was much lower than that of the experimental group.

5   Conclusions

The short-term network congestion active control system based on artificial intelligence aims to reduce the total congestion data and improve the distribution uniformity of the master node. It fully connects hardware execution devices such as motor modules, and determines the packet burst mechanism based on the definition of control nodes. The following conclusions can be drawn from the experiment.

(1)   When the experimental time is 80 min, the total congestion data of the experimental group is 3.24×10^9T, and the total congestion data of the control group is 7.55×10^9T. The total congestion data of the design system is significantly lower than that of the control group, which shows that the system in this paper can effectively reduce the total congestion data.

(2)   The distribution uniformity of the main control nodes of the system in this paper always keeps a high level, the maximum value of 76.26% exceeds the ideal extreme value of 70%, indicating that the distribution uniformity of the master nodes in this system is high.

From the perspective of practical application, the system has higher coordination ability and higher application and promotion value.

References

1. Yingda, L., Na, C., Xiaoying, S.: Design of transmission channel congestion intelligent control system for wireless sensor networks. Mod. Electron. Tech. **41**(16), 159–162 (2018)
2. Bo, Z.: Research on noise reduction method of laser particle image in mixed noise environment. Electron. Des. Eng. **27**(13), 58–61 (2019)
3. Xinwei, L.: Design of network congestion control system based on random detection algorithm. Mod. Electron. Tech. **42**(9), 63–67 (2019)
4. Cunwu, H., Mengqi, L., Wei, T., et al.: Simulation of congestion control for computer networks with multiple time delays. Comput. Simul. **36**(5), 320–324 (2019)
5. Wenjing, W., Jiangtao, L.: An active congestion control strategy in named data networking. Comput. Eng. Appl. **54**(10), 115–120 (2018)
6. Shijie, Z., Wan, C.: Computer network congestion control research based on nonlinear active queue management algorithm. Bull. Sci. Technol. **34**(11), 160–163 (2018)
7. Xu, W., Hayat, T., Cao, J., Xiao, M.: Hopf bifurcation control for a fluid flow model of internet congestion control systems via state feedback. IMA J. Math. Control Inf. **33**(1), 69–93 (2018)
8. Yang, X., Chen, X.Y., Xia, R.T., Qian, Z.H.: Wireless sensor network congestion control based on standard particle swarm optimization and single neuron PID. Sensors **18**(4), 1265 (2018)
9. Abdullah, W.A.N.W., Yaakob, N., Ahmad, R.B., Yah, S.A.: Classification of importance data for congestion control in remote health monitoring system. Adv. Sci. Lett. **24**(3), 1767–1770 (2018)
10. Rezaee, A.A., Pasandideh, F.: A fuzzy congestion control protocol based on active queue management in wireless sensor networks with medical applications. Wirel. Pers. Commun. **98**(1), 815–842 (2017). https://doi.org/10.1007/s11277-017-4896-6
11. Yao, Y., Liu, J.B., Xu, D.L., Zhi, R., Qing, H.: Centralized congestion control routing protocol based on multi-metrics for low power and lossy networks. J. China Univ. Posts Telecommun. **24**(5), 39–47 (2017)
12. Zhang, J., Wen, J., Han, Y.: TCP-ACC: performance and analysis of an active congestion control algorithm for heterogeneous networks. Front. Comput. Sci. **21**(4), 1–14 (2017)
13. Liu, S., Li, Z., Zhang, Y., et al.: Introduction of key problems in long-distance learning and training. Mobile Netw. Appl. **24**(1), 1–4 (2019)
14. Liu, S., Glowatz, M., Zappatore, M., Gao, H., Gao, B., Bucciero, A.: E-Learning, E-Education, and Online Training, pp. 1–374. Springer, Heidelberg
15. Li, Y., Chen, X., Lin, Y., Srivastava, G., Liu, S.: Wireless transmitter identification based on device imperfections. IEEE Access **8**, 59305–59314 (2020)

Decentralized Control Method for UAV Arriving Simultaneously Based on Large Data Analysis

Jian-jun Zhu and Jun Zhang[✉]

College of Electrical Engineering, Zhengzhou University of Science
and Technology, Zhengzhou 450064, China
lhy181025@163.com, zhaoxia1812@163.com

Abstract. Due to the limitation of the control method during the conventional simultaneous arrival control drone, a certain deviation is caused. In order to solve such problems, a research on the decentralized control method of simultaneous arrival drone based on big data analysis is proposed. Fusion of decentralized information is achieved through information filtering. Based on this, decentralized coordination of formations is controlled. The optimal path and convergence speed are calculated by calculation. Decentralized control methods are implemented through communication delay constraints. The proposed method is used for simulation experiments and it is found that the method can effectively reduce the error, which fully proves the feasibility of the control method.

Keywords: Large data analysis · UAV · Decentralized control

1 Introduction

With the advent of the era of big data, big data analysis has also emerged. Big data analysis refers to the analysis of large-scale data. Big data has the characteristics of large amount of data, fast speed, multiple types, value, and truth. Big data is the hottest vocabulary of the IT industry, and the subsequent use of data warehouse, data security, data analysis, data mining, etc., around the commercial value of big data has gradually become the focus of profit sought by industry professionals [1]. Nowadays, big data analysis technology has been applied in many fields, and certain results have been obtained. For example, by predicting demand to help enterprises deal with practical problems, in recent years, enterprises need to not only acquire customers, but also understand customer needs in order to improve Customer experience and develop long-term relationships. By sharing data, customers lower the privacy level of data usage, expecting companies to understand them, form corresponding interactions, and provide a seamless experience at all touch points. With the popularization of big data analysis technology, it is gradually adopted by the country's work and business, and the application of drones is becoming more and more extensive. UAVs are unmanned aircraft operated by radio remote control equipment and self-provided program control devices, or operated completely or intermittently autonomously by onboard computers.

S. Liu and L. Xia (Eds.): ADHIP 2020, LNICST 347, pp. 193–203, 2021.
https://doi.org/10.1007/978-3-030-67871-5_18

When multiple drones are operating at the same time, many problems will arise due to improper coordination. For example, multiple drones often need to reach the same or different target locations at the same time when completing tasks in coordination, or multiple drones taking off from different locations. UAVs gather in a certain position to form a close formation. This problem has been studied in many fields. However, compared with other hardware facilities, UAVs have their unique characteristics and applications. They move in three-dimensional space. It has a positive speed limit and cannot stop waiting or retreat. You can increase the path length by circling flight, and you can also adjust the flight speed within its allowable range. Reference [2] proposes a UAV control method based on acceleration feedback enhancement. This method introduces angular velocity and linear velocity feedback control on the basis of the original controller structure of the UAV to improve the disturbance suppression capability of the UAV. However, the control effect of this control method still needs to be further improved.

In order to solve the problem of the simultaneous arrival of UAVs, a decentralized control method is introduced. It adopts the basic design ideas of decentralized control, centralized operation and management, and adopts a multi-level hierarchical, cooperative and autonomous structure. Its main feature is its centralized management and decentralized control. The simultaneous arrival of multiple drones is a typical collaborative control problem, which usually includes two aspects of research content: one is path planning, that is, the path is planned for each drone under the conditions of environmental constraints and collaborative constraints; the other is trajectory Control, that is, by controlling the heading and speed of the UAV to make the UAV reach the target position along the planned path at the same time. This method can obtain the global optimal solution, but it is essentially a centralized control method. When calculating coordination variables, the coordination function information of all UAVs must be obtained. When some UAVs are affected by sudden threats, they must re-coordinateroute plan.

2 Design of Decentralized Control Method for UAV

It is assumed that in a certain mission, n UAVs are expected to arrive at a predetermined k target positions at the same time, of which $n \leq k \leq 1$. The initial position of UAV is the actual position of its current time, which may be arbitrarily distributed in space. Each UAV has and only has a definite target position. The target positions of different UAVs may be the same or different. In addition, it is assumed that UAV can obtain information about threats and obstacles (no-fly zone) in advance or in real time, and can independently plan the path offline or online, and give real-time estimates of the path length, and can autonomously fly along the planned path. The goal of multi-UAV simultaneous arrival is to find a control method or strategy to achieve the above tasks, and try to avoid the impact of adverse factors, such as path errors, sudden threats and so on. Seven of them have to reach two targets at the same time in order to attack two targets at the same time. In the process, UAVs have to avoid threats and no-fly zones [3]. Because the flight path and speed of UAV are not fixed, it can be guaranteed to arrive at the same time by adjusting the path length and flight speed. The UAV is regarded as a particle moving in two-dimensional plane without considering the change of flight altitude. The simplified motion model of UAV is taken as follows:

$$
\begin{pmatrix} x_i \\ y_i \\ \theta_i \end{pmatrix} = \begin{pmatrix} \cos\theta_i \\ \sin\theta_i \\ 0 \end{pmatrix} v_i + \begin{pmatrix} 0 \\ 0 \\ 1 \end{pmatrix} \omega_i, i \in \{1,2,\ldots,n\} \tag{1}
$$

Among them, $i \in \{1,2,\ldots,n\}$ is the UAV number, (x_i, y_i) is the position of UAV i, and θ_i is the direction angle of UAV i. The control input variables are linear velocity v_i and angular velocity ω_i of UAV i. Let $r_i = (x_i, y_i)$, $u_i = (u_{xi}, u_{yi})$, the system kinematics equation can be linearized by feedback, and the system model in the form of first order integral can be obtained.

$$
\begin{bmatrix} x_i \\ y_i \end{bmatrix} = \begin{bmatrix} u_{xi} \\ u_{yi} \end{bmatrix} \tag{2}
$$

u_i is the virtual control input of UAV. The relationship between the virtual control input and the actual control input variables is as follows:

$$
v_i = \sqrt{u_{xi}^2 + u_{yi}^2} \tag{3}
$$

$$
\omega_i = \frac{u_{xi}\bar{u}_{yi} - u_{yi}\bar{u}_{xi}}{u_{xi}^2 + u_{yi}^2} \tag{4}
$$

The decentralized control structure of multiple UAVs relying only on local information interaction is shown in Fig. 1, where $\theta_i (i \in V)$ is the coordination variable of the i UAV, and each UAV receives only the coordination variable information of its neighbors.

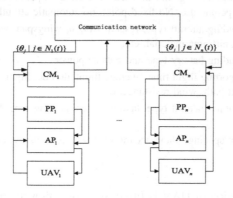

Fig. 1. Decentralized control structure

In the decentralized control structure shown in Fig. 1, all UAVs are equally positioned and connected by directed communication connections [4]. Among them, the lowest UAV module represents the UAV entity, which is the control object of the autopilot and outputs state information to the outside. The autopilot module AP represents a flight control system with heading and speed maintenance functions. It provides an instruction interface to the coordination module CCM and the path planner module PP. The path planner module can carry out path planning according to UAV status and environmental information, and output course angle instructions to the autopilot module to enable UAV to fly along the planned path, and output the estimated value of the remaining path length at the current time to the coordination module. Coordination module receives all neighbors' coordination variable information, then calculates speed instructions based on consistency algorithm and outputs them to autopilot module, and updates local coordination variables.

2.1 Decentralized Information Fusion

For the decentralized control of UAV, first of all, the flight information of all the target UAVs that need to arrive at the same time needs to be collected, counted and processed in a unified way. The UAV parameters to be counted are shown in Table 1.

Table 1. UAV information fusion parameters

Parameter	Specific information
Aircraft model	Wing Flying Kenong A6-160 Agricultural Plant Protection UAV, Phantom 3 SE, Mapping Eagle Trimble UX5, etc.
Flight Mode	GPS mode: suitable for beginners. Fixed-point positioning, attitude mode: NAZA is not equipped with GPS, can only fly in this mode. Usage rate is general, can not be fixed point but can be set high. The attitude needs to be revised to suit players with certain flight experience. Manual mode: can determine whether the center of gravity of the aircraft is appropriate, legend can rescue aircraft, or experienced experts for 3D flight, few people use. No fixed point, no automatic attitude correction
Fuselage parameter	Including aircraft type, reserve weight, wingspan, wing area, geometric size, airframe material, etc.
Battery	Including battery type and battery power
Flight parameters	Supporting flight time, average flight altitude, flight speed range, historical flight path, wind resistance, etc.
Take off and land	Including takeoff type, takeoff speed, landing type, landing speed, etc.
DTM	Point spacing, plane accuracy, elevation accuracy, etc.

The basic information of UAV is fused by information filtering. According to the type and characteristics of information, the information is divided into linear analysis and nonlinear analysis. The linear information is expressed as:

$$x(k) = F(k)x(k-1) + G(k)w(k) \tag{5}$$

Among them, $x(k)$ pseudo-weir state vector; $F(k)$ is state transfer matrix; $G(k)$ pseudo-noise input transfer matrix; $w(k)$ pseudo-zero mean uncorrelated Gauss white noise; $Q(k)$ pseudo-noise variance matrix. Nonlinear information is expressed as:

$$z(k) = h(kx(k)) + v(k) \tag{6}$$

Among them, $Z(k)$ is the measured value; $V(k)$ is zero, variance is $R(k)$ Gauss white noise; $h()$ pseudo-nonlinear measurement model.

Kalman filter obtains target state estimation χ^2 variance P, while information filter obtains information state y and Fisher information Y satisfying relation.

$$y = P^{-1}x \tag{7}$$

$$Y = P^{-1} \tag{8}$$

$$I(k) = H^T(k)B^{-1}(k)H(k) \tag{9}$$

$$i(k) = H^T(k)R^{-1}(k)(z(k) - h(x(k|k-1))) + H(k)x(k|k-1) \tag{10}$$

Among them, $i(k)$ is the information state distribution of the observation vector $Z(k)$, and $I(k)$ pseudo information matrix. $i(k)$ and $I(k)$ are only related to the target state dimension, but not to the sensor measurement dimension. When updating $y(k|k)$ and $Y(k|k)$, the decentralized information fusion algorithm considers not only the $I(k)$ and $i(k)$ of the UAV itself, but also the communication information of the nearby UAV. The decentralized information fusion algorithm based on information filtering can fuse the information of other UAVs through simple algebraic sum, so it has good scalability, heterogeneity and dynamic reconfigurability. The information filtering algorithm has good robustness to the selection of initial values. When the exact initial statistical characteristics of the system are not obtained, the small non-zero initial values can be selected to calculate iteratively, thus overcoming the problem that Kalman filtering algorithm is sensitive to the selection of initial values [5].

2.2 Decentralized Coordination Control of Formation

During the flight, UAV members will join at any time, and UAV exits due to single point failure. In this case, all UAV equipment need to be reconciled and formed in time to ensure that all UAVs can arrive at the same time. This decentralized coordination control is divided into the decentralized coordination within the formation and the decentralized coordination between the formation. Its principle is shown in Fig. 2.

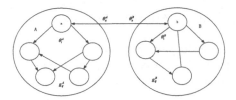

Fig. 2. Principle of decentralized

Suppose $\theta_t(t)(i = 1, 2, \cdots, n)$ is the coordinated information of the i platform at t-time, i.e. the coordinated variable, if satisfied:

$$\lim_{t \to \infty} \|\theta_i(t) - \theta_j(t)\| = 0(\forall \theta_i(0), \theta_j(0), i, j \in n) \tag{11}$$

Platform i and platform j are respectively called information pairs. The results are consistent. Decentralized coordination strategies are as follows:

$$\theta_i(t) = -\frac{1}{\sum\limits_{j \in \Omega_4(t)} \alpha_{ij}(t)g_{ij}(t)} \cdot \sum_{j \in \Omega_i(t)} \alpha_{ij}(t)g_{ij}(t)\{\theta_t(t) - [\theta_t(t - \tau_{ij}) + v_{ij}]\} \tag{12}$$

Among them, $\Omega_i(t)$ is the set of all platforms that send information to the platform i at t-time; $g_{ij}(t)$ is the communication topology at t-time; $\alpha_{ij}(t) > 0$ is the reliability or importance of transmitting information from the platform j to the platform i at t-time; and τ_{ij} is the communication noise between the platform j and the platform i. The distributed information coordination process of platform system is essentially an achievable condition of dynamic decentralized coordination consistency and convergence speed, which are related to communication topology, communication noise and so on [6]. Decentralized coordination among formations is the coordination between A and B in Fig. 2, which is deduced according to the mode of decentralized coordination within formations. In Fig. 1, A and B are the root nodes of the internal communication topology of Formation A and B, and the communication between Formation A and B is realized by the communication between Platform A and B.

2.3   Optimal Path Selection

After the formation of UAVs, the path of each UAV is planned separately. There are several principles to follow when planning the route: no overlap with other UAV flight paths; no roadblocks on the route. On the basis of these principles, the direction of flight is determined, and the path is calculated. On the basis of not violating the principle, the route with less energy consumption is chosen as far as possible, that is, the path with shorter distance. Considering the characteristics of fixed-wing UAV, a decentralized formation flight control strategy based on information consistency and relative position control is proposed under the assumption that the position, velocity and course of UAV are almost the same at the initial time.

$$v_i^c = \bar{v}_i^c \delta_i \Delta v_i^c \tag{13}$$

$$\psi_i^c = \bar{\psi}_i^c \delta_i \Delta \psi_i^c \tag{14}$$

The decentralized formation flight control strategy consists of two parts, one is v_i^c and ψ_i^c, which synchronize speed and course using relative velocity and course information between UAVs, the other is Δv_i^c and $\Delta \psi_i^c$ which use relative position information between UAVs to form and maintain formation. In the formula, $\delta_i \in \{0, 1\}$ is a binary switching variable and is defined:

When $\delta_i = 0$, only position and course are synchronized, and formation control is not started; when $\delta_i = 1$, both position and course are synchronized and formation control is started.

When $\delta_i = 0$, $v_i^c = \bar{v}_i^c$ $\psi_i^c = \bar{\psi}_i^c$ the aim is to make the flight speed and heading angle of all UAVs converge. For this purpose, the following decentralized heading synchronization control is given:

$$\bar{v}_i^c = v_i + \frac{1}{\alpha_{v,i}} \sum_{j \in N_i} a_{ij}(v_j - v_i) \tag{15}$$

$$\bar{\psi}_i^c = \psi_i + \frac{1}{1 + |N_i|} \sum_{j \in N_i} (\psi_j - \psi_i) \tag{16}$$

It can be seen from the formula that the velocity command $\bar{v}_i^c$ of the first UAV and the heading angle command $\bar{\psi}_i^c$ are only related to its neighbors, and can be obtained only by obtaining the relative speeds and heading angles of all neighbors. If the neighborhood set is empty, the speed and heading angle of the UAV will remain unchanged [7]. It can be seen that the formula is a decentralized control strategy using only local relative state information feedback.

Given course angle command v_i^c, the first UAV can turn from the current course angle to the command course angle clockwise along path 1 or counter-clockwise along path 2. In the graph, the deflection angle along path 1 is less than π, which is expected, while the deflection angle along path 2 is greater than π, which is not expected. In order to make UAV fly along the planned route in practical application, sometimes it is necessary to introduce external reference signal or virtual Leader. When UAV is far from the target position, try to adjust the path length through path planning so that it can fly at the appropriate speed, so as to retain a large margin of speed adjustment and better respond to path errors and sudden threats. When the UAV approaches the target position, the speed control is the main method, which can ensure the precise and simultaneous arrival.

2.4 Convergence Rate Calculation

In the minimum control speed in the air, both the longitudinal and lateral control of the aircraft are involved. The UAV must not only meet the conditions of simultaneous arrival, but also be higher than the minimum flight speed. The angular velocity of flight is expressed by formula (17):

$$v \geq \frac{V}{\sqrt{n_W}} \tag{17}$$

The V is the speed obtained at the first maximum when the acceleration of the elevator is less than 11 s per knot and the load coefficient of the aircraft is corrected to the speed obtained at the first maximum. n is found overload coefficient in track coordinate system at V. w is the weight of the aircraft, in units of N. S is the aerodynamic reference area of the wing. The initial trimming speed of the aircraft is not less than 1.13 V. The calculation method of convergence rate can be obtained. Preliminary estimates of the stall speed of the aircraft are as follows:

$$v = \sqrt{\frac{2W}{\rho SC}} \tag{18}$$

Among them, ρ is the air density and C is the critical lift coefficient of the aircraft. The trim angle of attack, throttle and elevator deflection of the UAV flying straight and flat at 1.13 V are calculated. Keeping the throttle unchanged, the final convergence speed can be obtained by deviating the elevator from the balancing position by one degree.

2.5   Communication Delay Constraints for Decentralized Control

From the time when UAV updates the coordination variable to the time when its neighbors use the coordination variable information to calculate the speed command, there must be a time difference, which is equivalent to introducing a time delay to the coordination variable [8, 9]. The time delay is determined by computing time, computing period and transmission time. The effect of time delay is equivalent to introducing deviation, which prevents UAV from arriving at the same time. Reducing the calculation period, predicting and compensating can weaken the effect of time delay [10].

UAV2 makes its flight speed far away from the limit by hovering flight. UAV also changes its flight speed and increases its speed adjustment margin [11, 12]. When a sudden threat occurs, four UAVs still reach the target position at the same time.

3   Simulation Experiment

In order to verify the effectiveness of the decentralized flight control method for UAV arriving at the same time, simulation experiments are carried out. In order to ensure the uniqueness of the test variables, assuming that all UAV models are the same, they have the same characteristic parameters and initial states as shown in Table 2.

Table 2. Characteristic parameters and initial states of a UAV

	UAV1	UAV2	UAV3	UAV4
Position (m, m)	(0, 0)	(−1000, 2000)	(1000, 1000)	(1000, 0)
Speed (m/s)	165	212	198	201
Heading angle rad	$\pi/4$	$\pi/6$	$3\pi/8$	$\pi/8$
Minimum flight speed	150	150	150	150
Maximum speed of flight	250	250	250	250
Minimum acceleration	−8	−8	-8	-8
Maximum acceleration	8	8	8	8

In order to ensure the rigor of the test, the error in the x direction is counted and compared with that in the experiment without the control method. The results are shown in Fig. 3.

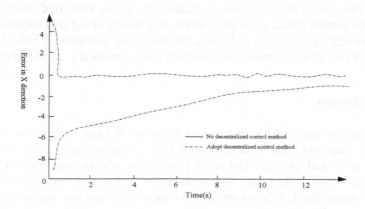

Fig. 3. Experimental comparison of decentralized control methods

It can be seen from the simulation results that the decentralized control method has a significant impact on the UAV target results and error convergence. The error in the x direction is divided into positive and negative directions. At the beginning of the flight, there will be a certain error value regardless of whether the method is used. The error value will gradually become smaller as the flight time passes, but the decentralized control method is adopted. The error can be adjusted to near the normal value within 2 s, and it can be maintained until the error value is approximately equal to 0.

In order to obtain more comprehensive experimental results, under the above experimental conditions, the convergence rate is used as the experimental comparison index to compare the method. The comparison results of the convergence rate of the proposed method and the comparison method are shown in Fig. 4.

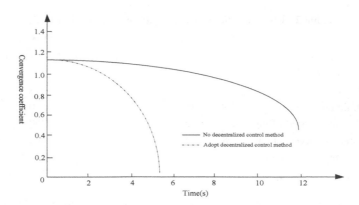

Fig. 4. Convergence rate comparison results

It can be seen from Fig. 4 that under the decentralized control method, the convergence rate control effect is better, indicating that the decentralized control method has a better control effect, while the convergence results of the decentralized control method are very unsatisfactory and time-consuming long. Therefore, it fully shows that the performance of the proposed decentralized control method is highly improved.

4   Conclusion

Only relying on the decentralized control structure of local information interaction reduces the need and difficulty of communication, and avoids single point failure. The combination of path planning and speed control can achieve complementary advantages, better cope with the impact of adverse factors such as path errors and unexpected threats to ensure that UAVs can arrive at the same time, and also has a certain feasibility in practical work.

Acknowledgments. The Henan Science and Technology Department Foundation of China under Grant No. 172102210114.

References

1. Tyushev, K., Amelin, K., Andrievsky, B.: The method of saving data integrity for decentralized network of group of UAV using quantized gossip algorithms. IfacPapersonline **49**(13), 259–264 (2016)
2. Dai, B., He, Y.Q., Gu, F., et al.: Acceleration feedback enhanced controller for wind disturbance rejection of rotor unmanned aerial vehicle. Robot **42**(1), 79–88 (2020)
3. Arbanas, B., Ivanovic, A., Car, M., et al.: Decentralized planning and control for UAV–UGV cooperative teams. Auton. Robots **1**, 1–18 (2018)
4. Wei, M., He, Z., Rong, S., et al.: Decentralized multi-UAV flight autonomy for moving convoys search and track. IEEE Trans. Control Syst. Technol. **25**(4), 1480–1487 (2017)

5. Conesa, A., Madrigal, P., Tarazona, S., et al.: A survey of best practices for RNA-seq data analysis. Genome Biol. **17**(1), 181 (2016)
6. Shuai, L., Weiling, B., Nianyin, Z., et al.: A fast fractal based compression for MRI images. IEEE Access **7**, 62412–62420 (2019)
7. Stergiopoulos, G., Kotzanikolaou, P., Theocharidou, M., et al.: Time-based critical infrastructure dependency analysis for large-scale and cross-sectoral failures. Int. J. Crit. Infrastruct. Prot. **12**(C), 46–60 (2016)
8. Liu, S., Fu, W., He, L., Zhou, J., Ma, M.: Distribution of primary additional errors in fractal encoding method. Multimed. Tools Appl. **76**(4), 5787–5802 (2014). https://doi.org/10.1007/s11042-014-2408-1
9. Cheng, Y.C., Wu, E.H., Chen, G.H.: A decentralized MAC protocol for unfairness problems in coexistent heterogeneous cognitive radio networks scenarios with collision-based primary users. IEEE Syst. J. **10**(1), 346–357 (2016)
10. Kamath, G., Shi, L., Song, W.Z., et al.: Distributed travel-time seismic tomography in large-scale sensor networks. J. Parallel Distrib. Comput. **89**(C), 50–64 (2016)
11. Hu, G.L., Gui, L., Quan, S.L., et al.: Design of four-rotor UAV control system based on series fuzzy PID. Exp. Technol. Manag. **36**(3), 132–135 (2019)
12. Liu, S., Glowatz, M., Zappatore, M., Gao, H., Gao, B., Bucciero, A.: E-Learning, E-Education, and Online Training, pp. 1–374. Springer, Cham (2018). https://doi.org/10.1007/978-3-319-93719-9

Hyperspectral Recognition and Early Warning of Rice Diseases and Insect Pests Based on Convolution Neural Network

Heng Xiao[✉] and Cao-Fang Long

Sanya University, Sanya 572022, China
xiaoheng564@163.com

Abstract. The traditional method of disease and pest recognition uses SVM to classify and recognize the image. Because of the large training convergence error, the recognition accuracy is not high. In view of the above problems, the paper studies the method of rice hyperspectral pest identification and early warning based on convolution neural network. By reducing the dimension of the collected spectral image, we can get more image information and extract image features. Based on alexnet, the structure of convolutional neural network is designed. The recognition database was established by collecting the spectrum images of rice diseases and insect pests, and the convolution neural network was trained by transfer learning, so as to realize the recognition and early warning of rice diseases and insect pests. The experimental results show that the convergence error of the method based on convolution neural network is small and the recognition accuracy is higher.

Keywords: Convolution neural network · Rice diseases and insect pests · Hyperspectral imaging · Recognition and early warning methods

1 Introduction

For agricultural production, the outbreak of crop diseases and insect pests has a great impact on the yield. However, the traditional detection methods mainly rely on personal experience and visual observation of the traditional methods of pest control, with high rate of misjudgment and poor real-time performance, which can not meet the needs of pest control. It is a new demand for the development of agriculture to find an accurate, fast, efficient and automatic method for identification and detection. With the development of information technology, image processing technology has been applied to the detection of crop diseases and insect pests, which has become one of the main ways to replace artificial identification. In addition, through the combination of artificial intelligence and other emerging knowledge, the detection of crop diseases and insect pests has the accuracy, convenience and intelligence, which brings a lot of convenience for the detection of crop diseases and insect pests, and increases the timeliness [1].

Hyperspectral resolution remote sensing is a technology to obtain a lot of very narrow spectral continuous image data in the visible, near infrared, mid infrared and thermal infrared band of electromagnetic spectrum, which can be used for ground

S. Liu and L. Xia (Eds.): ADHIP 2020, LNICST 347, pp. 204–214, 2021.
https://doi.org/10.1007/978-3-030-67871-5_19

object recognition. As a new technology, hyperspectral remote sensing develops rapidly, and its application in agriculture has also been studied deeply. Plant reflectance is related to physiological and biochemical characteristics, light, background and surrounding environment. The spectral characteristics of plants in visible and near-infrared bands are related to the growth and morphological characteristics of specific plant species [2]. After crops are damaged by diseases and insect pests, the leaf color, leaf and plant properties, physical structure, pigment content, etc. will change, which may lead to changes in spectral information. Hyperspectral technology provides a potential solution for the detection of the degree of pests.

Convolutional neural network is a variety of multi-layer perceptron. Compared with general neural network, convolutional neural network can directly process the two-dimensional matrix of image, and has outstanding performance in image processing [3]. Based on the above analysis, this paper will study the hyperspectral recognition and early warning method of rice diseases and insect pests based on convolution neural network.

2 A Method of Hyperspectral Recognition and Early Warning of Rice Diseases and Insect Pests Based on Convolution Neural Network

2.1 Rice Hyperspectral Dimensionality Reduction

In order to realize the hyperspectral recognition and early warning of rice diseases and insect pests, this paper uses the high-resolution spectral imaging system of UAV to collect the rice spectral image of the object area. After the rice spectral image is collected, because of the large amount of spectral data information collected, each spectral curve can be regarded as high-dimensional vector information, and multiple spectral curves constitute a high-dimensional vector space. In order to obtain the detail data of hyperspectral image, the dimension of hyperspectral image is reduced.

The multiscatter correction of hyperspectral image is carried out before dimension reduction. Multiple scattering correction is mainly to eliminate the scattering effect caused by the uneven distribution of particles and particle size. It can effectively eliminate the shift and shift of a line caused by the scattering effect between samples and improve the signal-to-noise ratio of the original absorbance spectrum. The calculation formula of multiple scattering correction is as follows [4]:

$$\begin{cases} \bar{x} = \frac{1}{n} \sum_{i=1}^{n} x_i \\ x_i = a_i + \bar{x} b_i \\ (x_i, MSC) = \frac{x_i - a_i}{b_i} \end{cases} \tag{1}$$

In the above formula, $\bar{x}$ is the average spectrum,x_i is the collected hyperspectral image, n is the number of images, a_i is the translation amount of hyperspectral curve, b_i is the relative offset coefficient, (x_i, MSC) is the corrected spectrum. After the hyperspectral correction of rice, in order to eliminate the measurement impact caused by spectral rotation, derivative algorithm is used to deal with Hyperspectral [5]. The derivative is calculated as follows:

$$\frac{dy}{d\lambda} = \frac{q_j - q_i}{\Delta\lambda} = \frac{q_j - q_i}{j - 1} \tag{2}$$

In the above formula, q is fluorescence intensity, spectral wavelength difference $\Delta\lambda$, and hyperspectral wavelength i,j. After the hyperspectral dimension reduction processing, through the dimension reduction processing can synthesize all aspects of rice Hyperspectral Information, and make the information not overlapping. In this paper, wavelet transform is used to reduce the dimension of rice hyperspectral.

The function $\psi(t)$ of the basic wavelet is shifted ε, and the inner product operation is ξ performed with the signal $x(t)$ to be analyzed at different scales [6]:

$$WT_x(\xi, \varepsilon) = \frac{1}{\sqrt{\xi}} \int_{-\infty}^{+\infty} x(t)\psi * \left(\frac{t - \varepsilon}{\xi}\right) dt \tag{3}$$

In the above formula, $WT_x(\xi, \varepsilon)$ is the hyperspectral image after dimension reduction processing. After the hyperspectral processing of rice, the characteristics of the hyperspectral skin image are extracted.

2.2 Extraction of Hyperspectral Image Features

In this paper, vegetation index is used to extract spectral features. Vegetation index is a combination of two or more wavelengths of surface reflectance to enhance a specific feature or detail of vegetation. According to the information of spectral characteristics of plants, for the main chemical components of plant leaves, pigment, water, carbon and nitrogen, we can extract broad-band green degree, narrow-band green degree, light utilization rate, canopy nitrogen, drought or carbon decay, leaf pigment, canopy water content. These vegetation indexes can simply measure the number and growth status of green vegetation, chlorophyll content, leaf surface canopy, leaf cluster, canopy structure, the utilization efficiency of vegetation for incident light in photosynthesis, the relative content of nitrogen in vegetation canopy, the carbon content of cellulose and lignin in dry state, the pigment related to stress in vegetation, and the canopy The real-time diagnosis of vegetation growth can be realized by monitoring the content of water content and so on. The available vegetation index is shown in the table below [7] (Table 1).

Table 1. Practical vegetation index

Serial number	Name	Descriptions
1	NVDI (Normalized Difference Vegetation Index)	The difference between the scattering of green leaves in the near-infrared band and the absorption of chlorophyll in the red band was increased
2	Simple Ratio Index	The ratio of the scattering of green leaves in the near-infrared band to the absorption of chlorophyll in the red band
3	Enhanced Vegetation Index	Enhanced NVDI to address the effects of soil background and atmospheric aerosols on dense vegetation
4	Atmospherically Resistant Vegetation Index	Enhance NVDI to better solve the effects of atmospheric scattering
5	Sum Green Index	Sensitivity of global light scattering in the green band range to canopy clearance
6	Photochemical Reflectance Index	It is useful for estimating leaf carotenoids (especially yellow pigment), leaf stress and carbon absorption efficiency

The convolution neural network structure is designed after extracting the curve features of rice hyperspectral image.

2.3 Structure Design of Convolution Neural Network

The structure of convolutional neural network designed in this paper is similar to the classical model alexnet, which is a very typical convolutional neural network model. Alexnet is an eight layer network model, mainly composed of one input layer, five volume layers, three full connection layers and one classification layer. In addition, the first five convolution layers and the first two full connection layers will be followed by the activation layer, using the activation function relu. The activation function is as follows [8].

$$f(x) = \max(0, x) \tag{4}$$

In the process of feature extraction of relu function, the amount of calculation is less, and as an activation function, it will affect some neurons, enhance the sparsity of neural network, speed up the convergence of neural network, and reduce the dependence between parameters. In the convolution neural network designed in this paper, the first activation layer and the second activation layer will be followed by the normalization layer. The specific parameters are shown in the table below [9, 10] (Table 2).

Table 2. Structure parameters of convolutional neural network

The sequence number of the layer	The type of layer	The parameters of the layer
1	Convolution layer	11 * 11 * 3 * 64
2	Convolution layer	5 * 5 * 64 * 256
3	Convolution layer	3 * 3 * 256 * 256
4	Convolution layer	3 * 3 * 256 * 256
5	Convolution layer	3 * 3 * 256 * 256
6	Fully connected layer	4096
7	Fully connected layer	4096
8	Fully connected layer	1000
9	Classification layer	—

In the above table, the parameters of the accumulation layer have 4 dimensions, which are expressed as $n_1 * n_2 * n_3 * n_4$, where $n_1 * n_2 * n_3$ is the core size, and n_4 is the number of cores. For the all connected layer n, the number of neurons in its parameters represents the number of neurons. The classification layer here is of softmax type and has no parameters.

In the convolution neural network designed in this paper, the size of the input image is 224 * 224 pixels. The size of convolution core of the first layer convolution layer is 11 * 11 pixels, the step length is 4, and the number of convolution cores is 64. Therefore, after calculation, $\frac{224-11}{4} + 1 \approx 55$, the size of the characteristic image obtained by the first layer convolution layer of 224 * 224 pixels is 55 * 55 pixels, and the number of characteristic images is 64. After the processing of the activation layer and the normalization layer, the characteristic image of 55 * 55 pixels is input to the first pooling layer. $\frac{55-3}{2} + 1 = 27$, the size of the first pooling layer is: 3 * 3 pixels, step size is 2, so after calculation: the size of the feature map obtained through the pooling layer is 27 * 27 pixels. Next, the characteristic image of the 27 * 27 pixel is input into the second convolution layer. The convolution kernel size of the second convolution layer is: 5 * 5 pixels, the step size is 1, two zeros are added around, and the number of convolution kernels is 256. Therefore, $\frac{27-5+4}{1} + 1 = 26$, after calculation, the size of the characteristic graph obtained through the second layer is 26 * 26 pixels, and the number of characteristic graphs is 256. The convolution kernel of the second convolution layer is mainly used to extract the texture information of the visible image. After that, the feature map with the size of 26 * 26 pixels is input to the second pooling layer. The size of the second pooling layer is 3 * 3 and the step size is 2, so after calculation, $\frac{26-3}{2} + 1 \approx 13$, the output size of the second pooling layer is 13 * 13 pixels. Similarly, the third, fourth and fifth convolution layers represent the high-dimensional information of the image. Finally, the feature map with the size of 6 * 6 pixels and the number of 256 is input into three fully connected layers, and finally through the softmax classification layer, the classification results can be obtained. After the structure of convolution neural network is determined, the convolution neural network is trained by using the data set of diseases and insect pests [11, 12].

In this paper, migration learning is used to train the designed convolutional neural network. The training process is as follows:

(1) Select the convolution network which has been trained on the Imagenet dataset;
(2) The mature network is used to extract the features of the image samples, and the 4096 dimension features are extracted from the full connection layer of the 7th layer as the input of the small-scale network to train the small-scale network;
(3) The mature network is used to extract the features of the image samples, and the 4096 dimension features are extracted from the full connection layer of the sixth layer as the input of the small-scale network to complete the small-scale network training;

The trained convolution neural network outputs the recognition results of rice diseases and insect pests according to the fitting results of Gauss function according to the rice diseases and insect pests collection in the recognition database. After the convolution neural network is designed and trained, the rice disease and pest identification database is established.

2.4 Establish the Database of Disease and Pest Identification

In order to accurately identify rice diseases and insect pests with hyperspectral data, it is necessary to establish a database of rice diseases and insect pests identification. Before the establishment of the database, the healthy and pest affected rice should be identified first.

Qualitative and quantitative description of "healthy leaves" standard:

(1) There is no insect bite, crawling trace and mechanical damage on the surface of leaves;
(2) The shape of leaves is complete, and there is no sudden change and wilting of leaves caused by external environmental factors such as overheating and super-cooling and internal physiological structure mutation;
(3) The SPAD value of chlorophyll was between 35.0 and 50.0;
(4) The net photosynthetic rate was 10.0–30.0, $\mu mol^{-2} \cdot s^{-1}$, CO_2;
(5) The F_v/F_m value of chlorophyll fluorescence parameter is between 0.80 and 0.84;
(6) The growth and development of the whole plant is neat, without excessive growth, and the leaf has no large or small protuberances.

"Leaves with diseases and insect pests" refers to leaves with obvious symptoms of diseases and insect pests or artificially inoculated with diseases and insect pests. After the difference between the rice affected by diseases and pests and the healthy rice is clarified, a discrimination database with the structure shown in the figure below is established (Fig. 1).

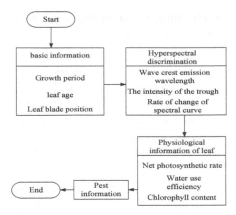

Fig. 1. Identifies the database structure

After the establishment of the discrimination database, the hyperspectral images of rice samples with different diseases and insect pests are collected and input into the database as the recognition standard to realize the recognition and early warning of rice diseases and insect pests.

2.5 Recognition and Early Warning of Rice Diseases and Insect Pests

After dimension reduction and other preprocessing, the collected hyperspectral image of rice is used in the classification layer of convolution neural network to judge whether the rice corresponding to the image is affected by diseases and pests by Gaussian function fitting.

If the hyperspectral point set obtained after dimension reduction is (x_i, y_i), the spectral curve of rice diseases and insect pests is fitted according to Gaussian function.

$$y = \frac{A_0}{\sqrt{\frac{\pi}{2} \bullet \omega_0}} e^{-2\frac{(x-\mu_0)^2}{\omega_0^2}} \tag{5}$$

In the above formula, A_0 is the area of the peak, μ_0 is the position of the peak and ω_0 is the parameter representing the half peak width. The convolution neural network outputs the fitting result of Gauss function after several data iterations. The fitting threshold value of Gaussian function is set as ± 0.1, and the curve whose fitting value is within the threshold value is determined to the rice disease corresponding to the hyperspectral curve of the sample. To complete the identification of rice diseases and insect pests. According to the results of rice diseases and insect pests identified, early warning of rice diseases and insect pests was carried out. So far, the research on Hyperspectral recognition and early warning method of rice diseases and insect pests based on convolution neural network has been completed.

3   Test Experiment

Accurate recognition and early warning of rice diseases and insect pests is an important prerequisite to ensure rice yield. Therefore, this paper studies the hyperspectral recognition and early warning method of rice diseases and insect pests based on convolution neural network. In order to test the performance of this method, this section will design a comparative experiment to complete the performance test of this research method.

3.1   Experiment Content

The experimental area is divided according to different growth periods of rice, and the hyperspectral images of rice collected in the corresponding area are used as the experimental data of this experiment. In order to ensure the authenticity and reliability of the experimental results, the diseases and insect pests of rice in the experimental area were checked manually. The test results are taken as the reference for the experimental test data.

The experimental group is based on support vector machine, and the experimental group is based on convolution neural network. There were two groups in the experiment. The convergence errors of the two methods were compared when using the same sample training set, and the recognition accuracy of healthy rice and pest rice was compared in the practical application. Through the comparison of the above two indicators to evaluate the advantages and disadvantages of the comparison group and the experimental group. Record the experimental data of two experiments, and analyze the experimental results.

3.2   Experimental Spectral Image Preprocessing

In view of the large spatial resolution of the original hyperspectral image collected by UAV, before rice pest identification, the original image is simply rough cut. After cutting, the spatial resolution of the spectral image is reduced to 400–900 pixels in width and 200–800 pixels in height. The spatial resolution histogram of the spectral image after rough cutting is shown in the figure below. According to the histogram, the spectral image width is mainly concentrated near 200 pixels, and the height is concentrated near 600 pixels. Therefore, the average hyperspectral image size is normalized to 200 × 600 pixels to achieve clearer experimental results (Fig. 2).

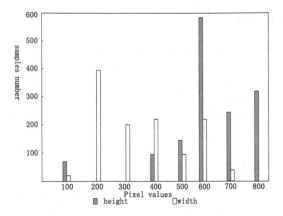

Fig. 2. Width and height histogram of hyperspectral image after rough cropping

After image preprocessing, two experiments were carried out. The data of the two experiments are analyzed and the experimental conclusion is obtained.

3.3   Experimental Results

The convergence error test results of the two methods of the experimental group and the comparison group using the same training set are shown in the figure below.

Compared with the support vector machine and convolution neural network used in the experimental group, the smaller the convergence error in training, the higher the classification accuracy in practical application, and the more stable the performance of the applied method.

It can be seen from the analysis in Fig. 3 that with the increase of training time, the curve decline speed of the control group gradually decreases, and finally tends to be

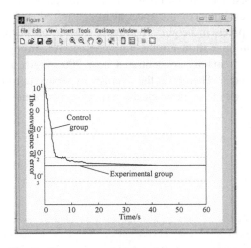

Fig. 3. Error curve during training convergence

flat. E. The results show that the convergence error of SVM is reduced and the classification accuracy tends to be stable. But compared with the experimental group, the convergence speed of the error curve o is slower than that of the control group. According to the theoretical results of training convergence error experiment, the classification accuracy is tested.

The identification results of the two groups of rice pest identification and early warning methods of the comparison group and the experimental group are shown in the table below (Table 3).

Table 3. Identification results of the two methods/strain

Rice growing period	Actual number of plants	Experimental group identification results	Contrast group recognition results
Seedling stage	46	46	53
Tillering stage	15	15	29
Jointing stage	34	34	21
Booting stage	22	21	9
Filling stage	42	40	24

It can be seen from the above table that in different rice growth periods, the number of plants identified by the comparison group method is significantly different from the real number of plants, while the number identified by the experiment group method is less different from the real number. Combined with the error curve in Fig. 3, The application of convolution neural network can improve the accuracy of pest identification, and the results of each stage have strong advantages. In conclusion, the aluminum network based on convolution neural network is better.

4 Concluding Remarks

It is very important to find rice diseases and insect pests in time and accurately for rice planting and development. The traditional identification method of rice diseases and insect pests relies too much on artificial and has low efficiency. Therefore, this paper studies the method of hyperspectral recognition and early-warning of rice diseases and insect pests based on convolution neural network. Through the contrast experiment with the method of recognition and early-warning of rice diseases and insect pests based on support vector machine, the performance of the method studied in this paper is better.

Fund Projects
1. Supported by Hainan Natural Science Foundation (619QN243).
2. Supported by Hainan Provincial Natural science Foundation of China (2019RC257).

References

1. Wang, R., Zhuang, Z., Wang, H., et al.: HRRP classification and recognition method of radar target based on convolutional neural network. Mod. Radar **41**(5), 33–38 (2019)
2. Jiang, P., Fu, H., Tao, H., et al.: Feature characterization based on convolution neural networks for speech emotion recognition. Chin. J. Electron Devices **42**(4), 998–1001 (2019)
3. Liu, Z., Sun, H., Ma, C., et al.: Vehicle recognition model based on multi-feature combination in convolutional neural network. Comput. Sci. **46**(1), 254–258 (2019)
4. Lin, L., Wang, S., Tang, Z.: Point target detection in infrared over-sampling scanning images using deep convolutional neural networks. J. Infrared Millim. Waves **37**(2), 219–226 (2018)
5. Tang, S., Yang, M., Bai, J.: Lung nodules detection and recognition based on deep convolutional neural networks. Sci. Technol. Eng. **19**(22), 241–248 (2019)
6. Liu, S., Liu, G., Zhou, H.: A robust parallel object tracking method for illumination variations. Mob. Netw. Appl. **24**(1), 5–17 (2018). https://doi.org/10.1007/s11036-018-1134-8
7. Xie, Z., Xu, H., Huang, Q., et al.: Spinach freshness detection based on hyperspectral image and deep learning method. Trans. Chin. Soc. Agric. Eng. **35**(13), 277–284 (2019)
8. Wang, C., He, Z., Wu, L., et al.: Multi-bands recognition of beef breeds with hyperspectral technology combined with characteristic wavelengths selection methods. Chin. J. Lumin. **40**(4), 520–527 (2019)
9. Liu, S., Fu, W., He, L., Zhou, J., Ma, M.: Distribution of primary additional errors in fractal encoding method. Multimed. Tools Appl. **76**(4), 5787–5802 (2014). https://doi.org/10.1007/s11042-014-2408-1
10. Song, W.: Linear contour recognition in layered pixel-level image fusion. Comput. Simul. **35**(6), 408–411+431 (2018)
11. Fu, W., Liu, S., Srivastava, G.: Optimization of big data scheduling in social networks. Entropy **21**(9), 902 (2019)
12. Liu, S., Bai, W., Zeng, N., et al.: A fast fractal based compression for MRI images. IEEE Access **7**, 62412–62420 (2019)

Industrialized Big Data Processing

Research on Abnormal Data Detection Method of Power Measurement Automation System

Ming-fei Qu$^{(\boxtimes)}$ and Nan Chen

College of Mechatronic Engineering, Beijing Polytechnic, Beijing 100176, China
qmf4528@163.com

Abstract. Aiming at the problems of long time consuming and low accuracy in traditional methods of abnormal data detection in power measurement automation system, this paper studies the methods of abnormal data detection in power measurement automation system. Design the data storage structure table of the electric power metering automation system database, and repair the missing data and denoise the data in the data table. Perform PAA calculation on the data to get the data feature sequence. After the P clustering algorithm pre-clusters the data, the iForest model is used to detect abnormal data to complete the research on the method. The experimental results show that the proposed detection method has the advantages of short detection time and high precision of 91.26–95.67%.

Keywords: Power measurement automation system · Abnormal data · Detection method · Iforest

1 Introduction

Electric power metering automation system refers to a system that can collect, monitor and analyze electric data on power generation side, power supply side, power distribution side and power sale side of power plants, substations, public transformers, special transformers and low-voltage customers, including metering automation master station, communication channel and metering automation terminal. The metering automation system realizes remote automatic real-time meter reading, abnormal alarm of electricity consumption information, voltage and power quality monitoring, line loss analysis, and prepayment by collecting, monitoring, analyzing and processing information such as data, voltage, current, load and other information of each monitoring terminal Functions such as fee management, electricity consumption inspection, load management and control, and power outage statistics provide data support for grid operation and management [1]. But before the realization of these advanced application research, the most basic premise is to ensure the timeliness, integrity and reliability of the data collected by the system (Fig. 1).

At present, the data content involved in the measurement automation system mainly includes various indicators of the terminal such as online rate, automatic meter reading rate, etc., collected data such as table codes, electricity, and power factor, and these basic data are subjected to secondary calculations such as voltage divider advanced application data such as loss, line loss, line loss in sub-stations, etc. Faced with such a

S. Liu and L. Xia (Eds.): ADHIP 2020, LNICST 347, pp. 217–228, 2021.
https://doi.org/10.1007/978-3-030-67871-5_20

large amount of data, relying on manual or traditional database software tools for daily data quality checks has been unable to effectively guarantee the reliability of data quality. The traditional method of outlier data detection is to use RBF classifier to detect outlier data, and the time cost of this detection method for large system data detection is too high, and the error of detection results is also large [2]. Therefore, based on the above analysis, this paper studies the abnormal data detection method of power measurement automation system.

2   Abnormal Data Detection Method of Electric Power Measurement Automation System

2.1   Establish Data Structure Table of Electric Power Measurement Automation System

The data used to detect abnormal data in the power metering automation system is data in tables such as the communication flow table of the system terminal and the field operation and maintenance record table. Therefore, the system data structure table needs to be established in the terminal database.

The communication flow table is used to record the communication status between each terminal and the master station, mainly including communication flow, number of reconnections and online time, etc. the specific definition is shown in Table 1. The data in this table is collected and counted by the master station at zero every day, reflecting the previous day's communication. If the terminal fails, the master station will not be able to collect current data, so there is no corresponding record in the database. The sending and receiving bytes in the table refer to the master station. The sending bytes represent the number of bytes sent from the master station to the terminal, and the received bytes represent the number of bytes received by the master station [3]. The data flow in the table indicates the flow used to transmit electric energy data among all flows. In addition to the above flow, the table also includes other control information of the terminal, including the number of reconnections, alarm flow and heartbeat flow. The online time indicates the number of heartbeat signals received by the master station, and the terminal transmits once per minute. If the value is 1440, it indicates that the terminal is online for 24 h (Table 2).

Table 1. Communication flow table

Project	Database field name	Data
Terminal code	rtuid	——
Data date	datatime	——
Send (downstream) byte	sendbytes	——
Receive (uplink) bytes	recvbytes	——
Reconnect times	logintimes	——
Data flow	databytes	——
Alarm flow	alarmbytes	——
Heartbeat flow	linkbytes	——
online time	onlinetimes	——

The on-site operation and maintenance table records the maintenance and inspection results of the operation and maintenance personnel on the fault end, including the terminal information and fault information. The table is generated by the system to generate terminal related information, including all kinds of numbers, data time and location from the master station and fault related information, which are filled in by the operation and maintenance personnel. The fault type and fault description are manually filled in by the operation and maintenance personnel and then input into the system. Therefore, for the same fault, the names filled in by different operation and maintenance personnel may be different, which needs further processing.

Line loss refers to the loss of electrical energy during power transmission and transformation. The line loss table is used to record the input and output power and line loss rate of each line, and contains basic information related to line loss for each line daily, as shown in the following table:

Table 2. Line loss table

Numbering	Project	Database field name
1	Line number	LINEID
2	Line name	DISC
3	Date of data	DATATIME
4	Input power	ENERGY_IN
5	Output power	ENERGY_OUT
6	Line loss rate	LINELOSS_RATE
7	Date data sent	SENDDATE

After the above data table is established in the database of the electric power metering automation system, the data collected and stored in the database by the system is processed.

2.2 Processing Data of Power Metering Automation System

Due to various reasons, the data will be incomplete and inconsistent. These data are called error data, which has a great impact on subsequent anomaly detection. Therefore, data cleaning is very important for abnormal data detection.

Data cleaning must first delete the redundant data in the data set. Redundant data is the only characteristic that destroys every record in the data set. When multiple identical records appear, the redundant data must be deleted. Every user must have electricity readings for every hour of the day, and the serious absence is defined as:

1) 20% of the reading points are missing from the curve;

2) The curve is continuously missing more than 2 consecutive reading points. If the data is missing to a serious extent, the user is excluded from the research scope, and the multi-level Lagrangian interpolation method is used to repair the missing value. The missing value repair formula is as follows [4]:

$$P_t = \frac{\sum_{k=1}^{m_1} P_{t-k} + \sum_{i=1}^{m_2} P_{t+i}}{m_1 + m_2} \tag{1}$$

In formula (1): m_1 is the number of forward periods, m_2 is the number of backward periods, t is the time when the system data is missing, P_t is the missing value after repair, P_{t-k} is the system data at time k before t System data, k is the at time i after time t. After the data is patched, the data is denoised by smoothing the system timing relationship curve.

Set the total operation time T of the system for a period of time t' has been in the abnormal state recorded by the observation equipment, the starting point of the abnormal state is recorded as P_{start}, the ending point of the abnormal state is recorded as P_{end}, the starting point of the abnormal state is recorded as $time_{Pstart}$, and the ending point of the abnormal state is recorded as $time_{Pend}$. Suppose that the data set $P_{t'}$ collected by the power metering automation system in t' is expressed as $P_{t'} = \{P_{start}, P_2, \cdots, P_{n-1}, P_{end}\}$, and the data point $P_{i'}$ in t' after processing is recorded as [5, 6]:

$$P_{i'} = \frac{P_{i'-1} \pm (P_{start} - P_{end})}{time_{Pend} - time_{Pstart}} \tag{2}$$

After processing, the data points in the time series relation curve can only be divided into two types, normal data and abnormal data. After processing the data of electric power measurement automation system, the abnormal data features are extracted.

2.3   Feature Extraction of Abnormal Data

In the process of abnormal data detection in the electric power measurement automation system, the time series of each data in the network database is first obtained, which takes the average value of the time series as the element. The specific steps are as follows:

Suppose that $Q = \{q_1, q_2, \cdots, q_m\}$ and $C = \{c_1, c_2, \cdots, c_n\}$ represent the two data time series, and w_q and w_c represent the time series of the two data time series. Use the following formula to calculate the feature average of all data elements in each time series [7]:

$$q_i = \frac{1}{w}(m) \frac{Q}{C} \bullet \frac{\{q_1, q_2, \cdots, q_m\}}{\{c_1, c_2, \cdots, c_n\}} \bullet \frac{[w_q, w_c]}{Q} \tag{3}$$

In formula (4), w represents the eigenvalue to form a new data sequence, m represents the piecewise aggregation approximation, and w_q and w_c represent the mean value of data elements. Assuming that $d(i,j)$ represents the dynamic time bending distance, ξ represents the eigenvalue to form a new data series, and N represents the data change form of the data time series, then use Eq. (4) to obtain the data characteristic series with the average value of the time series as the element [8, 9].

$$r(i,j) = \frac{L_{DTW} \times D_{w_q \times w_c} \bullet d(i,j)}{[\xi \bullet N] \bullet q_i}[\theta] \tag{4}$$

In formula (4), L_{DTW} represents the process in which the two time series are first converted into feature sequences, $D_{w_q \times w_c}$ represents the original time series data, and $[\theta]$ represents the distance accumulation matrix. After extracting the abnormal data features of the power metering automation system, the abnormal data is detected.

2.4 Implementation of Abnormal Data Detection

The clustering analysis of the AP algorithm uses an abnormal data feature similarity matrix, and the similarity between data points is expressed by the square of the negative Euclidean distance. If there are n data, then these data points constitute the similarity matrix S of $n \times n$, $S(i,j)$ represents the similarity between data points i and j, the calculation formula is as follows:

$$\begin{cases} S(i,j) = -\|x_i - x_j\|^2 \\ S(i,j) \in (-\infty, 0] \end{cases} \tag{5}$$

The element value $S(i,j)$ on the diagonal of the similarity matrix is used to judge the cluster center. $S(i,j)$ is called the preference parameter, which indicates the suitability of data point i as the cluster center of the class. If its value is larger, it means that the point is more suitable to be the cluster center. Set the mean value of similarity matrix as *preference*:

$$preference = \frac{\sum\limits_{i,j=1,i\neq j}^{n} S(i,j)}{n \times (n-1)} \tag{6}$$

The AP algorithm continuously updates the attraction matrix R and the attribution matrix A during the iteration process. The attraction information $R(i,k)$ is the information sent by the sample point i to the possible clustering center k, which indicates the degree of attraction of the sample point i to k. If the value is larger, it indicates that k is more likely to become the center of i; the attribution degree information $A(i,k)$ is potentially. The information sent by the clustering center k to the sample data i expresses the degree of attribution of k as the center of the sample point i. If the value is larger, i is more likely to belong to the cluster with the center k. In the iterative process, the above two information matrices depend on each other and are updated alternately. The update process is as follows [10, 11]:

then,

$$R(i,k) \leftarrow S(i,k) - \max\{A(i,k') + S(i,k')\} \tag{7}$$

When $i = k$,

$$R(k,k) \leftarrow S(k,k) - \max\{A(k,k') + S(k,k')\} \tag{8}$$

When the number of iterations of attraction matrix R and attribution matrix A exceeds the maximum number of iterations given in advance or the change of $R(i,k) + A(i,k)$ is lower than a given threshold, the iterative process will stop. After the AP clustering algorithm clusters the data sets, iforest detects the outliers of the clustered data.

Assume that the whole data set after AP algorithm classification is Ψ, $\Psi = (\psi_1, \psi_2, \cdots, \psi_n)$, where ψ_q represents the q-th cluster.

iForests is composed of f iTree isolated trees, each iTree is a binary tree structure. An iTree training procedure is as follows [12, 13]:

1) Put the m values of dataset $D = \{x_1, x_2, \cdots, x_m\}$ into the root node of the tree.
2) Randomly specify an attribute r, and randomly generate a split value p in the current data. The size of the split value p is the number between the maximum and minimum values of the specified attribute r in the current data.
3) The partition value p divides the current data space into two sub spaces, divides the data less than p in the specified attribute r into the left sub tree, and divides the data greater than or equal to p into the right sub tree.
4) Recursion steps 2 and 3 generate new nodes until there is only one data in the node.

The following figure is a schematic diagram of the construction of iTree.

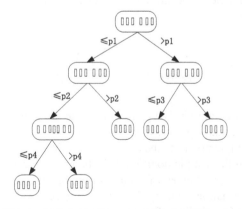

Fig. 1. Schematic diagram of iTree construction

The training steps for f iTree are as follows:

Randomly select ς samples from cluster ψ_q, and use these ς samples to establish an iTree according to the above process. Perform the above steps f times to get f isolated trees, which form an iTree set. When the itrees collection is built, the next step is the exception evaluation phase of the data.

First, calculate the path degree $h(x)$ of the data x to be detected. Path degree $h(x)$ refers to the number of iterations from the root node to the end of the leaf node. For an itrees tree, data x is moved down the partition condition corresponding to the creation until the leaf node is accessed, and the path length $h(x)$ is recorded. Because the structure of itrees is the same as that of the binary search tree, the path length of the leaf node containing data x is the same as the path length of the failed query in the binary search tree, that is, from the root node to the middle node, reach the leaf node, the number of edges traversed. Traverse the itrees set and calculate the exception score for data x. According to the abnormal score, judge the abnormal data. So far, the research on abnormal data detection method of power measurement automation system has been completed.

3 Experimental Research

In order to verify the effectiveness of this method, the following design of comparative experiments.

3.1 Experimental Content

This experiment is a simulation experiment. The experimental group is the abnormal data detection method of the electric power metering automation system studied in this paper, and the experimental group is the traditional abnormal data detection method of the electric power metering automation system. A total of two sets of experiments were conducted in the experiment. The first set of experiments compared the iForest model of the experimental group and the RBF model of the comparative group when detecting data sets containing the same abnormal data; the second group of experimental comparison indicators were abnormalities of the experimental group and the comparative group Data detection method When the abnormal data is detected in the experimental data set, the recall rate and precision rate of the method.

3.2 Experiment Preparation

The power metering data selected in this experiment includes the power consumption information of some users. The details of the data source are as follows (Table 3):

Table 3. Experimental data set details

Serial number	Features	Type	Explain
1	CJ_MP_ID	Integer	Measurement point identification
2	DATA_DATE	Datetime	Data timescale
3	DATA_SORCE	Integer	data sources
4	WRITE_DATE	Datetime	Write date
5	PZ	Float	Total active power
6	PA	Float	Phase A active power
7	PB	Float	Phase B active power
8	PC	Float	Phase C active power
9	QZ	Float	Total reactive power
10	QA	Float	Phase A reactive power
11	QB	Float	Phase B reactive power
12	QC	Float	Phase C reactive power
13	MID_I	Float	Zero sequence current

The calculation formula of the accuracy rate of the abnormal data detection method is as follows:

$$\text{Precision} = \frac{TP}{TP + FP} \tag{9}$$

The calculation formula of the recall rate of the abnormal data detection method is as follows:

$$\text{Recall} = \frac{TP}{TP + FN} \tag{10}$$

In formulas (9) and (10), TP is normal data judged as positive by the detection method. FP is the data judged as positive by the detection method, but is actually abnormal, and FN is the data judged as abnormal by the detection method, but is actually normal data.

Different data sets carry out anomaly data detection one by one, record the precision and recall, take the precision as the ordinate and recall as the abscissa, and draw the P-R curve. If the P-R curve of one detection method is included by another, the latter is better than the former. If the P-R curves of the two classifiers cross, we can choose the balance point, which is the value when p = R. the larger the balance point BEP, the better the performance of the detection method.

Record the experimental data of the two experiments, process and analyze the experimental data, and draw the corresponding experimental conclusion.

3.3   Experimental Results

In order to verify the effectiveness of this method, the abnormal data of iforest model and contrast group RBF model are detected, and the detection results are shown in the figure below.

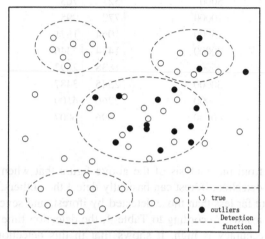

(A) RBF detection effect

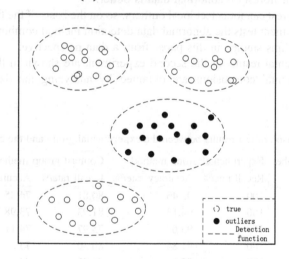

(B) iForest detection effect

Fig. 2. Comparison of detection effect

Analysis of the data in Fig. 2 shows that iforest can basically detect abnormal data when the dataset containing outliers is used for detection, while the abnormal data detected by RBF is far less than that detected by iforest, indicating that the abnormal data detection effect of iforest model in power metering automation system is better.

The abnormal data detection time of iforest model and contrast group RBF model are compared and analyzed. The comparison results are shown in Table 4.

Table 4. Comparison of detection operation time of two methods/MS

The amount of data	iForest	RBF
5000	521	763
10000	772	995
15000	1064	1571
20000	1474	2346
25000	1838	2892
30000	2345	3487
35000	2760	4109
40000	3596	6007

It can be seen from the analysis of the above figure that when the data set with outliers is used for detection, iforest can basically detect the outliers, while the outliers detected by RBF are far less than those detected by iforest, and several normal data in the RBF detection results. According to Table 4, the detection time of iforest is short and the detection accuracy is high. It shows that in this detection experiment, the detection effect of iforest on abnormal data is better.

However, iForest cannot detect local outliers, so on the basis of the first experiment, the second experiment tests the abnormal data detection method combining iForest and clustering algorithms studied in this paper from a data perspective.

The experimental results of the second experiment are shown in the table below, and the experimental conclusions are obtained by analyzing the data in the table (Table 5).

Table 5. Comparison of test results between the experimental group and the comparison group

Data set number	Experimental group method		Contrast group method	
	Recall rate/%	Accuracy rate/%	Recall rate/%	Accuracy rate/%
1	100	95.45	80.81	74.35
2	100	94.13	81.93	73.98
3	100	95.67	75.97	74.11
4	100	91.84	80.30	73.27
5	100	92.63	78.97	74.65
6	100	92.01	79.23	74.00
7	100	94.75	75.25	73.93
8	100	94.75	82.04	73.88
9	100	91.26	77.71	73.33
10	100	94.42	77.66	74.69
11	100	94.69	79.42	73.69
12	100	95.11	78.45	73.83

According to the above table, the recall rate of the test methods in the experimental group is 100%, and the recall rate interval of the test methods in the comparative group is seventy-four point three five ~ 82.04% The results show that the experimental group can accurately judge the normal data. The precision interval of the test method in the experimental group is ninety-one point two six ~ 95.67%. The precision interval of the control group method is seventy-three point two seven ~ 74.69%, which is far lower than the detection method of experimental group, indicating that the accuracy of abnormal data detection of experimental group is high and the number of false detection is small.

In summary, the automatic abnormal data detection method of the power metering system studied in this paper takes less time, has higher detection accuracy, and has higher practicability.

4 Conclusion

In the power metering automation system, the detection of abnormal data may obtain more useful value than the analysis of normal data. This paper proposes to use a combination of clustering algorithm and isolated forest as an anomaly detection method. Through comparison experiments with traditional methods, it proves that the research method takes less time to detect, and the detection results are more reliable, and the method is feasible.Fund projects

Design and implementation of new energy inverter based on MCU control (CJGX2016-KY-YZK034)

References

1. Yang, X., Qu, Y., Pang, H., et al.: Power metering pipeline fault warning technology based on deep learning algorithm. Electron. Des. Eng. **28**(04), 153–157 (2020)
2. Gao, S., Li, C.: An improved spectral clustering algorithm for anomaly detection of power data. Comput. Simul. **36**(11), 239–242 + 304 (2019)
3. Yang, J., Zeng, X., Yao, L., et al.: Research on abnormal electricity monitoring based on large data mining. Autom. Instrum. **08**, 219–222 (2019)
4. Tong, X., Yu, S.: Fault detection algorithm for transmission lines based on random matrix spectrum analysis. Autom. Electr. Power Syst. **43**(10), 101–115 (2019)
5. Xu, G., Ning, B., Zhong, Y.: Automatic matching of voltage blackout events in metering automation system. Electron. Test **04**, 111–112 (2019)
6. Huang, J., Dai, B., Zhang, L., et al.: Study on dynamic identification of abnormal data of electric energy measurement device. Guangxi Electr. Power **41**(04), 53–55 + 64 (2018)
7. Chen, Q., Zheng, K., Kang, C., et al.: Detection methods of abnormal electricity consumption behaviors: review and prospect. Autom. Electri. Power Syst. **42**(17), 189–199 (2018)
8. Liu, S., Glowatz, M., Zappatore, M., et al. (eds.): e-Learning, e-Education, and Online Training, pp. 1–374. Springer, Heidelberg (2018)

9. Zhang, J., Chen, F., Li, B., et al.: Research on energy metering abnormity and fault analysis technology based on metrological automation system. Yunnan Electr. Power **46**(02), 63–65 (2018)
10. Fu, W., Liu, S., Srivastava, G.: Optimization of big data scheduling in social networks. Entropy. **21**(9), 902 (2019)
11. Liu, S., Li, Z., Zhang, Y., et al.: Introduction of key problems in long-distance learning and training. Mob. Netw. Appl. **24**(1), 1–4 (2019)
12. Ding, M., Li., C., Li, H., et al.: Fault detection method of photovoltaic inverter based on massive data mining. Electr. Autom. **40**(03), 30–32 (2018)
13. Shuai, L., Gelan, Y.: Advanced Hybrid Information Processing, pp. 1–594. Springer, Heidelberg

Research on Data Optimization Method of Software Knowledge Base Operation and Maintenance Based on Cloud Computing

Gang Qiu[1,2] and Shi-han Zhang[3(✉)]

[1] Department of Computer Engineering,
Changji College, Changji 831100, China
chougan1467@163.com
[2] College of Computer Science and Technology, Shandong University,
Jinan 250100, China
[3] Shenyang Institute of Technology, Fushun 113122, China
Zhangshihan1324@163.com

Abstract. To improve the accuracy of data matching in software knowledge base operation and maintenance, a data optimization method based on cloud computing is proposed. In order to achieve the goal of accurate detection, the steps of anomaly detection of software knowledge base operation and maintenance data are improved, and the optimization of software knowledge base operation and maintenance data is completed. Finally, the experiment proves that the matching accuracy of the software knowledge base operation and maintenance data optimization method based on cloud computing is significantly improved compared with the traditional operation and maintenance method.

Keywords: Cloud computing · Software · Knowledge base · Operation and maintenance data

1 Introduction

With the rapid development of modern technology, the scale of software knowledge base data cluster is gradually expanding. In order to better guarantee the stability of knowledge base operation, it is necessary to further improve the database operation and maintenance technology of software knowledge base [1]. Therefore, through the analysis and research of the current common methods of operation and maintenance data optimization, it is found that due to the unreasonable structure of software knowledge base and the lack of effective association matching module, it is difficult to distribute and mine the massive dimension data accumulating in the database in time, resulting in the poor quality of data operation and maintenance and other problems [2]. In order to solve the above problems, combined with cloud computing method to optimize the operation and maintenance data method of software knowledge base, through the effective collection of multi-dimensional information of knowledge base, and according to the results of collection, the dimension data association of software database is analyzed, and the dimension operation and maintenance data relationship of

S. Liu and L. Xia (Eds.): ADHIP 2020, LNICST 347, pp. 229–238, 2021.
https://doi.org/10.1007/978-3-030-67871-5_21

database is judged and matched scientifically and reasonably. Thus, the data feature management and potential value of software knowledge base operation and maintenance can be effectively excavated, and the quality and efficiency of data operation and maintenance can be effectively improved.

2 Data Optimization of Software Knowledge Base Operation and Maintenance Based on Cloud Computing

2.1 Cloud Computing-Based Knowledge Base Operation and Maintenance Data Association Algorithms

Operations and Maintenance Related Knowledge Base includes two parts: Operations and Maintenance Database and Data Retrieval System. In order to achieve the research goal of optimizing the Operations and Maintenance Data, the first step is to initialize the Operations and Maintenance Data Association in the Knowledge Base. In the construction of software knowledge base structure, the characteristic change period and stability of operation and maintenance data are mainly judged by the delay time and the quality of anti-interference factors in the operation and maintenance process [3]. Without considering the influence of interference factors for the time being, the longer the period of data feature change, the higher the accuracy of data extraction in software knowledge base.

According to the above principles, combined with the network optimization scheduling algorithm, the arbitrary change time of the data features in the knowledge base is set to α. If the data amount in the software knowledge base is n and the feature acquisition interference degree of the data amount is x, then using the feature distribution extraction principle, the initial feature change degree of the data can be effectively described by the algorithm. The specific calculation formula can be expressed as follows:

$$C_n(x) = \begin{cases} 1 - \left(\frac{x}{\alpha}\right)^{-\alpha}, x \geq \alpha \\ 0, 0 < x \end{cases} \tag{1}$$

According to the principle of the algorithm mentioned above, if the data stream ς in the software knowledge base DRC is processed by feature classification, If m is the common characteristic parameter of data, D is the data flow of data change in database HOL, and ϑ is the lowest tail parameter of data potential relevance. Then the probability algorithm for feature association of abnormal data can be described as:

$$S_m(x) = \begin{cases} \frac{[\vartheta*C(x)]^\varsigma}{[DRC\text{-}\text{HOL}_{n-1}]^\vartheta} - \vartheta \\ \exp D * \left(\frac{\alpha-1}{1+\sqrt{n}}\right) \cdot \frac{[DRCn*C(x)]^\varsigma}{[\text{HOL}_m]^\vartheta} \end{cases} \tag{2}$$

If the probability index of feature change of all data in knowledge base is t, the standard information association matching parameters of all data in data stream $\varsigma(x)$ and $\vartheta(x)$ can be obtained, which is expressed as [4, 5]:

$$A(n) = \begin{cases} S_m(x)[\alpha - t(n)] - 1, t_{\max}[n] > \alpha \\ S_m(x)[t(n) - 1] + \frac{t}{\alpha}, t_{\max}[n] < \alpha \end{cases} \tag{3}$$

According to the above algorithm, we can accurately grasp the relativity of operation and maintenance data in software knowledge base.

2.2 Software Knowledge Base Operation and Maintenance Data Anomaly Detection

In the process of operation and maintenance, a large number of abnormal data are easy to occur, which leads to the unrelated results of data operation and maintenance. This paper combines cloud computing method to optimize the database operation and maintenance management method [6, 7]. According to the acquisition of abnormal data features, the types and characteristics of abnormal data can be accurately judged and matched in the database, and the required knowledge data can be selected for comparison and correction, so as to complete the effective operation and maintenance of abnormal data. It is very important to collect the key parameters of abnormal data features in the process of abnormal data detection and correction. If there are errors in data acquisition, it will directly affect the effect of data association matching and operation and maintenance [8, 9]. Due to the relatively complex collection function of abnormal data, in order to facilitate subsequent operations, the spatial behavior of data is visualized, to promote the effective operation and maintenance of similar abnormal data in the future, and finally achieve the optimization of operation and maintenance of acid-base knowledge base data.

The spatial behavior of software knowledge base includes social behavior data, logical language behavior data and spatial movement behavior data. The steps of mining spatial behavior data are divided into four stages: spatial behavior data preparation, data mining, data presentation and data evaluation. The process is shown in Fig. 1.

In the process of mining the spatial behavior data of the operation and maintenance of the software knowledge base according to the above steps, the minimum support degree of data mining is assumed to be s and the minimum confidence degree to be C_0, and then the candidate item set will be obtained. If the support degree of the set is greater than or equal to the minimum support degree, it is called the frequent item set. The data in the database is scanned until no new candidate sets are generated.

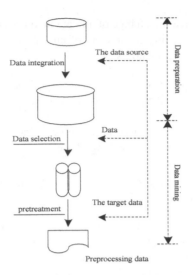

Fig. 1. Software knowledge base operation and maintenance data mining steps

According to the above software knowledge base operation and maintenance data mining steps, social behavior data is mined. Social data, topics, named entities and their associations are defined as hierarchical semantic model. Each message is defined as a node.

$$\prod = \{n{:}n \in V_T\} \tag{4}$$

Where n represents the message data in the spatial behavior data, and V_T is the message set of the same topic. The clustering graph obtained after partition traversal is represented by matrix vector, and the graph expression is shown in formula 2.

$$AG = \langle V_T, E_T \rangle \tag{5}$$

In Eq. (5), E_T is the classification relationship of named entities. According to the same mining method, the data of logical language behavior and spatial movement behavior in spatial behavior are mined, and the clustering fuses spatial behavior data to output the final mining results.

The spatial behavior data information in the operation and maintenance data of the software knowledge base is converted into the representation of the graph, so the spatial behavior trajectory needs to be transformed. The transformation process is divided into two steps, namely the generation of spatial behavior trajectory and the trajectory transformation. The generation of spatial behavior trajectory needs to calculate the distance of behavior and judge the direction of spatial behavior. When calculating the distance and direction, each node in the space needs to be traversed, and the buffer of the path intersects to obtain the set L of spatial behavior, then the total distance length is also L, where the length coefficient between each two nodes is κ, then the overall direction value of spatial behavior is calculated as follows:

$$L_\alpha = \kappa_1\alpha_1 + \kappa_2\alpha_2 + \cdots + \kappa_n\alpha_n \qquad (6)$$

The direction Angle of each space segment is α_n, the value of the direction Angle and the distance length value are calculated, and the result of the spatial behavior trajectory transformation of the operation and dimension data of the software knowledge base is finally obtained.

2.3 Implementation of Data Optimization for Operation and Maintenance of Software Knowledge Base

After completing the effective matching of massive data by using the knowledge base operation and maintenance data association algorithm, the abnormal data which failed to match is detected and corrected. After the correction of abnormal data is completed, the existing operation and maintenance system communication transmission technology is used to encrypt the operation and maintenance data and other aspects of management and transmission processing, so as to avoid the intrusion of aggressive data [10, 11]. Since the content of data operation and maintenance management is trivial, I will not make more statements here.

After data management and validation, the validation information is fed back to improve the accuracy, security and effectiveness of data operation and maintenance of software knowledge base. In order to effectively optimize the operation and maintenance data of software knowledge base, first of all, it is necessary to collect and analyze the data features, and add the corresponding functions of feature recognition, category analysis, data storage and retrieval of associated data into the structure of software knowledge base [12, 13]. By setting and optimizing data management module, feature classification change module (application layer) and data transmission and publishing management module (data support layer) in software knowledge base, the accurate feature extraction of different operation and maintenance data can be realized effectively. According to the acquisition characteristics to determine the correlation between data, and in the software database to complete the data transmission and management work. Integrating the previous ideas and algorithms, the data transmission process of operation and maintenance of software knowledge base is designed, and the following Fig. 2 is obtained.

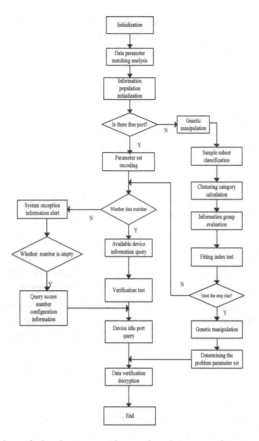

Fig. 2. Software knowledge base operation and maintenance data transmission process

As shown in the Fig. 2, combined with the previous method to optimize the software knowledge base operation and data transmission process, combined with data feature collection, data encryption and other technical principles, the process of data encryption transmission and decryption verification realizes the construction of IoT data movement operation and maintenance model Therefore, the abnormal data in the knowledge base is detected in time to achieve the goal of accurate collection and optimization of operation and maintenance data.

3   Analysis of Experimental Results

In order to verify the practical performance of the cloud computing-based software knowledge base operation and maintenance data optimization method, a simulation experiment was carried out, and the functional and non-functionality of the operation and maintenance system were compared and tested. In order to clarify the detection target, the problems in the test are corrected and perfected. Firstly, the data and information system of the operation and maintenance system is recorded, and the following Table 1 is obtained.

Table 1. Operation and maintenance data optimization test standards

Purpose	Testing standard
Testing requirements	Reasonable layout and complete functions
	Information matching
Test procedure	Register log in
	Data authentication upload
	Functional operation detection
	Information detection
	Data authenticity detection
Test results	The match is reasonable and the test passes
	The data detection is complete without defects, the data is in line with expectations, the abnormal data is reduced, the matching degree is improved, and the functional test is passed

Experiments were performed according to the standardized data in the above table. The software knowledge base operation and maintenance data optimization method and the traditional operation and maintenance method are compared and verified by combining the cloud computing method. The operation and maintenance effects are judged by comparing and analyzing the data matching degree. The specific detection results are as follows (Fig. 3).

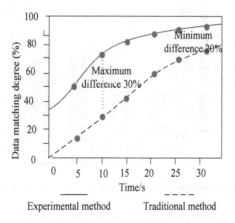

Fig. 3. Compare experimental test results

According to the above experimental results, it is not difficult to find that the data matching degree of the software knowledge base compared with the traditional method is improved by the cloud computing method, and the overall matching situation is improved by 20%–30% compared with the traditional method. In the operation and maintenance process, the higher the data matching degree, the less the abnormal data is, and the higher the operation and maintenance effect is. Therefore, it is confirmed that the cloud computing method fully satisfies the research requirements of the software knowledge base data operation and maintenance method.

The proposed method is compared with the traditional method for data operation and maintenance (s). The experimental results are shown in the Fig. 4 below.

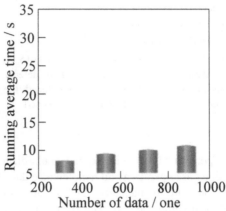

（a）The average maintenance time of the method presented in this paper

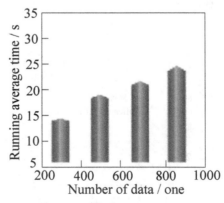

（b）Average operation and maintenance time of traditional methods

Fig. 4. Comparison of data operation and maintenance average time in different methods

Figure 4 shows the average data operation and maintenance time of the two methods under different data volumes. According to Fig. 4, with the increase of data volume, the running time of both methods increased. However, through comparison, it can be seen that the running time of the traditional method is much higher than that of the proposed method, and the running time increases more during the experiment. By comparison, this method is superior to the traditional method and has higher data operation and maintenance efficiency.

4 Conclusions

The cloud computing method is used to innovate and optimize the software knowledge base of the software knowledge base to ensure the safe and reliable operation of the knowledge base. The comparison experiment is carried out to compare the optimization effect of the method. The experiment proves that the combination of cloud computing method has a strong practical value for the software knowledge base data operation and maintenance method, which can process data information more quickly and accurately, and ensure knowledge. The security and stability of the library data processing capabilities.

Acknowledgements. This work was supported by the Research Project on Teaching Reform of Ordinary Colleges and Universities in Xinjiang Uygur Autonomous Region: Changji College - ZTE ICT Applied Talents Training Mechanism Research Project (2017JG117), as well as the South Xinjiang Education Development Research Center Project: Research on Xinjiang Bilingual Basic Education Curriculum Resources (XJEDU070116C11), and the 2019 Xinjiang Uygur Autonomous Region Higher Education Scientific Research Project (XJEDU2019Y057, XJEDU2019Y049).

References

1. Ding, Y.: Technical analysis of intelligent monitoring system for operation and maintenance data collection and business process of broadcasting system. Western Radio Telev. **426**(10), 189–190+193 (2018)
2. Jin, X., Yan, L., Liu, J., et al.: Fault analysis and operation research for enterprise database: taking Oracle Database as an example. Software **38**(10), 178–181 (2017)
3. Dou, J., Dai, F.: Research on operation and maintenance scheme of intelligent log analysis platform based on big data environment. J. Jiujiang Vocat. Tech. Coll. **25**(04), 96–98 (2017)
4. Simplicity, Jing, G., Hu, C., et al.: Software vulnerability detection algorithm for 8031 single chip microcomputer system based on vulnerability knowledge base. J. Beijing Inst. Technol. **34**(4), 371–375 (2017)
5. Zheng, P., Shuai, L., Arun, S., Khan, M.: Visual attention feature (VAF): a novel strategy for visual tracking based on cloud platform in intelligent surveillance systems. J. Parallel Distrib. Comput. **120**, 182–194 (2018)
6. Zhong, L., Guo, T., Zhang, M.: Design and implementation of online ceramic mineral resources management knowledge base based on LINGO. Intell. Comput. Appl. **36**(1), 68–71 (2018)
7. Liu, S., Li, Z., Zhang, Y., Cheng, X.: Introduction of key problems in long-distance learning and training. Mobile Netw. Appl. **24**(1), 1–4 (2018). https://doi.org/10.1007/s11036-018-1136-6

8. Wang, F.: Summary of technological innovation of broadcasting operation and maintenance data collection and process monitoring and processing system. Modern Telev. Technol. **34** (5), 138–141 (2017)

9. Zhao, C., Sun, L., Gao, X., et al.: Discussion on protection intelligent operation and maintenance technology based on source data maintenance mechanism. Power Grid Clean Energy **52**(09), 82–87 (2017)

10. Tan, L., Zhong, H.: Operation and maintenance practice of data center construction based on cloud computing model. Inf. Comput. (Theor. Ed.) **408**(14), 21–23 (2018)

11. Shuai, L., Weiling, B., Nianyin, Z., et al.: A fast fractal based compression for MRI images. IEEE Access **7**, 62412–62420 (2019)

12. Yang, Y., Zhang, S., Kang, Q.: Design of intelligent early warning system based on grid operation and maintenance data. Inner Mongolia Electric Power Technol. **35**(4), 20–23 (2017)

13. Shuai, L., Gelan, Y.: Advanced Hybrid Information Processing, pp. 1–594. Springer, New York

Dynamic Data Mining Method of Cold Chain Logistics in Drug Distribution Under the Background of Cloud Computing

Meng-li Ruan(✉)

Shandong Management University, Jinan 250357, China
ruanmengli988@163.com

Abstract. Because of the huge volume of cold chain logistics data, the traditional data dynamic mining method can not mine the whole local drug circulation data, resulting in the lack of a large number of data mining results, reducing the integrity of the data. Therefore, in the context of cloud computing, a new dynamic mining method is proposed for the cold chain logistics data of drug circulation. Under the cold chain logistics model, the method is further developed by defining the drug circulation mode. The data mining uses cloud computing technology to extract the target data; uses data cleaning, data elimination, data supplement and data conversion to preprocess the target data; according to the association rules between the acquired data, realizes the dynamic mining of cold chain logistics data information. Experiments show that, compared with the traditional methods, the proposed mining method can find the target data in the huge cold chain logistics data, and achieve all the data mining. It can be seen that the data mining method proposed in this paper has higher data integrity.

Keywords: Cloud computing · Drug circulation · Cold chain logistics data · Dynamic mining

1 Introduction

For a long time, high-performance computing (HPC), as an important topic in the field of computer research, has been concerned. Since the 1990 s, HPC has grown from 1 billion floating-point operations per second to 10 billion floating-point operations per second. However, with the increase of computing speed, the cost of supercomputers has also increased, and its high cost makes people rarely able to afford [1]. However, cloud computing technology has changed this situation, people began to use parallel and distributed computing technology to build high-throughput cloud computing system. The development style of high-end computing system is changing, and the computing strategy has changed from HPC to HTC, that is, to high-throughput computing. Generally, in order to make the results of data mining more accurate, people tend to choose larger data sets. Considering the problems of running efficiency and computing speed, the distributed computing and storage of cloud computing system have more advantages than single processing. Therefore, more and more technologies are put into the use of cloud computing technology [2].

S. Liu and L. Xia (Eds.): ADHIP 2020, LNICST 347, pp. 239–250, 2021.
https://doi.org/10.1007/978-3-030-67871-5_22

Data mining is a process of building models and discovering the relationships among data in large-scale data of database by using various analysis means and tools, and obtaining valuable information that has not been known before. It belongs to an interdisciplinary field, including statistics, database, pattern recognition, machine learning, artificial intelligence, high-performance computing, data visualization and other fields. Figuratively speaking, data mining looks for small valuable parts from large raw materials, just like extracting gold from soil or sand. The main task of data mining is to extract patterns from data sets. According to function classification, data mining can be roughly divided into two general characteristics: description and prediction. Description is mainly to find the data stored in the database through data mining, while prediction is the basic data knowledge of data mining, which can be divided into feature analysis, association analysis, classification prediction, clustering analysis, inference and prediction [3].

Cold chain logistics generally refers to the system engineering to reduce the loss of products in the production, storage, transportation, sales and all links before consumption, which is always in the specified low temperature environment to ensure product quality. Cold chain logistics is established with the progress of science and technology and the development of refrigeration technology. It is a low-temperature logistics process based on refrigeration technology and means of refrigeration technology [4]. Therefore, at this stage, some drugs circulation, as long as the cold chain logistics mode is adopted, while ensuring the quality of drugs, the information in each link is clear. In order to grasp the status of drugs in circulation in time, a dynamic mining method of cold chain logistics data is proposed under the background of cloud computing.

2 Drug Circulation Under Cold Chain Logistics Model

In a certain period of time, according to the change of drug information, drug information is divided into fixed drug information and changed drug information. Because the traffic network information changes all the time, in order to increase the anti-interference of the preset model, a coefficient of variation can be set. When the coefficient of variation is greater than a threshold, it is understood that the traffic network information has changed; when the coefficient of variation is less than the threshold, the traffic network has not changed. The changed drug information and the changed traffic network information can be collectively referred to as the changed information. Therefore, when the changed information is not generated, the cold chain logistics distribution network is static. According to the existing information, the distribution cloud platform plans the distribution path, sends the planned route map to the terminal of the cloud distribution system, and guides the driver to drive. With the implementation of refrigerated vehicle distribution task, once the distribution cloud platform receives the change information request, the distribution cloud platform will adjust and plan the original driving route according to the new data to respond to the change information.

Since the change information is generated based on the concept of time, the strategy of "time axis recording + cloud service platform delivering distribution information" is adopted. When the change information is generated, record the time t, analyze the unfinished, ongoing and newly generated drug information at the time t, so as to divide the dynamic distribution problem into static problems by using the time axis, and the cloud service platform will The distribution information is known and transferred to the intelligent vehicle scheduling system for global optimization [5], as shown in Fig. 1.

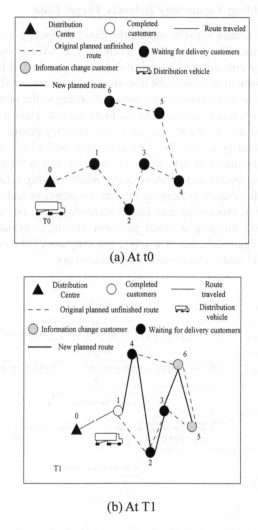

(a) At t0

(b) At T1

Fig. 1. Dynamic distribution path of cold chain under time axis

(a) It means that when the refrigerated vehicle has not started in the distribution center, it plans the distribution path according to the known customer information and real-time road condition information. It is known that the acceptable service time of customer 5 and customer 6 is adjusted at time t, and the vehicle just serves customer 1, so a new path planning is carried out for the remaining customers.

3 Dynamic Data Mining Method of Cold Chain Logistics

3.1 Cloud Computing Technology Extracts Target Data

According to the cold chain logistics model, the cloud computing technology is used to extract the target data. Cloud computing technology is widely used in the logistics industry. Through the collection of historical data, current data and follow-up new data in the cold chain information source, the historical data and current data are analyzed to explore the general law of follow-up new data. According to the new law, the status of drugs in the circulation link is analyzed to facilitate the cold chain logistics enterprises to formulate targeted development plans and lines Industry planning [6].

In the aspect of drug safety, the data collection and visual analysis under the background of cloud computing can obtain the logistics data in the circulation link; the distributed computing results and database query results of drug safety can be obtained by integrating the distributed processing and data warehouse technology; the sensor inspection information processing and fusion technology can be used to analyze the drug in each link by building a cloud platform Stream information and real-time environment data. Figure 2 below is a schematic diagram of the cold chain logistics data collection model under cloud computing technology.

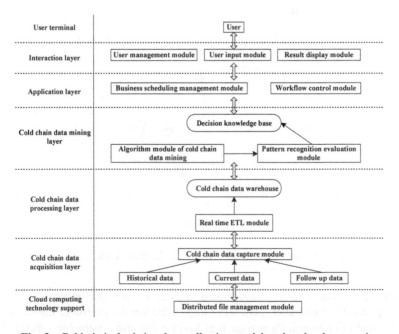

Fig. 2. Cold chain logistics data collection model under cloud computing

One of the key points of cold chain logistics data collection is temperature information, because the temperature is too high or too low, which has an impact on the quality of drugs. According to this characteristic, temperature and humidity recorder and acquisition standard are constructed. In order to ensure the temperature monitoring in cold chain logistics and provide the most timely data, cold chain logistics cloud computing center provides a non-contact temperature and humidity recorder that can realize the automatic collection and storage of temperature data.

The logic components of the temperature and humidity recorder for cold chain logistics include temperature data acquisition, data processing and data storage. The corresponding hardware is integrated digital temperature and humidity sensor, low-power microcontroller and data memory. According to the length of refrigerated transport car, install appropriate number of temperature and humidity probes, 3 below 10 m, 4 above 10 m. The single-chip controller suitable for industrial environment is selected, which has the design of anti disassembly and anti destruction. The internal integrated GPS module, six axis gyroscope, TF memory card interface, SIM card and standby power supply. After the temperature and humidity recorder is installed in the cold chain transport car, it can collect data in real time, record vehicle turning, braking, tilting and position information, and send it to the cloud computing center through 3G/4G mobile data network. Table 1 is the main collection index of temperature and humidity recorder [7].

Table 1. Main collection indexes

Index item	Parameter	Value
Temperature	Scope of work	−40 °C–80 °C
	Temperature error range	±0.5 °C
	Waterproof grade	IP65
	Working voltage	3–36 V
Humidity	Scope of work	0–100%
	Humidity error range	3%
	Waterproof grade	IP65
	Working voltage	3–36 V

The cloud computing technology is used to count these data and classify them according to the data sources, so as to provide reliable data sources for further data mining.

3.2 Target Data Preprocessing

The data extracted from cloud computing is often incomplete and noisy. The quality of data often determines the quality of data mining. In a mess of disordered data, it is impossible to carry out effective data mining and data analysis. The four processes of data cleaning, data integration, data conversion and data reduction are collectively referred to as data preprocessing. Before data mining, data preprocessing can improve data integrity, accuracy and availability, remove redundant data, reduce the amount of computation, and improve the quality of data mining.

Data cleaning is the processing of data noise and incomplete data cleaning; data integration is the unified storage and management of data of different data formats and different data sources; data transformation is to transform data of different structures into data input structures with applicable data mining algorithms on the premise of modifying the data structure input by data mining algorithms; data reduction In the face of massive data, data reduction can reduce the amount of data while keeping the integrity of data unchanged [8].

The main content of data cleaning includes two parts: noise removal and incomplete data processing. Noise removal refers to the treatment of random errors, deviations and redundancy in data, mainly including smoothing, clustering and de duplication. Smoothing is to smooth data values by deleting data that does not conform to the general data distribution curve. Clustering is a group of similar data, and outliers may be noise. The reason of de duplication is that there is no need to repeat data in many analyses, so de duplication is used to remove noise. Processing incomplete data refers to the processing method of treating incomplete data, including two processing methods: deleting data and supplementing complete data. Specifically, it includes data deletion, mean value supplement and neighbor supplement. Among them, deleting data is to delete all data with null attribute, and the number of data applicable to null attribute is far less than the total number of non null data; mean value supplement is to use the mean value of all non null values of attribute instead of null value, and the change degree between attribute values is small; nearest neighbor supplement is to make up for the missing data through proximity data.

If the attribute values of the known data a and b are a_i and b_i respectively, there is a formula:

$$f(a_i, b_i) = \begin{cases} 1 & if \quad a_i = b_i \\ 0 & else \end{cases} \tag{1}$$

According to the above formula, the nearest neighbor measures of a and b are calculated as follows:

$$\eta(a, b) = \sum f(a_i, b_i) \tag{2}$$

According to the above calculation results, find out the nearest m neighbor with the largest measure, that is, the most similar neighbor, use the attribute value of neighbor instead of null value, this step can deal with binary Boolean data. The threshold supplement first calculates the distance C between the incomplete data and all other data. If $C < Threshold$ is less than the threshold, the corresponding attribute value of the data is used instead of the null value, which C can be Euclidean distance, and its calculation equation is:

$$C = \left(\sum_i |a_i - b_j|^2 \right)^{1/2} \tag{3}$$

You can also default the above formula to Manhattan distance:

$$C = \sum_i |x_i - x_j| \tag{4}$$

The formula can deal with discrete and continuous data. Experience supplement is to supplement the vacancy value manually according to experience, which requires the processing personnel to have rich experience and low data complexity.

According to the above process of cleaning data, according to the results of cleaning data conversion processing. Data conversion includes four cases: unit conversion, data generalization, normalization and attribute construction: unit conversion is the unit of original data, which may need unit conversion. For mobile cold chain data, common unit conversions include: time and normal time conversion; time measurement second, minute and hour conversion. Data generalization is to layer the concept and replace the low-level or "original" data with the high-level concept. For mobile cold chain data, common data generalization includes: attributes of mobile user location can be generalized to higher-level concepts; time of a week can be generalized to workdays and weekends; time of a day can be generalized to morning, noon and evening. Normalization is to give equal weight to attributes of different dimensions so that they have the same influence on the analysis results. The normalization methods for data include attribute normalization and attribute standardization.

Data specification includes four main methods: dimension reduction, numerical reduction, data sampling and discretization: dimension reduction uses coding mechanism to reduce the size of data set and delete irrelevant or weakly related attributes in data. Numerical reduction is to replace a large amount of data with a small amount of data in an average sense. Data sampling is faced with a large number of data samples and can not be processed. It can sample the data and analyze the subset. Discretization is the discretization of continuous attribute values. The continuous attribute values in the original data are marked with discrete intervals to simplify the number of attribute values and simplify the original data. According to the above data preprocessing method, the cold chain logistics data in the drug circulation link is preprocessed to ensure the authenticity and integrity of drug related logistics data. Through dynamic mining method, the required logistics data is obtained.

3.3   Dynamic Mining of Logistics Data

The dynamic data mining task is implemented, which is based on the preprocessed data and finds out the same type of data characteristics according to the association 2^m rules between the data. And processed data sets, generally, there are m data clusters of different phases. There may be frequent s data clusters, and there will be a rule.

Therefore, in these complex data sets, all frequent data clusters that meet the minimum support threshold are found, and then association rules with high confidence are mined from these clusters. The possible rules between frequent data clusters can be expressed by the following formula (5):

$$\sigma = \sum_{\varphi=1}^{m} \left[EDI \begin{pmatrix} m \\ \varphi \end{pmatrix} \times \sum_{Z=1}^{m-\varphi} (m - \varphi)/Z \right] \tag{5}$$

In the formula: σ possible association rules obtained from cluster data cluster analysis; φ is number of candidate clusters; Z support count of each candidate cluster. According to these possible association rules, the association degree between these rules is measured [9–11].

According to these possible association rules, the association degree between these rules is measured [12]. Set $B = \{w_1, w_2, \ldots, w_m\}$ to represent the set of all items. If $X \subseteq B$ is a drug logistics data X mode, X is called the item set when there are h data information in the mode h.

It is assumed that data P is a set in drug circulation, in which each circulation data H is a set of items, and each data type is represented by HP. U and V are two cluster item sets, which exist in $U \subseteq B$, $V \subseteq B$ and $U \cap V \lambda$ are included U in the drug logistics data H. If and only then $U \subseteq H$, the association rule is $U \Rightarrow V[u, v]$. Therefore, the description equation of support and confidence is:

$$\begin{cases} support(U \Rightarrow V) = P(U \cup V) \\ confidence(U \Rightarrow V) = P(V/U) = P(U \cup V)/P(U) \end{cases} \tag{6}$$

$$\frac{|W - K_i|}{|W|} \leq \beta, 1 \leq i \leq m, \beta \in [0, 1] \tag{7}$$

In the formula: β indicates the specified parameter, which W is used to describe the division precision Y_i. It will be divided into data blocks. If the above formula $K_1, K_2, \ldots, K_n$ and W does not hold for all sums, it needs W to be divided into data blocks Y_n'. So far, under the background of cloud computing, the dynamic mining method of cold chain logistics data of drug circulation link can be realized.

4   Experimental Study

This paper proposes a comparative experiment, compares the data mining method of this study with the traditional data mining method, analyzes the differences between the two mining methods, and draws specific experimental conclusions according to the experimental test results.

4.1   Experiment Preparation Process

Set up the experimental test platform, select the experimental test software, and the following Fig. 3 is the operation page of the experimental test software.

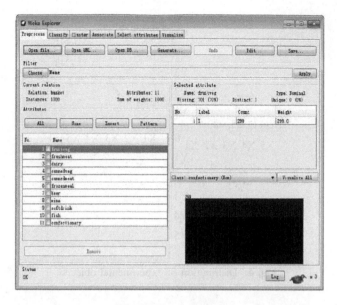

Fig. 3. Test software

Randomly select a pharmaceutical group in a certain city, take the cold chain logistics selected by its drug circulation as the experimental test object, randomly select the drug circulation data in three directions as the object to be mined, respectively recorded as N1, N2, N3. Table 2 below is the basic information of the object to be mined.

Table 2. Basic information of objects to be mined

	Data size	Data type
N1	45.5 GB	Distribution link
N2	28.63 GB	Transportation link
N3	104.27 GB	Transportation link

Figure 4 below is the distribution diagram of these drug circulation data in the cold chain logistics data storage database.

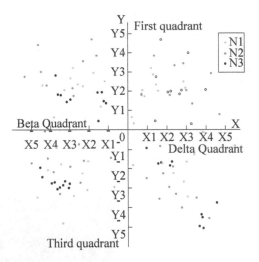

Fig. 4. Distribution of experimental objects

According to the above figure, the data in N1-N3 group are randomly distributed in four quadrants and do not have specific laws, which meet the requirements of this experiment. Therefore, two data mining methods are used to mine the data in Fig. 4 and analyze the mining effect of the two methods.

4.2   Experimental Results and Analysis

Taking the proposed data mining method as the experimental group and the latest data mining method proposed by other scholars as the control group, all data in N3 group are mined, as shown in Fig. 5.

After analyzing the two groups of test results, it is known that under the same experimental conditions, the mining method proposed in this paper is to mine all N3 data in four quadrants, and there is no problem of missing and mining errors. However, under the latest mining methods, for the remote N3 data, it is not mined at all. At the same time, there are some mining results of N2 and n3ex data in a small range. Based on the above experimental results, we can see that the proposed data mining method has already cleaned and compensated the data in the preprocessing stage, so as to get the complete mining results.

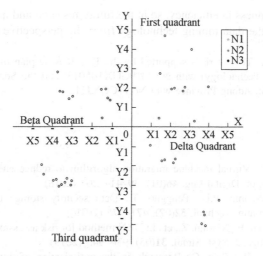

(a) Test results of experimental group

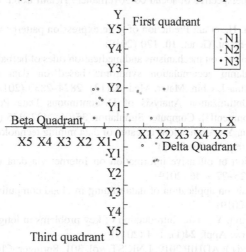

(b) Control group test results

Fig. 5. Experimental test results

5 Concluding Remarks

On the basis of combing the relevant research results at home and abroad, this paper improves the original traditional data mining methods by using the technology of cloud computing, data mining and logistics management. By means of cleaning and compensation, the repeated data and missing data in the circulation of drugs are preprocessed to achieve a higher degree of data mining integrity. However, the proposed mining method does not consider the mining efficiency, so there may be a certain degree of backwardness in efficiency, so in the future research and innovation tasks, we can also improve the data mining technology from the perspective of efficiency.

Acknowledgements. This work is supported by the Key R & D plan of Shandong Province (public science and Technology) with No.2019GGX105013, and the Social science planning research project of Shandong Province with No.17CQXJ11.

References

1. Shi, W., Liu, Z.: Virtual machine migration algorithm to reduce energy consumption in datacenter. Comput. Digital Eng. **46**(01), 39–41+ 203 (2018)
2. Zhigang, W., Wenbin, P.U., Pengguo, T.: Data security storage technology in cloud computing. Commun. Technol. **52**(02), 471–475 (2019)
3. Weijiong, C., Wen, F., Xiaolin, Z., et al.: DBN method for risk assessment of dairy products cold chain logistics. J. Syst. Simul. **31**(05), 936–945 (2019)
4. Zhou, L., Shang, R., Zhan, C.: Research on the optimization of drug safety supervision system under the perspective of social co-governance. Health Econ. Res. **36**(06), 54–57.52 (2019)
5. Liu, S., Lu, M., Li, H., et al.: Prediction of gene expression patterns with generalized linear regression model. Front. Genet. **10**, 120 (2019)
6. Yan, Y., et al.: Analysis on mechanisms and medication rules of herbal prescriptions for gout caused by heat-damp accumulation syndrome based on data mining and network pharmacology. China J. Chin. Mater. Med. **43**(13), 2824–2830 (2018)
7. Fuwen, X.I.E.: Optimization Analysis of Discontinuous Data Path Mining in Cloud Computing Environment[J]. Computer Simulation **35**(06), 432–435 (2018)
8. Liu, M., Lü, D., An, Y.: Applications research of data mining technology in big data era. Sci. Technol. Rev. **36**(09), 73–83 (2018)
9. Wang, L.: Detection of offensive information on Internet via data mining. J. Hunan Ind. Polytech. **19**(01), 25–27 + 46 (2019)
10. Huang, Q.: Analysis on application of data mining in cloud computing. wuxian hulian keji **16**(14), 145–146 (2019)
11. Liu, S., Li, Z., Zhang, Y., et al.: Introduction of key problems in long-distance learning and training. Mob. Netw. Appl. **24**(1), 1–4 (2019)
12. Gui, G., Yun, L. (eds.): ADHIP 2019. LNICST, vol. 301. Springer, Cham (2019). https://doi.org/10.1007/978-3-030-36402-1

Distributed Data Collaborative Fusion Method for Industry-University-Research Cooperation Innovation System Based on Machine Learning

Wen Li, Hai-li Xia, and Wen-hao Guo[✉]

Business School, Suzhou University of Science and Technology,
Suzhou 215009, China
ling531@hotmail.com

Abstract. Computer technology and the Internet industry are developing rapidly, and the amount of data is exploding, and people are entering the era of big data. Massive data contains a lot of knowledge value, and machine learning can extract useful key information from massive data. There are many short-comings in traditional fusion methods, in order to better process the data in machine learning, a distributed data collaborative fusion method based on machine learning and industry-university research cooperation innovation sys-tem is proposed. The method is analyzed by research method theory and method function. The method function mainly realizes the temporal and spatial fusion of data through time synchronization, delay and misalignment of uncertain data processing, data association and weighted fusion. The simulation experiment is carried out according to the design and implementation steps of the method, and the feasibility and use value of the method are verified by experiments, and the performance of this method is superior.

Keywords: Machine learning · Innovation system · Distributed data · Fusion method

1 Introduction

Machine learning is a multi-disciplinary subject involving many disciplines such as probability theory, statistics, approximation theory, convex analysis, and algorithm complexity theory. Specializing in how computers simulate or implement human learning behaviors to acquire new knowledge or skills and reorganize existing knowledge structures to continuously improve their performance [1]. In the process of machine learning, there is an innovation model of industry-university-research coop-eration to improve the efficiency of machine learning. The so-called industry-university-research cooperation is the synergy and integration of the division of labor in research, education and production in terms of functions and resources. It is the docking and coupling of technological innovations, middle and downstream. Due to the ref-erence of innovative forms, the data in the machine-learning industry-university cooperation innovation system presents a distributed construction model. And in order to better process and analyze the data, the data is processed by a cooperative fusion method. Data Collaborative Fusion is an automated information synthesis processing

© ICST Institute for Computer Sciences, Social Informatics and Telecommunications Engineering 2021
Published by Springer Nature Switzerland AG 2021. All Rights Reserved
S. Liu and L. Xia (Eds.): ADHIP 2020, LNICST 347, pp. 251–261, 2021.
https://doi.org/10.1007/978-3-030-67871-5_23

technology developed and developed in the 1980s. It makes full use of the complementarity of multiple data sources and the high-speed computing and intelligence of computers to improve the quality of the resulting information. Data Co-synthesis, also known as multi-sensor fusion, is based on the combination of multi-sensor information to obtain more accurate information than any single input data source to improve the effectiveness of the entire sensor system. This performance improvement comes at the cost of increased system complexity.

Nowadays, the traditional data collaborative fusion method has the method based on time series data correlation mining. This method analyzes the operation mode of the industry university research cooperation innovation system, evaluates the related lines among the three, and calculates the correlation coefficient between the three. According to the correlation coefficient, the distributed data collaborative fusion model of industry university research cooperative innovation system is constructed, and the function is solved by using generalized EM algorithm, and the effectiveness of the fusion result is analyzed considering the weight. However, the fusion effect of this method is poor, and the actual application effect is not ideal.

Below, we analyze the architecture and fusion of the fusion system, discuss the clock synchronization problem in the distributed fusion system, and propose solutions.

2 Distributed Data Cooperative Fusion Theory

This chapter introduces the theoretical basis of the distributed information fusion method for networked systems. According to the research content of this thesis, it includes the principle of level set method, the theoretical basis of clustering algorithm, and the theory of optimal estimation, which provides a theoretical basis for the research of subsequent methods. Since its introduction, DS evidence theory has been widely used in the fields of pattern recognition, multi-sensor information fusion, image recognition, and uncertainty decision-making because of its superiority in the description and processing of uncertain events. In DS evidence theory, the recognition framework X refers to the complete set of objects studied, and the elements in X are incompatible with each other and are discrete values. The horizontal method principle is a set of horizontal slices that express an n-dimensional function by using a higher level, i.e., a level set of n 1-dimensional functions. The zero level set is a curve composed of a set of functions having the same value on the surface $f(x,y)$ expressed by a three-dimensional continuous function, that is, a set of zero level sets of $f(x,y) = 0$ [2]. Distributed data co-fusion is analogous to the human brain to analyze the information conveyed by synthetic neuronal synapses, and thus this is an adaptive complex process. It comprehensively analyzes the environmental information that can be detected by different sensors and gives a reasonable and valuable processing result. Data fusion is a multi-level, all-round analysis process that detects, combines, correlates, estimates, and combines multiple sources of data. This results in accurate state estimation and timely, comprehensive situation assessment and threat estimation. Distributed data collaborative fusion is different from traditional data information processing in the processing of data information. The essence of data fusion is that the data processed by data fusion is more massive, bulky and highly complex. At the same time, its analysis of data is reflected in the fusion of different levels. Combined with the above definition of data fusion, target detection and tracking is a level-level fusion, which belongs to the state estimation process of the target; target recognition belongs to the estimation process. The distributed data cooperative fusion processing structure firstly processes the data source information locally, and transmits the processed data to the fusion center, thereby reducing the bandwidth required for communication and the processing pressure of the fusion center, data fusion efficiency and system reliability are relatively more concentrated.

3 Distributed Data Collaborative Fusion Method

The architecture of the distributed data collaborative fusion method is generally distributed hierarchical processing, distributed fusion processing, and completely centralized fusion. It distributes functions such as signal processing, interconnection, tracking, attribute combination, sensor management, and status assessment to multiple processors in different ways.

It can be seen from the figure that the sensor nodes are physically distributed, and each node has a local processor (NP), which performs local data fusion (level one); The information fused locally by each node NP is linked to the global processor (P), and then the hierarchical correlation and the level two and three processing are completed by him. The fusion method is the same as the fusion mode of other structures. For the distributed data cooperative fusion model, in order to obtain a comprehensive estimation of the target motion state, the existing time fusion and spatial fusion problems are processed and considered separately. That is, when a plurality of sensors distributed at different positions are used to observe moving targets, the observation values of the sensors at different times and in different spaces will be different, thereby forming a set of observations. If there is one sensor that observes the same target at n times, there may be: $s*z$ observations, represented by set V as $V = \{V_i\}$ (i = 0,1,...,s); $V_i = \{V_j(k)\}$ (k = 0,1,...,z), where Vi represents the set of observations of the i-th sensor, and Vj(k) represents the observation of the i-th sensor at time k. These observations have corresponding sequence numbers in time and space. In practical applications, in order to obtain the target state, these two fusions are often used in combination. Time/space fusion: Time fusion of each sensor's observation set is performed to obtain an estimate of the target state of each sensor, and then the estimation of each sensor is spatially fused to obtain a final estimate of the target state. For example, track fusion. Space/time fusion: firstly, the observation values of each sensor are fused at the same time, and the target position estimates at different times are obtained, and then time fusion is performed to obtain the final state; Time and space fusion: time fusion and spatial fusion are carried out at the same time. This method has good effect and does not lose information, but it is the most difficult and suitable for large data fusion systems [3]. The accuracy and synchronization of time is a key indicator of information accuracy, which will directly affect the accuracy of tasks such as target detection and track tracking. Obviously, for distributed data structure models, sensors are distributed in different spatial locations, and each node's local processor has its own clock. How to judge whether multiple information is collected at the same time and keep time synchronization is a key issue. With reference to the relevant theory of computer architecture, some algorithms can be used to solve the problem.

3.1 Distributed Data Clock Synchronization

In order to achieve clock synchronization, time fusion of distributed data is performed by referring to both active and passive algorithms. The active algorithm is the Berkeley algorithm, in which the time server is active, it periodically asks each machine for time. Then based on these answers, calculate the average and tell all machines to dial their clocks up or down to a new value. In this way, the time server has a time daemon and its time is maintained by the administrator. The process is as follows: First, the time daemon tells his time to other machines at a certain moment and asks them their respective time; The machines then tell the time daemon the difference between their time and the time of the daemon process; finally the daemon calculates their average and tells each machine how to adjust its time. The passive algorithm is the Cristian algorithm, which instructs a computer to be a time server, and each of the other machines periodically sends a message to the time server to ask for the current time. When the time server receives the message, it will reply to the message containing the current time value as soon as possible. When the sender gets the answer, he sets his clock to T. It is necessary to consider the time it takes to send a response from the time server to the sender. A simpler method is to accurately record the time interval from when the request is sent to the time server to when the response is received. Assuming that the start time is T0 and T1, they are all measured by the same clock. Even if the sender's clock has a certain difference from the time server clock, the time interval measured is relatively accurate. In the absence of any other information, the best estimate of the message transmission time is (T0-T1)/2. When the response message arrives, the time in the message plus this value gives the current time and the estimated time of the processed message. This estimate can be further refined if you know the time of the time server interrupt processing and the processing of the message time. Let the processing time be 1, then the transmission time interval is T0-T-I, so the one-way transmission time is half [4]. Both synchronization algorithms have a time server to provide standard time, which requires time servers to have precise time to achieve time-coherent integration on distributed data.

3.2 Uncertain Data Processing with Delay and Misordering

In the process of implementing time synchronization, the data may be affected by uncertain factors such as data transmission delay or misalignment. In order to ensure the synergistic integration of distributed data in the industry-university-research cooperation innovation system of machine learning, it is necessary to deal with such uncertain data factors. A delay compensation strategy is proposed for the delay problem. At the sampling instant k, the transmission delay is $q(k)$, and the measured output value $p(k)$ is stored. The received measurement output $u(k)$ is used for state estimation $b(k|t)$ at time $t + q(k)$. Since the delay reduces the performance of the model, the proposed linear time-delay compensation method based on estimation is to reduce the computational complexity and further reduce the negative impact of the delay on the model. Assuming that the current sampling time is k and the received data packet is $p(k)$, and the estimation state $b(k|t)$ is to be performed, the state prediction value $b(t + 1|t)$ is used for the proposed linear compensation method. Based on the maximum delay N and the current transmission delay q(k), the estimated value b(k|t)} is obtained by the compensation formula 1.

$$b\langle k|t\rangle = (1 - \frac{q(k) - 1}{N})b\langle t + 1|t\rangle \tag{1}$$

The delay is converted to a form without time lag through a series of operations. For the out-of-sequence data, the distributed data needs to be reordered, that is, reordered. When the received time-stamped data packet data is $z(k_1)$, the stored signal $p_z(k)$ is recombined into: $p_z(k) = z(k-z(k))$. Looping all the data in a distributed packet for reordering purposes, with time-stamped packets meaning in the communication network. Data is transmitted from the equipment to the filter, and the filter can obtain information on data delay and packet loss. In practical applications, the measured object is affected by the correlated noise. It is assumed that the process noise w_k and the measurement noise v_k have correlation at the same sampling time. The statistical attribute relationship is:

$$E\left(\begin{pmatrix} w_k \\ v_k \end{pmatrix}\left(w_l^T (v_l)^T\right)\right) = \begin{pmatrix} Q_k\theta_{k,l} & S_k\theta_{k,l} \\ (S_k)^T\theta_{k,l} & R_k\theta_{k,l} \end{pmatrix} \tag{2}$$

The covariance is a symmetric matrix. Find similarities in noise to perform similar processing. By using the formula to delay transform and reorder the data containing noise, the data can be removed by the method of minimum error covariance to obtain the cooperative data that can be cooperatively combined.

3.3   Data Association

Tracking distributed data and finally forming an effective trajectory, the first problem that should be solved is the problem of data source information attribution, that is, the uncertainty of data source [5]. Clustering similarity calculation is the first step in data association. The similarity measure is the amount used to describe the degree of similarity between data objects [6]. Commonly used are distance-based similarity measures, kernel-based similarity measures, connectivity-based similarity measures, cosine similarity measures, and conceptual similarity measures. Let R denote a non-empty set, if any two of the elements $i = \{M_{i1}, M_{i2}, ..., M_{in}\}$ and $j = \{M_{j1}, M_{j2}, ..., M_{jn}\}$ are bounded by a certain rule and a real number d{i,j} corresponds, and satisfies: d{i,j} is not negative; d{i,j} = 0 if and only if i and j are the same sample; d{i,j} = d{j,i};d{i,j} $\leq$ d{i,h} + d{h,j}. In cluster analysis, the similarity measure based on Euclidean distance is common in distance as a measure of similarity.

In some cases, in order to express the importance of certain variables, Weighted Euclidean distance-based similarity measures, weighted Manhattan distance-based similarity measures, and weighted Minkowski distance-based similarity measures are proposed [7]. Therefore, at the initial stage of the target track, it is necessary to perform a measurement and measurement correlation of the plurality of cycles for the measurement to ensure the correctness of the track start. The usual data association has three processes as shown in Fig. 1.

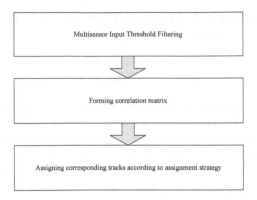

Fig. 1. Data association process

As can be seen from the figure, the data association process usually firstly filters the tracking threshold of the multi-sensor after pre-processing and data registration. Set thresholds based on prior statistical knowledge to filter out measurement data that should not appear [8]. Then, using the effective measurement of the tracking gate output to measure a track pair, an association matrix is formed, and the elements in the matrix represent the degree of association between the measurement point and the track prediction point. Finally, the measurement points with the highest degree of relevance to the predicted position are assigned corresponding tracks according to a certain assignment strategy. The assignment strategy is determined by various data association algorithms. The reasonable application of the data association algorithm is directly related to the pros and cons of the correlation effect [9].

3.4 Weighted Fusion Processing

The weighted fusion processing is performed after the correlation processing, and the distributed data weighted fusion processing flow is shown in Fig. 2.

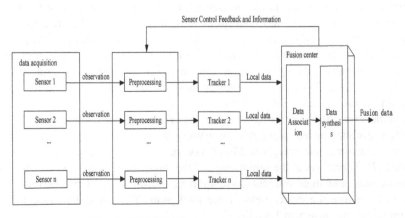

Fig. 2. Weighted fusion processing flow chart

The characteristic of the distributed fusion structure is that each branch has its own local processing system [10]. After the observation information acquired by the plurality of sensor branches is tracked in the local tracker, the target track formed by each tracker is fused at the fusion center [11]. The advantage of this configuration is that the tracking and compression processing of the measurement information is performed before the measurement information is transmitted to the fusion center, and the pressure of the communication data amount is reduced. Since local data information can be formed in each sensor branch, even if one of the roads fails, even the central system failure will not affect the formation of the track. Therefore, the reliability is high and the system survivability is strong. Increasing the tracking accuracy can be performed by time registration of the associated track and using the weighted probability to fuse the associated track [12]. Assuming that the covariance F_1 of the distributed data set 1 at the same time, and the filter value is G_1, the covariance of the data set N is F_n, and the filter value is G_n. The wave value is the covariance of the Ply radar N, and the filtered value is 1 JPN, then the covariance after the fusion and the fusion value are:

$$F = (G_1^{-1}F_1 + G_2^{-1}F_2 + \ldots + G_N^{-1}F_N)$$
$$G = (G_1^{-1} + G_2^{-1} + \ldots + G_N^{-1})^{-1} \tag{3}$$

3.5   Implementation of Data Cooperative Fusion Method

Realizing the distributed data collaborative fusion model of the industry-university-research cooperation innovation system for machine learning. The system is tested in a real network environment, and the Hadoop distributed data processing platform of the machine learning and research cooperation innovation system model is built [13]. The distributed big data collaborative fusion method of the industry-university-research cooperation innovation system of Mapkeduce machine learning written in Java (Fig. 3):

```
FIND-MAX-CROSSING-SUBARRAY (A, low, mid, high)
 1   left-sum = −∞
 2   sum = 0
 3   for i = mid downto low
 4       sum = sum + A[i]
 5       if sum > left-sum
 6           left-sum = sum
 7           max-left = i
 8   right-sum = −∞
 9   sum = 0
10   for j = mid + 1 to high
11       sum = sum + A[j]
12       if sum > right-sum
13           right-sum = sum
14           max-right = j
15   return (max-left, max-right, left-sum + right-sum)
```

Fig. 3. Algorithm pseudo code

4   Analysis of Experimental Results

The experimental demonstration uses all distributed data resources in the database of industry-university research cooperation innovation system of machine learning for data fusion processing. In order to ensure the rigor of the experiment, the traditional data fusion method, that is, the single station processing method is used as the experimental argumentation comparison, and the fusion results are statistically analyzed. The analysis results are shown in Fig. 4.

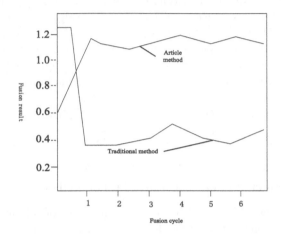

Fig. 4. Experimental results

It can be seen from the figure that in the previous cycle, the data processing mode of the single station is rapidly declining, and the data result of the data fusion mode is steadily increasing. After one cycle, the data results of the two methods are stable and the result of the fusion mode is much higher than the data result of the single station processing. Therefore, the method of transmitting the local fusion result and the processing method of the data fusion method have a large difference in the result. The system test results show that the model can guarantee the stable operation of all functions and achieve high-quality data fusion processing operations.

On the basis of the above experiments, compare the data fusion accuracy of this method and the traditional method, and the comparison results are shown in Fig. 5.

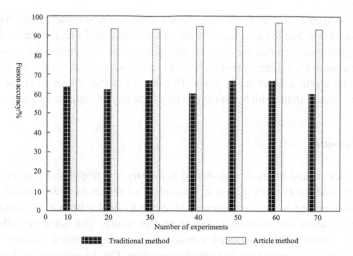

Fig. 5. Comparison of fusion accuracy

According to Fig. 5, the data fusion accuracy of traditional methods varies from 62% to 69%, while the data fusion accuracy of this method is always higher than 93%, which shows that this method can obtain accurate distributed data collaborative fusion results of industry university research cooperation innovation system. The reason is that the method mainly uses time synchronization, uncertain data processing with delay and wrong sequence, data association and weighted fusion to achieve data fusion in time and space, so it has high fusion accuracy.

On the basis of comparing the accuracy of data fusion of different methods, comparing the data fusion time of different methods, the lower the data fusion time, the higher the fusion efficiency and the better the data fusion effect. The comparison results are shown in Fig. 6.

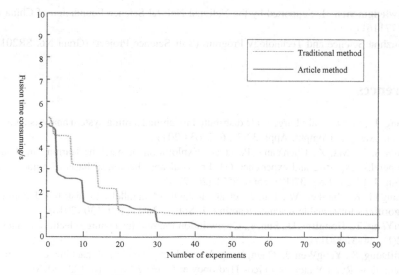

Fig. 6. Time consuming comparison of data fusion

As can be seen from Fig. 6, the data fusion time of the method in this paper varies from 0.5 s to 5.0 s, and the data fusion time of the traditional method varies from 1.0 s to 5.5 s, which indicates that the method can realize the distributed data collaborative fusion of the industry university research cooperative innovation system in a short time, and the actual application effect is good. The reason is that the method uses machine learning to process data, which improves the efficiency of data processing.

5 Conclusion

Under the background of knowledge-based economy, the traditional way of relying on factor driven and investment driven is becoming increasingly unsustainable. Relying on innovation driven is the fundamental strategy to deal with a new round of technological revolution and industrial change in the future. The intensive allocation of innovation resources and collaborative innovation among scientific research subjects will become the key support for industrial upgrading. On the premise of the difference in the goal orientation of cooperation subjects, to achieve the coupling interaction between different subjects in the cooperation network, and to maximize the limited resources into the competitive advantage, the collaborative role of industry university research cooperation subjects will become particularly prominent, It is of great significance to study the distributed data collaborative fusion method of regional industry university research cooperation innovation system. Through research, it provides a basic processing method of distributed data cooperative fusion, which provides some basic ideas for the entire data fusion strategy. The distributed data cooperative fusion method ensures the optimality of the decision result of the fusion center without increasing the amount of information transmitted and storage resources. The effectiveness of the fusion method is verified by simulation experiments and can be put into use in actual production and life.

Acknowledgments. 1. Supported by the National Natural Science Foundation of China (Grant No. 71771161);

2. Suzhou Science and Technology Program (Soft Science Project) (Grant No. SR201817).

References

1. Tang, L., Li, D., et al.: Large-scale distributed machine learning system analysis taking LDA as an example. Comput. Appl. **37**(3), 628–634(2017)
2. YuWei, S., Ma, Z., ChenYang, P., et al.: Exploration of machine learning methods with embedded expertise and experience (1): Proposal and theoretical basis of guided learning. Chin. J. Electr. Eng. **37**(19), 5560–5571 (2017)
3. LiangYi, K., JianFei, W., Liu, J., et al.: Review of parallel and distributed optimization algorithms for scalable machine learning. J. Softw. **36**(1), 109–130 (2018)
4. XuYang, T., HongBin, D., Sun, J.: Co-evolution method for feature selection. J. Intell. Syst. **12**(1), 24–31 (2017)
5. HuiDong, S., YangWen, J., GuangHeng, N., et al.: Application of machine learning in runoff prediction. Rural Water Resources Hydropower China **37**(6), 116–123 (2018)

6. Meng, Z., Yan, Z., MingDi, L., et al.: Distributed multi-sensor cooperative tracking algorithm under communication constraints. Firepower Command Control **42**(6), 6–9 (2017)
7. ZhaoMing, L., WenGe, Y., Dan, D., et al.: Distributed cooperative navigation filtering algorithm for multi-satellite cooperative targets. J. Beijing Univ. Aeronaut. Astronaut. **44**(3), 462–469 (2018)
8. Lu, M., Liu, S.: Nucleosome positioning based on generalized relative entropy. Soft. Comput. **23**(19), 9175–9188 (2018). https://doi.org/10.1007/s00500-018-3602-2
9. Fu, W., Liu, S., Srivastava, G.: Optimization of big data scheduling in social networks. Entropy **21**(9), 902 (2019)
10. Ma, H., Xie, H., Brown, D.: Eco-driving assistance system for a manual transmission bus based on machine learning. IEEE Trans. Intell. Transp. Syst. **19**(2), 572–581 (2018)
11. Xiao, F.: Multi-sensor data fusion based on the belief divergence measure of evidences and the belief entropy. Inf. Fusion **46**(2), 45–57 (2018)
12. Bader, K., Lussier, B., Sch, N.W.: A fault tolerant architecture for data fusion: a real application of Kalman filters for mobile robot localization. Robot. Auton. Syst. **88**(10), 11–23 (2018)
13. Chen, F.C., Jahanshahi, R.M.R.: NB-CNN: deep learning-based crack detection using convolutional neural network and Naïve Bayes data fusion. IEEE Trans. Industr. Electron. **65**(99), 4392–4400 (2018)

Research on Automatic Defense Network Active Attack Data Location and Early Warning Method

Jian-zhong Huang[1(⊠)] and Wen-da Xie[2]

[1] Modern Education Technology Center, Jiangmen Polytechnic,
Jiangmen 529030, China
baihuahualai2018@163.com
[2] Computer Engineering Technical College (Artificial Intelligence College),
Guangdong Polytechnic of Science and Technology, Zhuhai 519090, China

Abstract. The traditional automatic defense network active attack data location and early warning method has the shortcoming of poor localization performance, so the research of automatic defense network active attack data location and warning method is put forward. The active attack data is detected by the space distance of the network node data, and the active attack data is judged whether there is active attack data in the network, which is based on the detected active attack data. The multi-objective binary particle swarm optimization (BPSO) algorithm is used to obtain the optimal task allocation scheme for active attack data location. Based on it, the algorithm of extreme learning machine is used to realize the location and early warning of active attack data. Through the experiment, put forward the automatic Compared with traditional methods, the convergence value of active attack data location and early warning method of defensive network increases 23.61 and the error rate of location decreases by 15. It is fully explained that the proposed automatic defense network active attack data location and early warning method has better localization performance.

Keywords: Automatic defense · Network · Active attack data · Location · Early warning

1 Introduction

With the continuous development of Internet technology, the network usage rate is gradually increasing. Nowadays, the network has reached a popular state, but due to the openness of the network, the security of the network cannot be well protected. To ensure the security of the network, it is necessary to carry out corresponding positioning warning and processing of its active attack data to ensure the security of the network environment [1].

In recent years, network security has gradually been valued by many scholars, and more research has been done on the automatic defense, location, and early warning methods of network attack data. Automatic defense network active attack data location and early warning is the premise of network attack blocking and service recovery, and is the key link to ensure optical network security. The characteristics of the network itself make it face more security threats than traditional wired and wireless networks, and active attacks are one of the challenges that must be overcome in the security field [2]. In recent years, domestic and foreign scholars have studied the data and early warning of active attack data, and obtained some research results, but the active attack research in the network is still a challenging problem, which needs further research [3]. The traditional automatic defense network active attack data location and early warning method has the defect of poor positioning performance, which can not meet the needs of more and more network active attack data location and early warning. Therefore, the research on automatic defense network active attack data location and early warning method is proposed [4].

2 Design of Automatic Defense Network Active Attack Data Location and Early Warning Method

2.1 Active Attack Data Detection

In order to locate and alert the active attack data, it is first necessary to detect whether there is active attack data in the network data. In this paper, the active attack data is detected by the spatial distance of the network node data. The specific process is as follows.

Suppose there are n beacon nodes and m unknown nodes in the network. For an unknown node, its node i data can be represented by a vector as

$$RSS_i = \{s_{i1}, s_{i2}, \cdots, s_{ij}, \cdots, s_{in}\}, i = 1, 2, \cdots, m \tag{1}$$

Where RSS_i represents the i th node data; s_{ij} represents the signal strength of the unknown node i sent to the j th beacon node, Then the node data of each unknown node is equivalent to a point in an n-dimensional space. The node data of the same physical location node is close in the signal space, and will fluctuate up and down at a mean value, and there is a certain distance between the node data of different physical location nodes [5].

According to the "distance-loss" model, the calculation formula of s_{ij} is:

$$s_{ij}(d_{ij})[dBm] = P(d_o)[dBm] - 10\gamma \log\left(\frac{d_{bj}}{d_{aj}}\right) + \Delta F \tag{2}$$

Where $P(d_o)$ represents the transmit power of node i; d_o indicates the reference distance; d_{ij} represents the ranging distance between the unknown node i and the beacon node j; γ represents the path loss factor. For beacon nodes j. The same $P(d_o)$ of any two unknown nodes a, b.

The node data difference is

$$\Delta s_j^{ab} = s_{aj} - s_{bj} = 10\gamma \log\left(\frac{d_{bj}}{d_{aj}}\right) + \Delta F \tag{3}$$

Where d_{aj}, d_{bj} represents the ranging distance between the node a, b and the beacon node j, respectively.

For n beacon nodes, the node data of Node B is equivalent to two points in the n-dimensional space, and the spatial distance between the two nodes is:

$$D^{ab} = \|RSS_a - RSS_b\| = \sqrt{\sum_{j=1}^{n}\left(\Delta s_j^{ab}\right)^2} \tag{4}$$

Where D^{ab} represents the Euclidean distance of the node a, b.

According to the principle of non-central χ^2 distribution, when two nodes are in different physical locations, $\Delta s_j^{ab}, j = 1, 2, \cdots, n$ obey $N\left(10\gamma \log\left(\frac{d_{bj}}{d_{aj}}\right), 2\sigma^2\right)$ distribution $N\left(10\gamma \log\left(\frac{d_{bj}}{d_{aj}}\right), 2\sigma^2\right)$. Then the probability density function of the node data is

$$f\left(\frac{X}{same}\right) = \frac{1}{2^{\frac{n}{2}}\Gamma\left(\frac{n}{2}\right)} X^{\frac{m}{2}-1} \cdot D^{ab} \tag{5}$$

Where $\Gamma()$ represents the Gamma function; X represents random node data.

The detection of active attack data is implemented according to the comparison of the probability density function of the node data with the threshold. The detection performance depends on the selection of the threshold. If the threshold is too large, the probability that the attacking node is misjudged as a legitimate node increases; if the threshold is too small, the probability that the legitimate node is misjudged as an attacking node increases. After selecting the appropriate threshold, the active attack data is detected. If the probability density function is greater than the threshold, it means that there is no active attack data in the network. If the probability density function is less than the threshold, it means that the network contains active attack data [6]. Through the above process, the detection of the active attack data is realized, and the following is prepared for the positioning and early warning of the active attack data.

2.2   Active Attack Data Location Task Assignment

Based on the above-mentioned detected active attack data, the positioning tasks are allocated correspondingly, and multiple active attack data positioning is realized at the same time, which can greatly improve the efficiency of the automatic defense network active attack data positioning and early warning method. The multi-target binary particle swarm task assignment algorithm is used to allocate active attack data location tasks. The specific process is as follows [7]. The active attack data location tasks in the network are allocated to obtain a long-lived network life cycle, lower network energy consumption and balanced network load. Therefore, this chapter has designed three objective functions: total task completion time, total energy consumption, and load balancing. The multi-objective binary particle swarm task assignment algorithm is used to determine the optimal allocation scheme [8]. The multi-target binary particle swarm task assignment algorithm is expressed as

$$Minimize \quad f(X_i) = \{f_1(X_i), f_2(X_i), f_3(X_i)\} \tag{6}$$

Among them, $X_i = \{x_{l1}, x_{l2}, \cdots, x_{lj}, \cdots, x_{ln}\}, l = 1, 2, \cdots, q$ represents the decision variable, which corresponds to a distribution scheme of tasks in the particle. f_1 is the total time of the target function task completion, f_2 is the total energy consumption of the objective function, f_3 is the objective function load balance. Therefore, the decision space is q-dimensional and the target space is 3 dimensions,That is, the task allocation scheme is a q-dimensional particle, and an optimal solution set under the f_1, f_2, f_3 3 dimensional objective function.

In the multi-objective binary particle swarm task assignment algorithm, the position X of the particle represents a task assignment scheme, expressed as

$$X = \begin{bmatrix} x_{11} & x_{12} & \cdots & x_{1j} & \cdots & x_{1n} \\ x_{21} & x_{22} & \cdots & x_{2j} & \cdots & x_{2n} \\ \vdots & \vdots & \vdots & \vdots & \vdots & \vdots \\ x_{q1} & x_{q2} & \cdots & x_{qj} & \cdots & x_{qn} \end{bmatrix} \tag{7}$$

Among them, $x_{lj} = \{0, 1\}, l = 1, 2, \cdots, q, j = 1, 2, \cdots, n$, x_{lj} indicates the state in which the beacon node joins the task, 0 means unselected, and 1 means selected.

The velocity of a particle indicates the probability of a change in particle position. If the velocity of the particle is large, the probability that the particle position is taken as 1 is large. If the velocity value of the particle is small, the probability that the particle position is taken 1 is small.

The individual extremum of each particle is denoted as $pb = (pb_{ij})_{q \times n}$, and the global extremum of the entire particle swarm is denoted as $pg = (pg_{ij})_{q \times n}$.

The update formula for particle position A is

$$x_{lj}(t+1) = \begin{cases} 0, & if \quad rand() \geq S(v_{lj}(t+1)) \\ 1, & if \quad rand() < S(v_{lj}(t+1)) \end{cases} \tag{8}$$

Among them, $S(v_{lj}(t+1))$ represents the transfer function.

In order to solve the problem that the binary particle swarm optimization algorithm is easy to fall into the local optimum, the global optimization ability is strong and the local search ability is strong. Therefore, the inertia weight is improved, so that the inertia weight is nonlinearly reduced as the number of iterations t increases.

In the algorithm, the inertia weight decreases nonlinearly with the increase of the number of iterations t, which is beneficial to jump out of the local extremum point, and obtain a larger value at the initial stage of the iteration. The particles in the population are quickly scattered throughout the search area to determine the optimal value. Approximate range; as the iterative nonlinearity decreases, the search space of most particles gradually decreases and shrinks to the nearest neighbor range; at the end of the iteration, when the maximum iteration t_{max} is reached, the particles are optimal. The global optimal solution is searched in the neighborhood.

The specific steps of the multi-target binary particle swarm task allocation algorithm are:

(1) Initialization algorithm: set the number of tasks to q, and generate the initial position of each particle and the individual extreme value pb under the constraint condition;

(2) Calculate the objective function: get the fitness value of the particle as $p \cdot f_1, p \cdot f_2, p \cdot f_3$;

(3) Particle update: first update the inertia weight, then update the speed and position of the particle;

(4) Constraint test: If the particle p satisfies both the workload constraint and the space constraint, go to step (5); otherwise, go to step (3) and re-update the particle;

(5) Individual extreme value selection: Calculate the fitness value of the particle. If the current position of the particle is better than the historical best position, update the individual extreme value pb ;

(6) Elite file strategy: delete duplicate members in the archive, sort the members in the file according to the descending distance, and get a better archive. At the same time, according to the dense distance, the global optimal pg is selected for each particle by the proportional selection method;

(7) If the number of iterations $t \geq t_{max}$, the algorithm ends and the optimal solution set S is output, otherwise it goes to step (3).

The flow of the multi-objective binary particle swarm assignment algorithm is shown in Fig. 1.

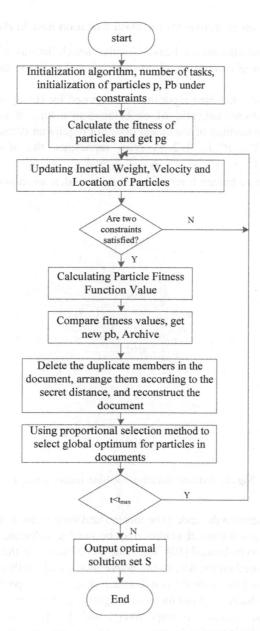

Fig. 1. Flow chart of multi-objective binary particle swarm optimization task assignment algorithm

Through the above process, the multi-target binary particle swarm task assignment algorithm is used to allocate the active attack data location task, and the optimal allocation scheme S is obtained, which provides support for the implementation of the following active attack data location warning [9].

2.3   Implementation of Active Attack Data Location and Early Warning

Based on the optimal allocation scheme of active attack data location tasks obtained above, the algorithm of extreme attack learning is used to locate and warn the active attack data.

The speed learning machine algorithm is proposed by Huang et al. It is a single hidden layer feedforward network, which belongs to a kind of neural network. Its advantage is that the learning of the model does not require an iterative process. Given N data $(x_i, t_i) \in R^n \times R^m, i = 1, 2, \cdots, N.x_i$ represents the $n \times 1$ input vector $x_i = [x_{i1}, x_{i2}, \cdots, x_{in}]^T$, t_i is a $m \times 1$ target vector $t_i = [t_{i1}, t_{i2}, \cdots, t_{im}]^T$. The speed learning machine has a hidden layer with L hidden nodes, as shown in Fig. 2.

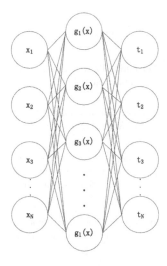

Fig. 2. Extreme learning machine model network

If the distance between the node to be located and the beacon node in the network is known, a trilateral positioning algorithm may be used to calculate the position coordinates of the node to be located [10]. The node to be located in the article is an active attack data node. Based on the data of the active attack node, a three-edge positioning algorithm can be used to locate the active attack data, and the position of the beacon node B_1, B_2, B_3 is obtained according to the algorithm of the extremely fast learning machine,$(x_1, t_1), (x_2, t_2), (x_3, t_3)$ is respectively, and the A_s coordinate of the active attack data node to be located is (x_s, t_s).

The real data value of the active attack data node A_s can be represented by a vector, The distance between the active attack data node A_s and the beacon node B_j is d_{sj}. The formula for calculating the ranging distance of the active attack data node A_s to the beacon node B_1, B_2, B_3 :

$$\begin{cases} d_{s1} = \sqrt{(x_s - x_1)^2 + (t_s - t_1)^2} \\ d_{s2} = \sqrt{(x_s - x_2)^2 + (t_s - t_2)^2} \\ d_{s3} = \sqrt{(x_s - x_3)^2 + (t_s - t_3)^2} \end{cases} \qquad (9)$$

Substitute d_{sj} into the above formula, solving the equations can calculate the coordinates (x_s, t_s) of the active attack data node A_s as:

$$\begin{bmatrix} x_s \\ t_s \end{bmatrix} = \begin{bmatrix} 2(x_1 - x_3) & 2(t_1 - t_3) \\ 2(x_2 - x_3) & 2(t_2 - t_3) \end{bmatrix}^{-1} \begin{bmatrix} x_1^2 - x_3^2 + t_1^2 - t_3^2 + d_{s3}^2 - d_{s1}^2 \\ x_2^2 - x_3^2 + t_2^2 - t_3^2 + d_{s3}^2 - d_{s2}^2 \end{bmatrix} \qquad (10)$$

Iteratively optimize the positioning results. After multiple iterations, the positioning result closest to the real coordinates of the active attack data node can be obtained, thereby achieving accurate positioning of the active attack data node [11–13]. The prompt sound device is used to prompt and display the above-mentioned active attack data, and the operator processes the data according to the corresponding result [14].

Through the above process, the application of automatic defense network active attack data location and early warning method is realized, which fully proves the feasibility of the method. The positioning performance of the method was analyzed by comparative experiments as described below.

3 Analysis of Location Performance of Active Attack Data Location and Warning Method

Through the above process, the application of automatic defense network active attack data location and early warning method is realized, which fully proves the feasibility of the method. The positioning performance of the method was analyzed by comparative experiments as described below.

Two evaluation indexes of convergence and positioning error are used to compare and analyze the performance of the proposed method and the traditional method. Using NS2.35 as the network simulation platform, The difference between a common node and a beacon node is that the beacon node can locate, while the ordinary node does not have the localization function, 101 sensor nodes are arranged in the network area, where the location of the aggregation node is (0, 0), and 80 common nodes and 20 beacon nodes are randomly generated uniformly using the NS2.35 scene generation tool covering the entire network. The network environment parameter values are shown in Table 1.

Table 1. Network environment parameter value

Parameter	Numerical value
Mac protocol	MAC/802_15-4
Routing protocol	AODV
Energy model	EncrgyModel
Initial energy of nodes	11520 J
Communication radius	200-250 m
Wireless propagation model	Shadowing
Path loss factor	2
Transmitting power	0.282 W
Simulation time	100 s

The multi-objective binary particle swarm task assignment algorithm has a population size of 20, a particle velocity of [−6,6], a maximum number of iterations of 200, a learning factor of 2, and an inertia weight of [0.4, 0.9], which can be set by algorithm parameters. Get a good solution in a short period of time。

3.1   Convergence Analysis

The convergence comparison is shown in Fig. 3

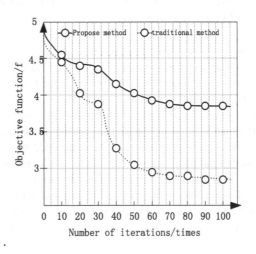

Fig. 3. Convergence contrast graph

As shown in Fig. 3, the proposed method converges to 3.78 and the traditional method converges to 2.89. The convergence value of the proposed method is 23.61% higher than that of the traditional method.

3.2   Positioning Error Analysis

The positioning error of the proposed method and the conventional method is shown in Table 2.

Table 2. Positioning error Table

Number of experiments	Proposed method error rate	Error rate of traditional methods
10	10%	23%
20	23%	30%
30	5%	25%
40	8%	28%
50	11%	21%
60	5%	25%
70	9%	29%
80	16%	26%
90	4%	24%
100	11%	21%

As shown in Table 2, the error rate of the proposed method is reduced by 15% compared with the conventional method.

Through the above two experiments, it can be seen that the proposed automatic defense network active attack data location and early warning method has better recognition performance.

4   Conclusion

The proposed automatic defense network active attack data location and early warning method improves the convergence, reduces the positioning error rate, and greatly improves the performance of the automatic defense network active attack data positioning and early warning method. However, the interference of the influencing factors is neglected in the experiment process. Therefore, further research and analysis on the automatic defense network active attack data location and early warning method are needed.

References

1. He, Y.: Research on active security defense system of campus network based on honeypot technology. Comput. Fans **12**(4), 21–22 (2016)
2. Zhu, C., Zhao, D.: A packet allocation protection method for network attack crime prevention. Sci. Technol. Bull. **32**(8), 203–206 (2016)
3. Dong, X., Lin, L., Zhang, X., et al.: Application of active defense technology in communication network security engineering. Inf. Secur. Technol. **7**(1), 80–84 (2016)

4. Pujiang, L.L.: Research on network security early warning and defense system based on three-domain model. Inf. Secur. Technol. **8**(8), 68–72 (2017)
5. Jiang, S., Luo, T.: Research on detection method of network small disturbance intrusion source location under non-uniform noise environment. Sci. Technol. Eng. **17**(5), 247–251 (2017)
6. Luo, X.W., Tao, H.: Research on Simulation of attack signal location and recognition in wireless communication networks. Comput. Simul. **33**(11), 320–323 (2016)
7. He, C.: Design of computer network security active defense model in big data era. J. Ningbo Vocat. Tech. Coll. **20**(4), 97–99 (2016)
8. Di, Z.: Research on the current situation and defense measures of network security management in Colleges and Universities. Electron. Technol. Softw. Eng. **56**(13), 221–221 (2016)
9. Liu, S., Li, Z., Zhang, Y., et al.: Introduction of key problems in long-distance learning and training. Mob. Networks Appl. **24**(1), 1–4 (2019)
10. Chaocheng, Q., Jianhong, Q.: Network security defense model of metal trading based on attack detection. World Nonferrous Metals **23**(7), 77–78 (2016)
11. Huang, R., Huang, R.: Simulation research on privacy information protection of network users. Comput. Simul. **51**(11), 319–322 + 423 (2017)
12. Liu, S., Bai, W., Srivastava, G., Machado, J.A.T.: Property of self-similarity between baseband and modulated signals. Mob. Networks Appl. **25**(4), 1537–1547 (2019). https://doi.org/10.1007/s11036-019-01358-9
13. Shuai, L., Weiling, B., Nianyin, Z., et al.: A fast fractal based compression for MRI images. IEEE Access **7**, 62412–62420 (2019)
14. Shuke, Yu.: Research on target intrusion detection based on mobile wireless sensor networks. Digital Commun. World **59**(12), 6–8 (2017)

Efficient Retrieval Method of Malicious Information in Multimedia Big Data Network Based on Human-Computer Interaction

Jia-ju Gong[1], Wen-da Xie[2], and Jing-hua Wang[3(✉)]

[1] Jiangmen Polytechnic, Electronic and Information Technology,
Jiangmen 529030, China
gongjiaju5866@163.com
[2] Guangdong Polytechinc of Science and Technology, Computer Engineering
Technical College, Guangdong 519090, Zhuhai, China
[3] College of Electronic and Optical Engineering & College of Microelectronics,
Nanjing University of Posts and Telecommunications, Nanjing 210003, China
xiewenda4376@163.com

Abstract. In order to solve the phenomenon that the malicious information retrieval in the traditional method is not comprehensive and the precision is not high, an efficient retrieval method of malicious information in multimedia big data network based on human-computer interaction is proposed and designed. On the basis of analyzing the principle of information retrieval, the human-computer interaction form is used to divide the metrics of malicious information. On this basis, the human-computer interaction retrieval logic is established, and the malicious information clustering method is used to realize the multimedia big data network. Efficient retrieval of malicious information. Through the method of experimental argumentation analysis, the validity of the human-computer interaction retrieval method is determined. The results show that the method has a high recall rate of 28.31% compared with the traditional method, and the retrieval accuracy is extremely high. In this paper, the method of network malicious information retrieval is effective and suitable for popularization.

Keywords: Human interaction · Big data network · Malicious information · Search processing · Information clustering

1 Introduction

Human-computer interaction is the study of people, computers and their mutual influence of technology. Man-machine interaction refers to the communication between the user and the computer system. "Conversation" here is defined as a communication, that is, an exchange of information, and a two-way exchange of information, which can be input by a person to a computer, or feedback by the computer to the user. This exchange of information takes the form of various symbols and actions, such as keystrokes, mouse movements, and displaying symbols or graphics on the screen. Systems can be a variety of machines, or they can be computerized systems and software. A human-computer interface is usually the part that is visible to the user. The

S. Liu and L. Xia (Eds.): ADHIP 2020, LNICST 347, pp. 273–282, 2021.
https://doi.org/10.1007/978-3-030-67871-5_25

user communicates with the system through the man-machine interaction interface and carries on the operation. The principles of interaction design need to include an understanding of the user, the user environment, and the design strategy. Different product designs require specific coping principles and strategies, but the principles of design are the same for all products. Although application principles are not a specific method to solve problems, they are a general principle that can be applied to the whole product design system to interact with users, and can be used as a part of the design strategy, as a guide to the design of interactive toy products. Based on user psychology and cognitive science, the following basic principles are proposed to guide interaction design:

(1) Usability principle

The principle of ease of use requires that interactive toy products be available to most people and have a wide range of applications. Users of different ages, cultures and regions can experience the convenience, efficiency and fun of products in the process of human-computer interaction. A good interactive product can ensure the user in the use of the process of availability and convenience, users in the experience and use of the product will not be unable to understand the manual or complex operation and lose interest.

(2) Adaptability principle

The user is the main body of human-computer interaction and should be in the control position. Therefore, the hardware and software design of the product should adapt to the user's needs in many aspects.

(3) Principle of participability

The main body of human-computer interaction is human and computer. The principle of participability emphasizes the spiritual experience of users when using the product. The product should not only meet the functional requirements of users, but also be easy, effective, interesting, fulfilling and satisfying in the process of interaction between users and the product.

Big data networks can handle multiple large-scale information at the same time, and when it comes to flexibility, it can copy data to multiple locations, and the size of the processed information will become larger and larger. But in fact, the most important attribute of a big data network is not its scale, but the ability to split big data into many small data. It can spread the resources of a task to multiple locations for parallel processing. The efficiency of the data processing process [1]. At this stage, the concept of big data network has become synonymous with cluster environment. According to the characteristics of different applications, the cluster requirements that meet the application are established, and the data load inside the functional partition is realized, and the quantitative relationship between each data load is correctly processed.

Human-computer interaction is a study of the interaction between the system and the user. The system mentioned here can be a variety of machines, or it can be a computerized system and software. The human-computer interaction interface is usually visible to the user part. The user communicates with the system through the human-computer interaction interface, and performs operations to realize the situational description of information communication, and interacts with the real environment to meet the functional requirements of the person.

Reference [2] proposed a network malicious information retrieval method based on big data network, which can obtain implicit feedback information from the network by observing the actions selected by users when browsing web pages, and establish user interest update model. Vector is used to describe the web documents that users browse, and the corresponding weight is given to each browsing behavior. Malicious information is extracted and implicit feedback information retrieval is carried out, which can effectively improve the detection time. But the recall rate is low and the detection efficiency is poor. Literature [3] proposed a Bayesian network malicious information retrieval method, through the Bayesian network technology to build a Bayesian network retrieval model, detailed analysis of its working principle, and the malicious information retrieval method is optimized, this method can improve the detection efficiency, but the detection accuracy is low.

Aiming at the above problems, this paper proposes and designs an efficient retrieval method of malicious information in multimedia big data network based on human-computer interaction, and validates the effectiveness of the method in the simulation platform. The results show that the human-computer interaction retrieval method can Improve the recall rate of malicious information, and thus improve the retrieval accuracy of malicious information, which has higher superiority than traditional methods.

2 Design of Efficient Retrieval Method for Malicious Information

2.1 Fragmentation of Malicious Information Metrics

The collection of malicious information in big data networks is the basis of retrieval. It has important significance [4, 5]. This paper uses metrics to collect and classify malicious information. Firstly, the Kullback-Liebler algorithm is used to cluster the malicious information in the big data network, namely:

$$KL = \sum_{w_i \in d} \frac{n(w_i, d)}{|d|} \log \frac{|d|}{nc/|c|} \tag{1}$$

Among them, KL represents the semantic distance between standard information and malicious information, obviously there is a situation of $KL = 0$; $n(w_i, d)$ is the number of times the malicious information w_i concept appears in the big data network d; $|d|$ represents the total number of malicious information; c represents the cluster of the algorithm Coefficient, this calculation does not do orientation analysis.

With the development of network technology, more and more big data networks adopt an open information distribution structure, which provides a development platform for the efficient retrieval process of computers [6, 7]. Therefore, this paper combines the human-computer interaction mode to simultaneously apply this efficient retrieval method. Applied to the human-computer interaction interface, the technician can feel the interference process of malicious information through the computer, and then accurately classify the retrieval criteria of the malicious information.

According to the classification form of the retrieval database, the browsing mode and the search engine are queried. During the query process, the document information in the entire big data network is sorted according to the probability of generating the malicious information model [8, 9]. Assume that in the big data network model based on human-computer interaction, each document information represents a polynomial of each term t in the information D vocabulary, then each document information is sorted according to its production probability when performing information retrieval. The probability of generating document information is calculated by the following formula:

$$sim(Q, D) = \frac{P(D|Q)}{KL/n(t|Q)} \tag{2}$$

Where $sim(Q, D)$ represents the frequency of occurrence of document information D and $Q.P(D|Q)$ represents the equivalent feature of all document data corresponding to the statistical item, and does not affect the sorting, and can be ignored in the actual calculation process. As can be seen from the formula, the hypothetical terms in each malicious information model are independent of each other. $n(t|Q)$ is the test probability of selecting document information, which can be used for the setting process of weights such as malicious document information retrieval parameters. All $P(D)$ values are the same in the calculation process of this paper, so it can be ignored in calculation.

2.2   Human-Computer Interaction Retrieval Logic Establishment

In the malicious information metric defined above, the spatial knowledge of all semantic concepts is concentrated on the level of words and sentences [10–14]. Firstly, a word-to-concept knowledge comparison table should be established based on the HNC concept node table to map and demap the malicious information features in the big data network. This paper introduces the concept of human-computer interaction and establishes the key information in the big data network information knowledge base. And its characteristics are shown in Table 1.

Table 1. Big Data Network Information Characteristics

Features	Type	Describe
frequency	Relevance	Frequency of Key Features of Malicious Information Query in Title and Text
BM25	Relevance	Query based on BM25 formula, including the relevance score of title, text and words of network information
N-gram BM25	Relevance	Query based on BM25 formula, including the relevance score of the title, text and meta-index items of network information
Edit distance	Relevance	Relevance Score of Title, Text and Editing Distance of Query Network Information
Number of incoming links	document information	Number of inbound links on Web pages
PageRank	document information	PageRank-based Web page importance is related to the number and quality of web pages'inbound links
Clicks	document information	Retrieve the number of clicks per page
BroseRank	document information	The importance of Web pages based on BroseRank score is related to the click probability of users when they browse the Web pages.
Malicious Information Assessment Value	document information	More than 80% of malicious information in web pages
Web Quality Assessment Value	document information	The Possibility of Web Pages as Low Quality Pages

Analysis Table 1 shows that the biggest advantage of introducing human-computer interaction retrieval logic is that there is no semantic ambiguity in the information concept. In the multimedia big data network, an information concept symbol corresponds to a certain semantic, which can fundamentally solve the inaccurate retrieval of malicious information. The problem. At the same time, according to the identification method of the information concept shown in Table 1, the establishment process of the retrieval logic can be realized by the clustering calculation of the concept. details as follows:

If an information concept is successfully identified, then the information retrieval process is considered as a problem of judging which document has malicious information, and only the reference value of the test data is considered, and the threshold clustering calculation is performed, and the interactive object of the malicious information is determined. You can get spatial information about any malicious information.

Assuming that the spatial coordinate of a malicious information is $N(x, y, z)$, the probability of generating the information $sim(Q, D)$ is introduced. According to the increasing arrangement of the distance values, the establishment basis of the retrieval logic is determined on the function curve, and then the spatial coordinates of any point on the curve are calculated. The spatial threshold of the malicious information is obtained, and the clustering result is automatically formed according to the threshold.

Finally, the matching process of the retrieval logic is completed through the query and clustering based on human-computer interaction. The function expression of the malicious information retrieval logic based on human-computer interaction calculated in this paper is as follows:

$$
\begin{cases}
N_x = \dfrac{c(sim(Q,D)+1)}{N} \\
N_y = \dfrac{1 - c \cdot sim(Q,D)}{N} \\
\quad N_z = 1 - \left(N_x/N_y\right)
\end{cases}
\tag{3}
$$

Where N_x, N_y and N_z represent the retrieval logic of the spatial abscissa, ordinate, and height coordinates of malicious information, respectively; M represents the semantic implied hint of malicious information in a big data network.

The human-computer interaction retrieval logic is established by referring to the HNC node concept, which provides a basis for realizing the retrieval and calculation of malicious information.

2.3   Efficient Retrieval of Malicious Information

In order to efficiently and accurately classify malicious information of big data networks, it is necessary to further search for malicious information based on the criteria of classification and the similarity threshold. Generally speaking, in a random text message in a big data network, the emergence distance of malicious information will continuously jump with the increase of density. These jumping points are the thresholds of malicious information we are looking for, and the distance between jump points is in ascending order, the corresponding threshold of malicious information can be determined by fitting the form of the incremental function. The incremental sequence of malicious information is shown in Fig. 1.

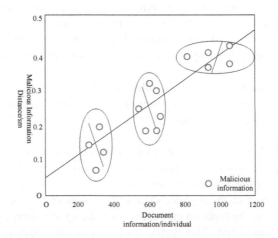

Fig. 1. Schematic diagram of the incremental sequence of malicious information

The curve of Fig. 1 is fitted by the idea of least squares method, and the problem of malicious information retrieval is transformed into the problem of finding the $f(x)$ value of the curve, so that the height fitting of the above malicious information N_z is minimized.

Suppose ξ_i is a malicious information point on $N_z | N_z \in f(x)$, where $f(x) = x_1 + x_2 + x_3 + \cdots + x_m$, then only the coefficient $x_1 + x_2 + x_3 + \cdots + x_m$ is determined, so that ξ_i is minimized, and the height fitting of N_z can be minimized.

Bring the expression of $f(x)$ into the calculation formula of ξ_i, and let the partial derivative of ξ_i to m_j be equal to 0, then get $m + 1$ equations and solve m_j from the equations.

According to the actual fitting effect, it is found that it is most suitable to fit the curve with the third-order polynomial. Combine the fitting equation of the curve to solve the second derivative, and make $f(x)' = 0$, then the x value is the curve inflection point. One inflection point is the curve coefficient, which is the minimum threshold of x corresponding to the N_z minimum value.

The minimum threshold obtained above includes the noise of malicious information. In order to improve the retrieval precision, the malicious information is also subjected to noise statistics, and the interference noise is eliminated to generate an absolute model of malicious information retrieval.

The noise reduction process is calculated as follows:

$$p = \frac{1 + \xi_i}{\log d_{\max}} \tag{4}$$

Where, p represents the malicious information retrieval coefficient without noise interference; $d_{\max}$ represents the maximum value of the malicious information cluster.

After the above definition, the efficient retrieval method based on human-computer interaction is used to retrieve and deduct malicious information in a random text of a big data network. By demarcating the metrics of malicious information, the method of establishing human-computer interaction logic is determined, and the retrieval coefficient is determined. The search coefficient is denoised to ensure that the method has an efficient and accurate retrieval advantage. In the process of retrieval, considering the existence of a single malicious information node, the method uses a set value to determine any information point. When the minimum value of the semantic distance ξ_i between the document information is greater than the retrieval coefficient, the information is determined to be an isolated point, which is consistent with the retrieval logic, and realizes efficient and accurate retrieval of malicious information in the multimedia big data network.

3 Simulation

In order to evaluate the efficient retrieval method based on human-computer interaction, the malicious information retrieval test was carried out on the method, and the malicious information retrieval test system based on Jelinek-Mercer smooth model and Bayesian model was obtained from the big data network. After proper processing and modification, the experiment is compared with the traditional malicious information retrieval method, and the malicious information retrieval effects of the two methods are counted.

3.1 Experimental Test Set

During the experiment, all the test data are the Chinese information retrieval test data set provided by TREC6. The test set includes a total of 164,811 big data network articles, all of which are from well-known big data networks, of which 1/10 are malicious information. The probability of 0.5% maliciously interferes with the multimedia big data network. All of these malicious information coexist in both Chinese and English. By changing the amount of malicious information and the length of interference, the comprehensiveness and accuracy of the search are analyzed.

3.2 Comprehensive Comparison of Malicious Information Retrieval

In order to analyze the effect of malicious information retrieval, taking recall rate as the experimental index, the traditional method and the method in this paper are used to compare the recall rate of malicious information retrieval of the two methods, and the results are shown in Fig. 2.

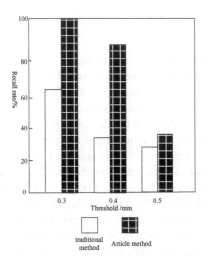

Fig. 2. Comprehensive comparison of malicious information retrieval results

The comparison results show that both methods can retrieve malicious information in the multimedia big data network, but the recall rate is different. When the threshold value is 0.3, the recall rate of traditional method is 65%, and that of this method is 100%. When the threshold is 0.4, the recall rate of traditional method is 35%, and that of this method is 88%. When the threshold value is 0.5, the recall rate of traditional method is 30%, and that of this method is 38%. Analysis of the reasons for this shows that with the increase of the threshold, the search process requirements for information matching are also improved, the number of malicious information that meets the requirements is reduced, and the recall rate is naturally reduced. When the threshold is between 0.3 and 0.4, the method has a better effect on the expansion of malicious search terms, and the correlation is greater. The result is that the recall rate is higher than the traditional method. Therefore, in order to ensure the comprehensiveness of malicious information retrieval, it is necessary to give a suitable threshold value for retrieval quantity control before retrieval.

3.3 Malicious Information Retrieval Accuracy Comparison

In order to analyze the effect of malicious information retrieval, taking the accuracy of malicious information retrieval as the experimental index, the traditional method and the method in this paper are used to compare the malicious information retrieval of the two methods, and the results are shown in Fig. 3.

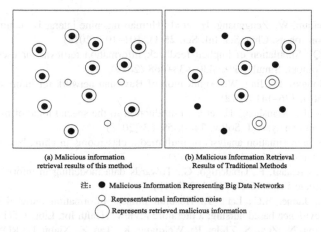

(a) Malicious information
retrieval results of this method

(b) Malicious Information Retrieval
Results of Traditional Methods

注: ● Malicious Information Representing Big Data Networks
○ Representational information noise
◎ Represents retrieved malicious information

Fig. 3. Comparison of malicious information retrieval accuracy

Through the analysis of the experimental results in Fig. 3, both methods can achieve the retrieval of malicious information in the multimedia big data network, but the retrieval accuracy is different. The analysis shows that this method can accurately retrieve and mark malicious information. However, the traditional methods have some defects in the retrieval of malicious information. It can be seen from the figure that the traditional method does not retrieve edge information in the big data network, so there must be accurate retrieval of malicious information. There is also the phenomenon of false information retrieval noise, which affects the final retrieval result of malicious information. Through the above analysis, the validity of the human-computer interaction retrieval method can be explained. When searching, the malicious information in the big data network can be accurately found and marked, and the validity and relevance of the query index words are guaranteed. Precision, superior to traditional methods

4   Conclusion

This paper has carried out a comprehensive and systematic research on the malicious information retrieval method based on human-computer interaction, but there are still many shortcomings. In the future research, we will further consider the in-depth development of human-computer interaction, with different modes. Human-computer interaction behavior improves the retrieval performance of malicious information. Also consider the semantic relationship between human-computer interaction and big data network, and use these relationships to enhance the semantic representation of malicious information, and then use semantic representation relationship to comprehensively analyze and compare malicious information retrieval methods, and further improve the efficiency of retrieval methods. Improve the information environment of big data networks.

References

1. Xu, L., Weiqun, W., Zengguang, H., et al.: Human-machine interactive control method for rehabilitation robots. Chin. Sci. Inf. Sci. **25**(1), 101–103 (2018)
2. Juanjuan, Q.: Simulation of implicit feedback information retrieval for users of big data network. Comput. Simul. **36**(9), 430–433+468 (2019)
3. Wei, Z., Hongxu, H., Jing, W.: Application of Bayesian network for information retrieval. Inf. Sci. **36**(6), 136–141 (2018)
4. Yao, Z., Bo, L., Xiansheng, H., et al.: Introduction to the special topic of multimedia data processing and analysis. J. Softw. **74**(4), 59–63 (2018)
5. Qi, Z.: Research situation analysis of multimedia classrooms in china based on big Data. Econ. Res. Guide **26**(18), 165–166 (2017)
6. Agosti, M., Crestani, F., Gradenigo, G.: Towards data modelling in information retrieval. Compr. Remote Sens. **15**(6), 143–162 (2018)
7. Berit, E.A., Janne, B.C., Lars, T.: Hospital nurses' information retrieval behaviours in relation to evidence based nursing: a literature review. Health Inf. Libr. J. **5**(1), 3–23 (2018)
8. He, J., Liming, N., Zeyi, S., Zhilei, R., Weiqiang, K., Tao, Z., Xiapu, L.: ROSF: leveraging information retrieval and supervised learning for recommending code snippets. IEEE Trans. Serv. Comput. **12**(1), 34–46 (2019)
9. Lu, Mengye, Liu, Shuai: Nucleosome positioning based on generalized relative entropy. Soft. Comput. **23**(19), 9175–9188 (2018). https://doi.org/10.1007/s00500-018-3602-2
10. Jihong, L., Weiguang, Z.: Study on the innovation of network information retrieval for university library in big data era. Agric. Network Inf. **1**(4), 30–32 (2017)
11. Liu, S., Liu, D., Srivastava, G., et al.: Overview and methods of correlation filter algorithms in object tracking. Complex Intell. Syst. (2020). https://doi.org/10.1007/s40747-020-00161-4
12. Li, J., Li, N., Afsari, K., et al.: Integration of Building Information Modeling and Web Service Application Programming Interface for assessing building surroundings in early design stages. Build. Environ. **153**(1), 91–100 (2019)
13. Stanton, L.: Eighth Circuit Says VoIP Is Information Service, Preempts Minnesota PUC. Telecommun. Rep. **84**(17), 1,35–36 (2018)
14. Tang, Z., Srivastava, G., Liu, S.: Swarm intelligence and ant colony optimization in accounting model choices. J. Intell. Fuzzy Syst. **38**(2),1–9 (2019)

Null Value Estimation of Uncertainty Database Based on Artificial Intelligence

Shuang-cheng Jia[(✉)] and Feng-ping Yang

Alibaba Network Technology Co., Ltd., Beijing 100102, China
xindine30@163.com

Abstract. Due to the complexity of the objective world, information loss and uncertainty are common. As a tool to express the real world, database uses null values to express the problem of information missing. Aiming at the problem of null value in uncertain database, an artificial intelligence based null value estimation algorithm is proposed. Firstly, the characteristics of uncertain database are analyzed, then the lost information retrieval model is constructed, and the empty value estimation of database is completed by feature selection and data transformation, artificial intelligence clustering, influence degree calculation, empty value step estimation and other methods. Finally, it analyses the time complexity of the algorithm, and improves the problem of poor evaluation effect of traditional algorithms. Supported by experimental data and environment, the results show that the proposed algorithm has higher accuracy than the traditional algorithm. It shows that this algorithm can effectively estimate the null value in the uncertain database, and has high practical application value, and can provide theoretical reference value for related research.

Keywords: Artificial intelligence · Uncertainty · Database · Null value estimation · Time complexity

1 Introduction

In recent years, with the vigorous promotion of the rapid development of computer technology and network information technology, various information systems have been growing, carrying more and more data storage and collection tasks. Especially in the big data environment, with the rapid growth of business data of organizations, all kinds of data are generated and processed at an unprecedented speed [1]. Therefore, how to mine and extract effective information from accumulated massive data has become a hot issue in academic circles.

With the maturity of relational database theory model, various relational database systems are widely used in various fields of social life, especially in the field of data mining. However, data in real databases often contain noise, default and ambiguity, which will affect the validity of data mining. Therefore, how to accurately estimate the null value in the process of data preprocessing is an important research topic [2]. In the face of this problem, there are usually several ways to deal with it: (1) Discarding records with null values; (2) Replace null value with a constant value; (3) Take an average value instead of the null value in the range of null value; (4) In the range of null

S. Liu and L. Xia (Eds.): ADHIP 2020, LNICST 347, pp. 283–294, 2021.
https://doi.org/10.1007/978-3-030-67871-5_26

value, a random value is used instead of null value. (5) Statistical distribution function of the original data, and then according to the distribution function to generate the replacement value of null value. However, the above methods can not deal with all the null value problems perfectly, the calculation process is complex, and the tendency of original data clustering is neglected, and the null value estimation effect can not be given very well [3, 4].

To this end, relevant personnel have proposed some database null value estimation methods, Reference [5] proposes a general boundary value estimation method for uncertain data model indicators. According to the characteristics of uncertain transaction databases with weights, a general boundary value estimation framework for commonly used model indicators is first designed, and then a quick estimation method for the upper bounds of model indicators under this framework is presented. Finally, the upper bounds of two typical model indicators are estimated to illustrate their feasibility. Experimental results show that although this method can realize the estimation of data null values, but the estimation effect is not good when facing the same local search time and the same risk measurement time. Reference [6] proposes an effective method for estimating the null value in relational databases. The method firstly mines the data in the data table to find the attribute set associated with the estimated attribute. This process only uses the data provided by the data itself. Information, avoiding the error caused by subjectivity when the expert determines the conditional attributes. Secondly, fuzzy clustering is performed according to the obtained attribute set to obtain a division of the original data, and then an estimated empty value in the relationship table is given based on the scored cluster and linear regression The method. Finally, the average absolute error rate is used to measure the accuracy of the algorithm estimation. The experimental results show that the result of this method has a high accuracy rate, but there is a problem of poor effect.

To solve the above problems, an artificial intelligence-based null value estimation algorithm for uncertain databases is proposed. This method introduces the principle of error to determine the order of estimating null values for each column. Through data mining, the attribute set associated with the estimated attributes is found. The original data is divided into fuzzy clusters by artificial intelligence method, and the null value is estimated by linear regression method within each cluster.

2   Characteristic Analysis of Uncertain Database

Uncertainty data is a general term for data that does not have complete confidence in data model. In reality, data is deterministic. The reason for producing uncertain data is due to its own knowledge limitation, which leads to the existence of uncertain data in data model. As a result, the following factors will lead to uncertain data:

(1) When describing the real world in the data model;
(2) When modifying or transforming data in the data model;
(3) When manipulating the data in the data model.

The term "uncertain data" is used to represent data that do not have full confidence in all data models. Overall, uncertain data mainly include the following categories:

(1) Probabilistic data: those data which are judged to be true or false by a certain probability value are called probabilistic data.
(2) Inaccurate data: This kind of data is available in data models, but not very clear. For example, the data may be a range or a non-value.
(3) Fuzzy data: In a data model, such data is expressed as vague in quantity or unit.
(4) Inconsistent data: Data with different true and false attributes at different times may change over time.
(5) Ambiguity data: Some data in the data model may lead to ambiguity or ambiguity.

The purpose of database system is to provide users with the information they want, and the information they interact with is the result of operation transformation of the description of real world information [7]. That is to say, interactive information with users is the focus of database system.

Since its birth, relational database has been widely used in various fields because of its simple and clear data structure, flexibility, independence, integrity, less redundancy and convenient application. But in practical application, it also reflects some short-comings of relational database, such as the introduction of null value to solve the problem of uncertain data processing in the real world [8].

A null value in a relational database represents an unknown value, which is neither a number 0, nor an empty string, nor any other meaningful value. In the early database, there was no concept of null value. All values were determined and knowable. With null value, we can express data that we don't know, undefined data and, of course, human error data. The introduction of null value makes the data representation of relational database more complete, and to some extent, it deals with the uncertainty problem [9].

3 Building Lost Data Retrieval Model

Building the lost data retrieval model is mainly to describe the simulation and abstract process of lost data retrieval. First of all, TIR technology is used to obtain database information, and the retrieval target is set to obtain information closely related to keywords in a certain period of time.

In order to better achieve the needs of lost information retrieval, the main objectives of lost data retrieval model are to define lost data retrieval, define retrieval results, calculate the relevance of retrieval results, etc. According to the characteristics of database, the lost data retrieval model is defined as four tuple form, which is represented by $[Q, D, R, S]$, where, Q represents query information; D represents data model; R represents lost data retrieval; S represents the scoring mechanism of query information and retrieval results.

With the development of network technology, data is mainly stored in the database, but for the data, it has the time attribute. In order to better express the time attribute of the data, the data in the database is represented by the temporal data graph.

The temporal data graph is represented by $G = (V_t, E_t)$, where, V_t represents the set of temporal nodes and E_t represents the set of temporal edges.

The temporal node is v_t, expressed as $v_t = [v, (ts_{vt}, te_{vt})]$, where, v represents the identification of the temporal node, $[ts_{vt}, te_{vt}]$ represents the semi open time interval, and E is the effective time of the data.

Temporal edge e_t is expressed as $e_t = [u_t, v_t, (ts', te')]$; $[ts', te']$ is the retrieval effective time.

The temporal data figure is shown in Fig. 1. As shown in Fig. 1, information has effective time and transaction time. In the process of lost data retrieval, the effective time of information is mainly considered.

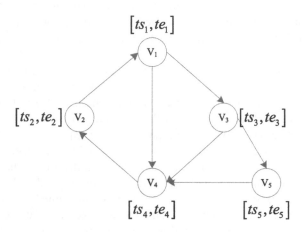

Fig. 1. Temporal data graph.

Based on the lost data retrieval model, according to the time constraints and key words in the query information, the result subtree is retrieved in the temporal data graph. In general, in the case of missing data query, we will get a lot of query result subtrees. In order to get the most similar missing data information, we sort the query result subtrees by similarity calculation, and the first one is similar missing data retrieval subtrees. Therefore, it is very important to calculate the similarity of lost data information.

In general, the smaller the temporal edge weight is, the greater the similarity is. The specific calculation method of temporal edge weight is as follows.

In order to ensure that as many keyword nodes are retrieved as possible, and ensure that the retrieval results are highly related to the lost data information of the query, calculate the node structure weight. The weight of node structure represents the importance of node in temporal data graph, which is introduced into the calculation formula of edge weight, and the calculation formula of edge weight is obtained as follows:

$$W(Q, e_t) = \frac{1}{IR_{(k,u)} + IR_{(k,v)}} \times W_e(u, v) \tag{1}$$

Where, $W_e(u, v)$ represents the node structure weight.

The temporal edge has timeliness, and the weights of different temporal edges are also different. Therefore, in the process of lost data retrieval, the temporal edge finite time needs to be consistent with the lost data query time. The temporal edge weight is set according to the lost data query time, and its calculation formula is:

$$W(Q, e_t)' = 1 - \frac{|I_c \cap I_e|}{|I_c|} \tag{2}$$

Where, I_c represents the query time of lost data; I_e represents the effective time of temporal edge.

Through the above formula, the temporal edge comprehensive weight value is obtained, so as to judge the similarity of the lost data and provide data support for the following uncertainty database null value estimation.

4 Space Value Estimation Algorithms Based on Artificial Intelligence

In the actual application of database, the problem of data missing is almost unavoidable, resulting in the problem of null value. Null value estimation has become the mainstream research direction of null value processing, and a large number of database null value estimation methods emerge [10]. Most of these methods use part of the complete data in the database table as training set, learn knowledge from the training set through machine learning or some theory of soft computing, derive decision rules or models, and finally estimate the null value according to the rules or models.

There are many commonly used null value estimation algorithms, such as rough set method, cloud model method and genetic algorithm based method. These algorithms have their own advantages, but there are also some obvious shortcomings. Rough set method is mainly based on the compatibility relationship between data, which is filled by compatible tuple values. However, if a tuple is not compatible with other tuples or the attribute values corresponding to compatible tuples are missing, then an estimate can not be given. The method of cloud model is mainly based on the generation of random points near the equilibrium position by the subordinate cloud generator to fit the original distribution of data, which will cause some "randomness" of the estimated value and affect the results of the algorithm [11]. The main disadvantage of null value estimation based on genetic algorithm is that it needs to analyze natural language semantics into effective coding, and the algorithm needs a long iteration time, and has poor scalability when the amount of data is large. In order to estimate the empty value in the relational database more accurately, based on the lost data retrieval model and the lost data similarity, a method of estimating the empty value in the uncertain database based on artificial intelligence is proposed. The specific implementation process is as follows:

Step 1: Feature selection and data conversion.

(1) Feature selection, attribute reduction algorithm based on rough set is used to reduce the attributes of the original data table and get the key attribute set after reduction.

(2) Data conversion, mainly refers to data preprocessing, making it easy to use data form. Firstly, the natural language semantic attributes are numeralized, so that the attributes can be conveniently used for data mining. Then the formula of fuzzy number is used to normalize the numerical information and simplify the calculation [12].

Step 2: Artificial intelligence clustering.

The non-null attribute sets associated with the attributes with null values obtained in step 1 are used for clustering. Make similar data together, different data are divided into different clusters. Figure 2 shows the compatibility of objects in different attribute sets.

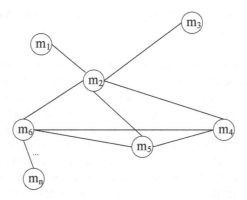

Fig. 2. Compatibility relationship among objects in different attribute sets

As shown in Fig. 2, considering that different attributes have different influence weights on columns with null values, the relevant weights are introduced:

$$w = \frac{r^2 - W(Q, e_t)'}{\sum_{k=1}^{m} r_1^2} \qquad (3)$$

In the formula, m is the number of attributes in the set of non-empty attributes related to attributes with null values; r represents the correlation coefficients of attributes with null values; w is the ratio of the correlation coefficients of attributes with null values and the sum of the correlation coefficients of all related attributes and attributes with null values, which reflects the weight of the influence of attributes with null values. After artificial intelligence clustering, the clustering center is obtained.

Step 3: Calculate the impact [13, 14].

After clustering the data into several clusters, for each cluster, the influence of different independent variables on dependent variables is different. Artificial intelligence regression coefficient is used to calculate the influence of different independent variables on dependent variables [15]. Firstly, the fuzzy correlation coefficient is used to represent the correlation degree between attributes, then the independent variable coefficient is determined, and finally the influence degree of attributes is obtained.

The formula for calculating the degree of correlation is as follows:

$$z_{a,b} = \frac{\sum_{i=1}^{n} (a_i - \bar{a}) \cdot (b_i - \bar{b})}{\sqrt{\sum_{i=1}^{n} (a_i - \bar{a})^2 \cdot \sum_{i=1}^{n} (b_i - \bar{b})^2}} \tag{4}$$

In the formula, $\bar{a}$ and $\bar{b}$ represent the sample mean of a, b of the fuzzy set.
The formula for determining the coefficient of independent variable is as follows:

$$COD = \pm \frac{r^2}{\sum_{k=1}^{m} r_1^2} \tag{5}$$

Step 4: Estimate null values.

Firstly, the Euclidean distance between the tuple and each cluster center is calculated, and the null value estimation algorithm is used to obtain the estimated value [16].

The null value estimation algorithm of uncertain database based on artificial intelligence is described. If the number of records in the data table is N, the number of records containing null values is N_{null}. The key attribute number after attribute reduction is m and the clustering number after partition is C. Then the time complexity of the algorithm is analyzed as follows:

(1) Feature selection and data conversion. In this step, an attribute reduction algorithm based on discernible matrix in rough set is used, and the time complexity of the algorithm is $O(N^2)$.
(2) Clustering using artificial intelligence algorithm of null value estimation in uncertain database, the time complexity of the algorithm is $O(N)$;
(3) Calculate the correlation degree to obtain the overall time consumption, i.e. linear complexity [17];
(4) Estimating null values and evaluating. This step estimates a small number of null values contained in the database tables with a time complexity of $O(N_{null})$, it is a high order infinitesimal of $O(N)$, which can be neglected [18].

In summary, the algorithm of null value estimation in uncertain database of artificial intelligence has high estimation accuracy.

5   Experimental Analysis

In order to verify the rationality of null value estimation algorithm in uncertain database based on artificial intelligence, experimental verification and analysis are carried out.

5.1   Experimental Data and Environment

Experimental environment: The operating system is Windows 7, using 3.40 GHz Intel Core i7-3770 CPU, using 8G memory bars, using C+ language on Microsoft Visual Studio 2012 development platform to realize the uncertainty database null value estimation algorithm based on artificial intelligence.

The algae dataset, a classical data set in data mining, is used in the experiment. The attributes are shown in Table 1.

Table 1. Independent attributes of Algae datasets.

Variable	Range
Season	{Spring, Summer, Autumn, Winter}
Size	{Small, medium, large}
Speed	{Low, medium, high}
PH	Positive real number
CL	Positive real number
NO_3	Positive real number
NH_4	Positive real number
PO_4	Positive real number

The table describes that under the influence of independent variables, the data set contains 184 pieces of data, 164 of which are used as training data set, and the remaining 20 as test set.

5.2   Experimental Steps

Experiments are carried out on algae datasets. The specific process is as follows:

Step 1: Data conversion and feature selection: because season, size and speed are text variables, they can not be directly processed, changing season's "spring, summer, autumn, winter" to "1, 2, 3, 4"; size's "small, medium, large" to "1, 2, 3"; speed's "low, medium, high" to "1, 2, 3". Then we use feature selection algorithm to eliminate the four attributes of season, CL, NH4 and PO4. The remaining attributes represent more than 95% of the data features.

Step 2: Clustering the remaining attributes, using clustering indicators to get a better clustering effect, dividing the original data set, and finding the clustering center.

Step 3: For each type of data, calculate the influence of independent variables on dependent variables.

Step 4: Estimate the null value.

Figure 3 is the flow chart of the experimental steps.

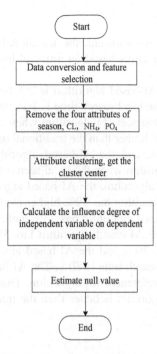

Fig. 3. Experimental flow chart

5.3 Experimental Results and Analysis

In order to study the validity of AI-based null value estimation algorithm for uncertain databases, local search time and risk measurement time are taken as criteria to compare the evaluation results of traditional references [5] algorithms and AI-based algorithms.

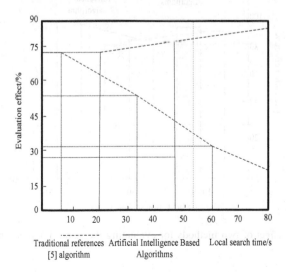

Fig. 4. Evaluating the effect of two methods using the same local search time.

(1) Local search time

When the local search time is consistent, the traditional references [5] algorithm is compared with the evaluation effect based on artificial intelligence algorithm, and the results are shown in Fig. 4.

Figure 4 shows that the AI-based algorithm is 2% better than the traditional references [5] algorithm when the local search time is 10 s; the AI-based algorithm is 8% higher than the traditional references [5] algorithm when the local search time is 20 s; the AI-based algorithm is 20% higher than the traditional references [5] algorithm when the local search time is 30 s; and the AI-based algorithm is 40 s higher than the traditional references [5] algorithm when the local search time is 40 s. Compared with the traditional references [5] algorithm, the AI-based algorithm has 32% higher evaluation effect; the AI-based algorithm has 39% higher evaluation effect when the local search time is 50 s; the AI-based algorithm has 53% higher evaluation effect when the local search time is 60 s; the AI-based algorithm has 58% higher evaluation effect when the local search time is 70 s; and the AI-based algorithm has 80 s higher evaluation effect when the local search time is 80 s. The AI-based algorithm is 60% more effective than the traditional references [5] algorithm. Therefore, under the same local search time, the AI-based algorithm is better than the traditional references [5] algorithm in evaluating the effect.

(2) Risk measurement time

When the time of risk measurement is consistent, the traditional references [5] algorithm is compared with the evaluation effect based on artificial intelligence algorithm, and the result is shown in Fig. 5.

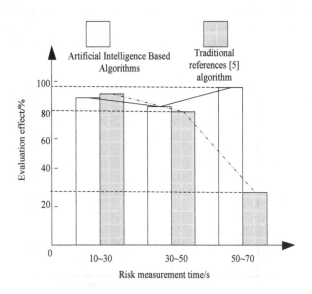

Fig. 5. The effect of two methods for evaluating the same risk measurement time.

Figure 5 shows that when the time of risk measurement is 10–30 s, the evaluation effect of traditional references [5] algorithm is 90%, and that of AI-based algorithm is 86%. When the time of risk measurement is 30–50 s, the evaluation effect of traditional references [5] algorithm is 80%, and that of AI-based algorithm is 83%. When the time of risk measurement is 50–70 s, the evaluation effect of traditional references [5] algorithm is 28%, and that of AI-based algorithm is 97%. It can be concluded that under the same risk measurement time condition, the AI-based algorithm is better than the traditional references [5] algorithm in evaluating the effect. This is because the method through the analysis of the characteristics of the uncertain database, data conversion, and the use of artificial intelligence technology for data clustering, and calculate the impact degree, and ultimately achieve the null value estimation of the database, through the above steps to enhance the effect of null value estimation.

To sum up, the null estimation effect of the proposed algorithm is better than that of the traditional references [5] algorithm under the same local search time or the same risk measurement condition, indicating that the proposed algorithm has high application value.

5.4 Experimental Conclusions

To sum up, the null value estimation algorithm of uncertain database based on artificial intelligence is effective. Under the same local search time, the maximum evaluation effect of artificial intelligence algorithm is 86%, and under the same risk measurement time, the maximum evaluation effect of artificial intelligence algorithm is 97%.

6 Concluding Remarks

Non-deterministic database is based on strict mathematical concepts. It has a single concept and simple and clear data structure. Its greatest advantage is that the relationship between entities can be expressed by relationship, that is, the indefinite database can describe itself. Many advantages make the indefinite database occupy the dominant position in the market and has been widely used.

With the gradual expansion of the application field of uncertain database, the data processing scope and ability of uncertain database are demanded in various fields. For example, in the fields of scientific computing, sensor application and knowledge learning system, the database is required to deal with uncertain data. However, most uncertain databases can only deal with accurate data, lacking a comprehensive method for dealing with uncertain data. Now the only way to solve this problem is to use the artificial intelligence algorithm for estimating the null value of uncertain databases.

This method uses artificial intelligence algorithm to classify the initial data, taking into account the fuzzy nature of data classification, and introduces weighted values according to the different dimensions of sample data to the contribution of clustering, which makes the clustering results more accurate. Then, the null value estimation model is constructed by multiple linear regression, which makes the null value estimation method more effective and accurate.

References

1. Alam, M.K., Aziz, A.A., Latif, S.A., et al.: Error-aware data clustering for in-network data reduction in wireless sensor networks. Sensors **20**(4), 1011 (2020)
2. Zhang, T.A.: Dynamic threats assessment based on intuitionistic fuzzy set under missing data condition. Fire Control Command Control **43**(8), 93–97 (2018)
3. Kim, K.: Identifying the structure of cities by clustering using a new similarity measure based on smart card data. IEEE Trans. Intell. Transp. Syst. **21**(5), 2002–2011 (2020)
4. Wang, J., Liu, F.X., Jin, C.J.: General bound estimation method for pattern measures over uncertain datasets. J. Comput. Appl. **38**(01), 165–170 (2018)
5. Liu, L., Wang, L.S., Wu, F.: An efficient method for estimating null values in relational database. Comput. Technol. Autom. **35**(03), 110–114 (2016)
6. Li, H.: RFID tag number estimation algorithm based on sequential linear Bayes method. J. Comput. Appl. **38**(11), 3287–3292 (2018)
7. Gao, J.M.: Adaptive deduplication simulation in privacy protection database. Comput. Simul. **36**(01), 239–242 (2019)
8. Song, X.P.: Global estimates of ecosystem service value and change: taking into account uncertainties in satellite-based land cover data. Ecol. Econ. **143**(1), 227–235 (2018)
9. Faghih, M., Mirzaei, M., Adamowski, J., et al.: Uncertainty estimation in flood inundation mapping: an application of non-parametric bootstrapping. River Res. Appl. **33**(4), 611–619 (2017)
10. Wang, Y., Tao, W., Yan, Z., Wei, R.: Uncertainty analysis of dynamic thermal rating based on environmental parameter estimation. EURASIP J. Wirel. Commun. Netw. **2018**(1), 1–10 (2018). https://doi.org/10.1186/s13638-018-1181-7
11. Ju, H., Zhang, G., Cui, J., et al.: A novel algorithm for pose estimation based on generalized orthogonal iteration with uncertainty-weighted measuring error of feature points. J. Mod. Opt. **65**(3), 331–341 (2018)
12. Frédérique, S., Bernard, C., Paolo, D.G.: High-resolution humidity profiles retrieved from wind profiler radar measurements. Atmos. Meas. Tech. **11**(3), 1669–1688 (2018)
13. Fu, W., Liu, S., Srivastava, G.: Optimization of big data scheduling in social networks. Entropy **21**(9), 902 (2019)
14. Forsberg, E.M., Huan, T., Rinehart, D., et al.: Data processing, multi-omic pathway mapping, and metabolite activity analysis using XCMS Online. Nat. Protoc. **13**(4), 633–651 (2018)
15. Liu, S., Liu, D., Srivastava, G., et al.: Overview and methods of correlation filter algorithms in object tracking. Complex Intell. Syst. (2020). https://doi.org/10.1007/s40747-020-00161-4
16. Koch, D.C.L., Jean-Paul, G., Xue, M., et al.: Terahertz frequency modulated continuous wave imaging advanced data processing for art painting analysis. Opt. Express **26**(5), 5358 (2018)
17. Yang, Y.G., Guo, X.P., Xu, G., et al.: Reducing the communication complexity of quantum private database queries by subtle classical post-processing with relaxed quantum ability. Comput. Secur. **81**(3), 15–24 (2019)
18. Liu, S., Li, Z., Zhang, Y., et al.: Introduction of key problems in long-distance learning and training. Mob. Netw. Appl. **24**(1), 1–4 (2019)

Research on Fuzzy Clustering Algorithms for Large Dimensional Data Sets Under Cloud Computing

Shuang-cheng Jia[✉] and Feng-ping Yang

Alibaba Network Technology Co., Ltd., Beijing 100102, China
xindine30@163.com

Abstract. With the rapid development of modern computer technology, the internet technology based on computer technology has also been remarkably developed, which has made great progress in modern information technology. The management and effective application of these large-scale data has become the main trend of the development of modern society. In order to improve the ability of fast processing and recognition of large data, data clustering analysis is needed. Aiming at the problems existing in traditional fuzzy clustering algorithm, this paper proposes a design of fuzzy clustering algorithm for high-dimensional and large data sets under cloud computing. Through data classification and data processing classification, the design of fuzzy clustering algorithm for high-dimensional large data sets is realized, and compared with the traditional algorithm through experiments. The simulation results show that the method is easy to calculate and fast in the fuzzy clustering of high-dimensional and large data sets. The clustering effect is good, and the clustering algorithm of high-dimensional and large data sets is well realized.

Keywords: Computer technology · Cluster analysis · Fuzzy clustering algorithm · High-dimensional large data sets

1 Introduction

With the rapid development of Internet technology, data storage and data compression technology, the emergence of interactive applications such as micro-blog, micro-messaging, social networks, the rise of cloud applications, and the wide use of various forms of digital devices, the data is explosive growth. All sectors of society, such as academia, enterprises and government departments, have paid close attention to the issue of big data. Data has become a new asset, which can bring endless social and economic benefits [1]. Big data is gradually becoming a powerful tool for people to understand and transform the world, making it easier for people to grasp the laws of things and accurately predict the future.

The resources in cloud computing are usually virtualized resources provided through the Internet [2]. There are many high-dimensional data sets in cloud computing. At present, there are still many doubts and disputes about the basic concept and key technologies of high-dimensional large data. Traditional relational database technology is not competent for the processing of these data, because the starting point of

© ICST Institute for Computer Sciences, Social Informatics and Telecommunications Engineering 2021
Published by Springer Nature Switzerland AG 2021. All Rights Reserved
S. Liu and L. Xia (Eds.): ADHIP 2020, LNICST 347, pp. 295–305, 2021.
https://doi.org/10.1007/978-3-030-67871-5_27

relational database system is to pursue data consistency and fault tolerance. Diversity is an important feature of large data, which means the universality of data sources and the complexity of data types. In this complex data environment, processing large data is a great challenge. To deal with large data, it is necessary to integrate data from all data sources, extract relationships and entities from them, and store these data in a unified structure after association and integration. Therefore, the pretreatment of large data is one of the key factors to determine the value of large data. In order to reflect the value of large data, the pretreatment of large data needs further processing [3]. Fuzzy clustering is one of the key technologies of large data analysis and an important method to reflect the value of large data [4].

At present, relevant personnel have proposed a large number of data clustering methods, among which, reference [5] proposes an algorithm to increase the confidence threshold. The algorithm repeats the constant confidence threshold algorithm several times while increasing the confidence threshold. The algorithm can almost certainly converge to a certain equilibrium point. A numerical example is given to illustrate the applicability of this method to clustering problems, but the calculation time of the method is slow. Reference [6] proposed a new error aware data clustering (EDC) technology in cluster head (CHS) application. Histogram based data clustering (HDC) module groups time-related data into clusters and eliminates relevant data from each cluster. Recursive outlier detection and smoothing (rods) of HDC module provides error aware data clustering, which uses temporal correlation of data to detect random anomalies. The experimental results show that the algorithm has a small amount of calculation and can reduce a large number of redundant data with the minimum error, but the data coverage is not high.

To solve the above problems, this paper proposes a fuzzy clustering algorithm for high-dimensional large data sets in cloud computing.

2 Design of Fuzzy Clustering Algorithms for High Dimensional and Large Data Sets

2.1 Data Object Classification

Before realizing the fuzzy clustering algorithm, the data obtained should be classified, the original data set should be divided into several data fragments, and these data fragments should be copied to the node of the task, and the data at the node should be processed separately. The main task of this process is to calculate the degree of membership of data objects to clustering centers in nodes. In the process of data classification, the data objects and initial clustering centers (or the clustering centers updated in the last iteration process) at this node are input, the index numbers of the clustering centers are output, and the membership degree of all data objects and data objects in the node are also given. Because the data generated in the classification process are stored on local disks, and these data can be merged locally, in order to reduce the communication overhead between the nodes and the calculation of the process. Map process is used in classification, which is carried out at the node, and can play the role of merging data at this node. The main task of map process is to calculate the sum of the product of all the corresponding subordinate degrees and the corresponding data objects, and to represent the sum of all subordinate degrees (fuzzy factors) for each cluster center. The map process outputs in the form of seu, which represents the index of the clustering center. The output is the data set composed of the index of the data center, the data object and the membership degree. The data set cooperates as the input of the map process. The input form is to compose a data set with the same values in the data set generated in the stage, so the essence of the process is to correspond to each value. A clustering center is indexed, so a clustering result can be obtained every time a process is executed.

After the above work is completed, the data are classified. In the correlation analysis of data classification, the corresponding data classification methods are different when the types of attribute values of data objects are different. The values of attributes are divided into continuous variables and classification variables. Interval variables or ratio variables are classified by the values of continuous variables. If the data object has the largest degree of membership to a clustering center, then the data object belongs to the corresponding class of the clustering center and classifies the data object according to the data parameters obtained above. The structure of the data object is shown in Fig. 1.

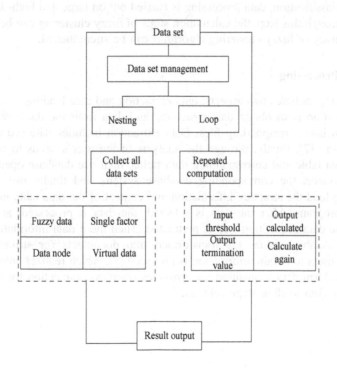

Fig. 1. Data object classification structural diagram.

The structure chart of data object classification shown in Fig. 1, when classifying data, classifies the data according to the process shown in the figure to ensure the accuracy of classification. The classification formula is as follows:

$$f_i = \sum_d^2 vb \frac{x_i - 1}{\max(x_i) - \min(x_m)} \tag{1}$$

In the formula, f_i represents the marked data; $x_i - 1$ represents the standard value of data evaluation; max represents x_i types of data; min represents m types of data, this calculation does not do directional analysis.

$$j = \frac{F_0 \cdot \omega(f_i)}{F \cdot v(j_0 + 1)} \tag{2}$$

In the formula, j represents the classified data after discretization; f represents the standard value of classification; F_0 represents the detected data; j_0 represents the quality level of the data being tested, this calculation does not do directional analysis.

After the above formulas, data object classification is realized. On the basis of object data classification, data processing is carried out on large and high-dimensional data sets. Through this step, the calculation steps of fuzzy clustering can be simplified and the accuracy of fuzzy clustering algorithm can be strengthened.

2.2 Data Processing

Data processing includes two aspects: data extraction and data loading. The main task of data extraction is to obtain data structures and data from the database, and then convert them into corresponding files. Data extraction includes data acquisition and data processing [7]. Firstly, through the analysis of the user's needs to obtain data, select the data table and corresponding data fields, then use database operation technology to connect the corresponding database system, and finally use appropriate statements to load the structure information and content of the data table into the data processing program. After the data is selected, the data is processed, and the data loaded by the data acquisition step is pretreated. Then these data information and the structure of database tables are transformed into documents (or strings) and the interface for users to obtain information is provided [8], which realizes cross-language and cross-platform data acquisition and provides seamless connection for data source acquisition in data loading stage (Fig. 2).

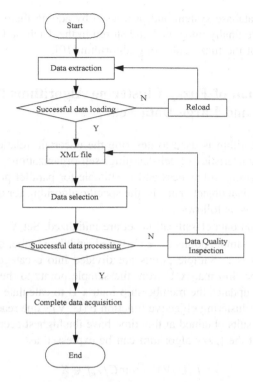

Fig. 2. Data processing diagram.

Data extraction is the basis of realizing the fuzzy clustering algorithm. Through data extraction, we can prepare for the next data loading. After obtaining the file, we need to analyze the XML file, and then create the appropriate parsing method according to the description of the XML and the table structure we need to store. According to the research of document parsing based on technology used by nodes in the platform, data can be processed in parallel. The acquired files are very small and can be read from the table as the input path of the whole job or read in parallel as the process input at the stage of setting up the job.

After data extraction, data is loaded. In this stage, the main task is to access different Web Service interfaces, parse the acquired XML data and load it to the target data source. In this paper, Base is used as the target data source [9]. The address and service method of each data source are loaded into it. By distributing these information to different nodes, each node accesses different data sources according to the information received by each node, and then stores these documents into the file system. At the same time, the information of these files (including storage address and file) is stored. Size, etc.) is stored in the database for subsequent management of these files. When data acquisition is completed, these documents need to be parsed, so we need to look at the preamble of these documents to understand the storage mode of each document and the format of the original data, and then design the corresponding conversion rules and store them. The paths and corresponding transformation rules are obtained from the database system and processed by adding them to different nodes. The acquired data are finally integrated and stored in the database to provide the basis for the completion of the fuzzy clustering algorithm [10].

3 Implementation of Fuzzy Clustering Algorithms for High Dimensional and Large Data Sets

Fuzzy clustering algorithm is used to describe the uncertain relationship among patterns. It has the characteristics of self-learning, self-organization and self-adaptation. This algorithm has strong robustness and is suitable for parallel processing. Through data preparation and data object analysis, the specific algorithm for designing the fuzzy clustering algorithm is as follows:

Detection data and data classification set are initialized. Set $X = (x_1, x_2, x_3 \ldots x_m)$ is assumed to be all the sample points to be clustered, and the number of clustering results of all sample points is c. Sample points are divided into c categories. The data set calculates the membership matrix U from the sample points to the clustering centers iteratively, and then updates the membership matrix U to calculate the new clustering center V, so that the clustering objective function F (U, V) value reaches the minimum, and the clustering results obtained at this time have the highest accuracy [11–13]. The objective function of the fuzzy algorithm can be expressed as:

$$F(U, V) = \min C / \lambda_n f \in X \tag{3}$$

In the formula, λ_n represents the sample size; C represents the fuzzy index; $\in$ represents the clustering effect; X represents the number of iterations of clustering data, this calculation does not do directional analysis.

According to the membership matrix U, which can be obtained from the above formula, the new clustering center is calculated according to the membership matrix U:

$$F = f_i + \sum_{i=1}^{k} A + M \cdot K \tag{4}$$

The detailed steps of the fuzzy clustering algorithm are as follows:

Initialization algorithm stop threshold, the number of iterations is $i = 1$, the maximum number of iterations is K;

Randomly select k sample points as the initial clustering center, calculate the membership degree of each sample point to each initial clustering center, and obtain a membership matrix F;

The clustering centers are recalculated according to the membership degree F, and all kinds of clustering centers are updated [14]. The fuzzy clustering algorithm terminates and outputs the clustering results.

From the implementation steps of the fuzzy algorithm, it can be seen that calculating the membership degree from a sample point to each clustering center is the most important work, and also the most time-consuming step. Since the membership degree of each sample point is independent of each other, the parallel execution of this process can be considered [15, 16]. It is possible to compute the membership degree of cluster centers of sample points in parallel. By calculating the membership degree of each sample point in parallel on distributed cloud platform with high-dimensional large database, the time required to compute the membership matrix U can be greatly shortened, thus greatly improving the operation efficiency of the fuzzy algorithm.

4 Experimental Demonstration

In order to verify the validity and feasibility of the proposed fuzzy clustering algorithm, a simulation platform is built and experimental analysis is carried out. At the same time, the algorithm is debugged in the context of cloud computing to maximize the accuracy of the experimental results. In order to ensure the rigor of the experiment, the traditional fuzzy clustering algorithm is compared with the fuzzy clustering algorithm designed in this paper. Because there are many kinds of high-dimensional and large data, the elements to be detected are complex, and the data obtained are generated in a short period and high frequency, the data stability is not high. In order to evaluate the quality of clustering results, it is necessary to define the evaluation criteria of clustering and the objective function or method of clustering by checking the clustering results of the algorithm to determine whether the algorithm needs further iteration.

4.1 Experimental Environment and Data Setting

Taking My-sea database as the data source, 10 data sets are selected from the database for experiment, and data analysis is conducted through online data analysis software MOA (an experimental tool for massive online analysis). Table 1 shows the data set used in the experiment.

Table 1. Parameters of experimental data set.

Dataset number	Data contained	Data dimension	Subseries
1	512	4	15
2	1024	12	10
3	261	8	12
4	258	6	16
5	637	10	10
6	298	8	8
7	186	6	12
8	254	12	14
9	301	10	15
10	209	4	5

Table 2 shows the parameters of computer virtual simulation platform.

Table 2. Parameters of computer virtual simulation platform.

Name	Parameter
CPU	Intel i7-8700, 3.75 GHz
Hard disk	512G Solid state disk
Graphics card	NVIDIA
Memory	16G
Operating system	Windows 8.1

According to the above experimental conditions, the data clustering results of different methods are compared, and the experimental conclusions are obtained.

4.2   Clustering Effect Comparison

In the implementation of clustering algorithm, the sum of squares of errors is used as the condition of clustering convergence, and the precision and recall rate are used to evaluate the quality of clustering results. The precision ratio refers to the ratio of the number of data objects correctly clustered into the current cluster and all objects clustered into the current cluster after clustering. Recall ratio refers to the ratio of the number of objects accurately clustered into the current cluster to the number of actual objects in the cluster after clustering.

1000 feature vectors are extracted from high-dimensional large database. In order to reflect the effectiveness of the proposed fuzzy data algorithm and traditional algorithm in computing, the same cluster and the same data set are used for clustering. The results are shown in the Table 3.

Table 3. Traditional algorithm result table.

Data set	Average precision	Average recall rate	Iteration times	Execution time	pH
64	0.39	4.5	12	90	8.1
83	0.42	3.6	10	89	8.3
52	0.52	2.1	11	74	8.2
178	0.69	5.7	12	98	8.2
124	0.78	6.1	14	100	8.1
281	0.96	2.5	11	114	8.1

Setting the same data set, the average precision rate is set to 0.5, the average recall rate is 5.5, the normal number of iterations is 5, the standard value of execution time is set to 60, and the pH is set to 7.

Cloud computing is used to record the results of the fuzzy clustering algorithm, and the average precision, recall, iteration times, execution time and PH of the two algorithms are compared according to the experimental data. The analysis of Table 3 and Table 4 shows that the average precision of the fuzzy clustering algorithm designed in this paper is similar to the standard values, and the average recall rate is higher than that of the traditional algorithm, which has obvious advantages. In terms of execution time, it costs less time and saves a lot of manpower and financial resources. Therefore, we can confirm the validity of the fuzzy clustering algorithm designed in this paper. When clustering high-dimensional large data sets, the number of detection iterations is low, and the operation time is short. It can accurately and quickly cluster high-dimensional large data sets. To a certain extent, it improves the clustering effect of large data, and in a sense, it has the promotion effect. This is because the fuzzy clustering algorithm is used to describe the uncertain relationship between patterns. The algorithm has the characteristics of self-learning, self-organization and self-adaptive. It has strong robustness and is suitable for parallel processing, thus improving the clustering effect of high-dimensional large data sets.

Table 4. The result table of the algorithm in this paper.

Data set	Average precision	Average recall rate	Iteration times	Execution time	pH
64	0.42	4.6	1	52	8.1
83	0.5	5.1	2	45	8.3
52	0.51	5.2	2	56	8.2
178	0.58	5.6	3	60	8.2
124	0.48	5.9	4	61	8.1
281	0.52	6.0	1	63	8.1

In order to compare the data clustering results of different methods more intuitively, the clustering results of different methods are presented in the form of image comparison. The results are shown in Fig. 3.

Analysis of Fig. 3 shows that the data coverage of the algorithm proposed in this paper is significantly higher than that of the traditional algorithm, the maximum data coverage of traditional algorithm is less than 40%, while the maximum data coverage of this algorithm is more than 70%, indicating that the clustering effect of the algorithm in this paper is better. This is because the algorithm realizes the design of fuzzy clustering algorithm for high-dimensional large data sets through data classification and data processing classification, and effectively improves the data coverage.

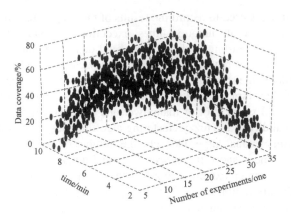

(a)The result table of the algorithm in this paper

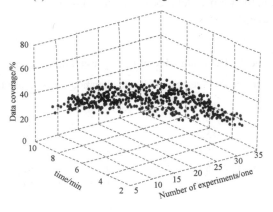

(b)Traditional algorithm result

Fig. 3. Data clustering results of different methods.

5   Concluding Remarks

With the rapid development of information science and technology and mobile Internet technology, people's life is full of various information data, which is undoubtedly exponential explosive growth. The era of big data has come quietly, and it promotes the development of all walks of life. In order to fully explore the development potential of various industries behind information and data, the high-dimensional and large data fuzzy clustering algorithm under cloud computing has become an indispensable tool. Many existing clustering methods have some problems. Through the analysis of the problems, the design of the fuzzy clustering algorithm is proposed. The experiment proves the effectiveness of the proposed algorithm. The algorithm can cluster high-dimensional large data sets accurately, reduce the residual rate of the fuzzy algorithm, and has very high effectiveness. But there is still room for further study.

References

1. Madadi, S., Mohammadi-Ivatloo, B., Tohidi, S.: A data clustering based probabilistic power flow method for AC/VSC-MTDC. IEEE Syst. J. **13**(4), 4324–4334 (2019)
2. Pengfei, L., Chunyu, L.N.: Real-time flow controllable clustering algorithm for large correlated data in cloud computing environment. Sci. Technol. Eng. **18**(7), 185–190 (2018)
3. Yecheng, G.: Prediction strength optimization of large data clustering algorithms based on distributed computing. J. Changchun Normal Univ. **37**(6), 71–73 (2018)
4. Kuwil, F.H., Shaar, F., Topcu, A.E., et al.: A new data clustering algorithm based on critical distance methodology. Expert Syst. Appl. **129**(9), 296–310 (2019)
5. Nguyen, L.T.H., Wada, T., Masubuchi, I., et al.: Bounded confidence gossip algorithms for opinion formation and data clustering. IEEE Trans. Autom. Control **64**(3), 1150–1155 (2019)
6. Alam, M.K., Aziz, A.A., Latif, S.A., et al.: Error-aware data clustering for in-network data reduction in wireless sensor networks. Sensors **20**(4), 1011 (2020)
7. Tadie, A.T., Guo, Z.: Optimal planning of grid scale PHES through characteristics-based large scale data clustering and emission constrained optimization. Energies **12**(11), 2137 (2019)
8. Bacciu, D., Castellana, D.: Bayesian mixtures of hidden tree markov models for structured data clustering. Neurocomputing **342**(21), 49–59 (2019)
9. Xiaodong, Z.: Research on parallel scheduling algorithms for large data of ships based on cloud computing. Ship Science and Technology **23**(4), 31–33 (2018)
10. Zheng, Q., Diao, X., Cao, J., et al.: From whole to part: reference-based representation for clustering categorical data. IEEE Trans. Neural Netw. Learn. Syst. **31**(3), 927–937 (2020)
11. Kaimin, L., Zili, Z.: Research on the connotation and solutions of big data mining based on cloud computing. Comput. Fans **91**(04), 39 (2018)
12. Fu, W., Liu, S., Srivastava, G.: Optimization of big data scheduling in social networks. Entropy **21**(9), 902 (2019)
13. Liu, S., Glowatz, M., Zappatore, M., Gao, H., Jia, B., Bucciero, A. (eds.): eLEOT 2018. LNICST, vol. 243, pp. 1–374. Springer, Cham (2018). https://doi.org/10.1007/978-3-319-93719-9
14. Ben HajKacem, M.A., N'cir, C.-E.B., Essoussi, N.: One-pass MapReduce-based clustering method for mixed large scale data. J. Intell. Inf. Syst. **52**(3), 619–636 (2017). https://doi.org/10.1007/s10844-017-0472-5
15. Liu, S., Lu, M., Li, H., et al.: Prediction of gene expression patterns with generalized linear regression model. Front. Genet. **10**, 120 (2019)
16. Wenying, Z., Junwei, Z., Binning, L., et al.: Research on higher vocational education resource sharing based on cloud computing and big data. Comput. Educ. **22**(04), 12–56 (2018)

Research on Big Data Classification Algorithm of Disease Gene Detection Based on Complex Network Technology

Yuan-yuan Gao[1(✉)], Ju Xiang[1], Yan-ni Tang[1], Miao He[1], and Wang Li[2]

[1] Changsha Medical College, Changsha 410219, China
[2] College of Big Data, TongRen University, Tongren 554300, China
lknm36500@126.com

Abstract. In order to improve the accuracy of the classification of the big data of disease gene detection, an algorithm for the classification of the big data of disease gene detection based on the complex network technology was proposed. On the basis of complex network technology, a distance-based membership function is first established. Considering the distance between the sample and the class center, the membership function of sample compactness is designed to complete the establishment of membership function of complex network. Combined with the design of the classification algorithm flow of the big data of disease gene detection, the design of the data classification algorithm was completed, and the classification of the big data of disease gene detection was realized. The experimental results show that the proposed algorithm is more accurate than the other two classification algorithms in the big data sets of different disease genes.

Keywords: Complex network technology · Disease genes · Big data · Classification algorithm

1 Introduction

The 21st century is the era of combining life science and information science. With the implementation of the human genome project, the sequence and structural data of nucleic acids and proteins have grown exponentially. At the end of 1980s, bioinformatics, a new interdisciplinary subject, emerged and developed rapidly. The main purpose of bioinformatics is to obtain, process, store, retrieve and analyze the biological experimental data, so as to reveal the biological significance of the data. The development of bioinformatics will bring revolutionary changes to the life sciences, and will also have a huge impact on the agriculture, medicine, health, food and other industries. Therefore, the governments and industries of all countries attach great importance to it and invest a lot of money and manpower [1]. After obtaining the complete sequence information of the genes, human beings must further understand what the functions of all these genes are and how they perform these functions, so that the genetic information of the genes can establish a direct link with the activities of life.

© ICST Institute for Computer Sciences, Social Informatics and Telecommunications Engineering 2021
Published by Springer Nature Switzerland AG 2021. All Rights Reserved
S. Liu and L. Xia (Eds.): ADHIP 2020, LNICST 347, pp. 306–319, 2021.
https://doi.org/10.1007/978-3-030-67871-5_28

Therefore, researchers are shifting their attention from the structural study of genes to the functional study of genes. However, how to get useful information from the massive genetic information to reveal the biological significance of these data has always been a difficult problem for researchers, and has important scientific significance and wide application value. Through gene classification, the massive gene expression data can be divided into a relatively small number of groups with biological significance, and valuable information can be extracted from them, so as to understand the functional relationship and working mode of different gene combinations. It is extremely useful to classify a large number of genes into relatively few classes to extract valuable information from the data [2]. In order to achieve this goal, people have applied many methods to solve the problem of gene classification.

Now, with the rapid development of science and technology, even for the most knowledgeable organisms, we only know a small part of the functional information of the genes, and this information is often incomplete and does not meet our scientific research needs. With the expansion of the application scale of gene chip and the development of gene expression database, in order to make full use of these data, more in-depth analysis of these gene data is needed to extract the implied information from a large number of data. For the gene chip data, through in-depth analysis, we can obtain other in-depth information, such as the evolution information of DNA sequence, the information of gene function, the synergistic relationship between genes, the spatial rule of gene expression, and the information related to diseases.

In gene expression studies, there is a basic assumption that information about when and where genes are expressed carries information about gene function. Thus, the first step in the analysis of gene expression data is to classify the tissue genes according to the similarity of the gene expression profile, or to classify the genes according to the pattern of gene expression. Gene classification is now one of the most widely used methods of expression data analysis. A single gene chip often contains hundreds of genes whose expression can be detected simultaneously [3]. The same chip was used to conduct gene expression experiments under different conditions (different time, different cells, different external effects), collect expression data, and put the original data together to form a data table. Each row in the table represents a gene, and each column represents the intensity of expression under different experimental conditions. Mathematically, a row or column of data in a table is a vector, and genetic classification is the grouping of these vectors according to their similarity.

The general process of pattern recognition of gene expression data is as follows: first, preprocess the original gene expression data, then select and extract the characteristics of the gene expression data, and finally classify the genes according to the characteristic information. However, in the actual environment, the distribution of the training sample set is usually unbalanced, that is, the number of samples of different categories in the training sample set is different or even quite different. Because of this imbalance, the training and prediction of the classifier tend to the categories with a large number of samples, which is not conducive to the accuracy of classification decision. For example, in a concentration of liver cancer samples containing 80 normal samples and 10 pathological samples, since the training and prediction output tend to be large samples, the prediction of a small number of pathological samples may result in "missed detection", in which the pathological samples are misjudged as normal samples. In practice, the risk of such misjudgment would be much greater than the risk of misinterpreting a normal sample as a diseased one [4]. Therefore, the bias of classifier training and prediction caused by the unbalanced distribution of sample sets should be considered in practical application.

However, most of the current research work focuses on how to improve the prediction accuracy as much as possible, but ignores the bias feature of the prediction, and seldom studies the risk cost of the prediction bias on the decision result from the theoretical perspective. In the diagnosis of fatal diseases, if the patient is misdiagnosed as a healthy person, it will affect the best chance for the patient to get timely treatment and pose a great threat to the safety of the patient's life. Therefore, it is necessary to design an effective algorithm for the above problems.

Although many traditional classification methods can also be used to classify gene expression data in gene chip technology, the classification of gene expression data in gene chip technology is far more challenging than the traditional classification. Overlap is because microarray gene expression data is very different from traditional data. Traditional data usually contain a large number of samples with relatively few attributes, while microarray gene expression data usually contain a large number of gene attributes with a small number of samples. The high-dimensional small sample characteristics of microarray gene expression data pose a severe challenge to traditional classification methods, causing "dimensional disaster" and "overadaptation" problems. In addition, of the thousands of genes in microarray gene expression data, only a few genes are actually closely related to the classification task, and most of the genes show different representations in different categories. There is no practical significance in the differentiation between liver cancer and normal tissue, and the operation of some genes with no practical significance will increase the classification cost and affect the classification accuracy. Therefore, in order to carry out a successful classification of genes, it is necessary to delete the genes that do not play a role in the classification task and to retain the genes closely related to the classification task [5].

2   Design of Big Data Classification Algorithm for Disease Gene Detection

2.1   The Membership Function of Complex Network Is Established

When using complex network technology. The design of the membership function is the key of the whole data classification algorithm, which requires that the membership function can objectively and accurately reflect the uncertainty of the samples in the system. At present, there are many methods to construct membership functions, but there is no general criterion to be followed. In dealing with the actual situation. Usually we need to determine the reasonable membership function according to the experience of specific problems. About membership functions. Many scholars have done some research in this field, but it is mainly based on the distance between the sample and the class center to measure the degree of membership [6]. In general. The basic principle to determine the membership size is based on the relative importance of the class in which the sample is located, or the size of the contribution to the class. The distance between the sample and the class center is one of the criteria to measure the contribution of the sample to the class.

In general, the basic principle to determine the membership degree is based on the relative importance of the class in which the sample is located, or the contribution to the class. The distance between the sample and the class center is one of the criteria to measure the contribution of the sample to the class. At present, in complex network technology, the determination of distance-based membership function is to regard the membership degree of the sample as a function of the distance between the sample and the class center in the feature space.

Let x' be the center of class and r be the radius of class, which is determined by Eq. (1):

$$r = \max_i \|x_i - x'\| \tag{1}$$

When the membership degree is determined according to the distance, the membership degree of each sample in the class is:

$$u(x_i) = 1 - \frac{\|x_i - x'\|}{r} + \delta \tag{2}$$

Where, $\delta > 0$ is a small constant preset to avoid $u(x_i) = 0$.

In the determination of distance-based membership function, the membership degree of the sample is regarded as a linear function of the distance between the sample and the center of the class. However, there is no simple linear relationship between the membership degree of the actual sample and the distance between the sample and the class center. The standard s-type function defined by Zadeh is modified to obtain the membership degree of samples. The membership function transformed from the standard s-type function is as follows:

$$u(d_i, a, b, c) = \begin{cases} 1 \\ 1 - 2\left[\dfrac{d_i - a}{c - a}\right]^2 \\ 2\left[\dfrac{d_i - a}{c - a}\right]^2 \\ 0 \end{cases} \tag{3}$$

Where, d_i is the distance between the sample and the center of the class in which it is located, and parameters a and b are predefined parameters, $b = \frac{a+c}{2}$, Now when $d_i = b$, $u(d_i, a, b, c) = 0.5$.

When determining the membership degree of the sample, not only the distance between the sample and the class center, but also the compactness between the samples should be considered. The tightness between the samples can be measured by how far the samples are from the origin [7]. Therefore, the membership degree of the sample should be determined according to the maximum distance p from the origin of the sample. For the samples distributed in and out of the region, two different methods were used to calculate their membership. The calculation formula of membership degree is as follows:

$$\mu(x_i) = \begin{cases} 0.6 * \left(\dfrac{1 - \rho/f(x_i)}{1 + \rho/f(x_i)} \right) + 0.4, f(x_i) \geq \rho \\ 0.4 * \left(\dfrac{1}{1 + (\rho - f(x_i))} \right), f(x_i) < \rho \end{cases} \tag{4}$$

Where, ρ represents the maximum distance of sample from the origin, and $f(x_i)$ represents the decision function of sample x_i, whose calculation formula is:

$$f(x_i) = \sum_{i=1}^{l} \alpha_i K(x_i, x_j), j = 1, 2, \cdots, l \tag{5}$$

As can be seen from the membership function $\mu(x_i)$ defined by Eq. (4), the farther the sample is from the origin, the greater the membership degree of the sample belongs to this class. At the same time, the influence of the position of the sample in the class was also considered. The membership degree of the sample in the class area was all greater than 0.5. However, the membership degree of the samples located outside the class area was all less than 0.5. As the noise and outfield samples are generally located outside the class region, their membership degree is less than 0.3. Therefore, by setting a threshold value less than 0.5, noise and outfield samples can be removed from the sample set, leaving samples that are beneficial to the construction of the optimal classification hyperplane [8, 9]; Next, the classification of the big data of disease gene detection is realized through the design of the algorithm process of classification of the big data of disease gene detection.

2.2   Design of Big Data Classification Algorithm for Disease Gene Detection

Since the number of genes in the gene expression data of diseases is usually much larger than the number of samples, a general classification algorithm for the gene expression data of diseases is proposed in this paper, which applies the complex network technology to the classification algorithm. In this algorithm, the classification algorithm is firstly screened based on the output inconsistency measure, then the data of the screened classification algorithm are classified by majority voting method, and finally the gene expression data are classified by the classified algorithm [10, 11]. In summary, the steps of the big data classification algorithm for disease gene detection based on complex network technology are described as follows:

Step 1: In order to avoid the generation of overfitting classification model of big data of disease gene detection, the big data samples of disease gene detection were divided by 50% cross-validation (one fold for testing, two thirds for training, and the remaining one third for validation);

Step 2: Initializes the big data set of disease gene detection, generates the initial solution randomly and encodes it;

Step 3: Initialize the classification algorithm parameters, including the number of large data sources SN for disease gene detection, the position of the i largest data source for disease gene detection, the maximum number of iterations MCN, and judge whether the solution falls into the local minimum value (threshold value limit) and reaches the tolerance rate ε;

Step 4: The solution obtained by Step 2 is decoded to obtain the input weight, hidden layer bias and training sample weight of the classification algorithm. The complex network is trained on the training set by these parameters, and the corresponding fitness value and the corresponding network output weight norm are calculated on the verification set according to formula (6) and (7) after obtaining the output weight;

$$AUC = \frac{1}{m \cdot n} \sum_{i=1}^{m} \sum_{k=1}^{n} I(x_i > y_k) \tag{6}$$

Type in the,$I(\cdot)$ is an index function, $I(true) = 1$ and $I(false) = 0$, m. n is the sample number of small class and large class respectively, x_i and y_k respectively represent the output of the algorithm for the i th small class sample and the k th large class sample.

$$v'_{i,j} = \begin{cases} v'_{i,j}, iff\left(v'_{i,j}\right) - f(x_{i,j}) > \varepsilon \cdot f(x_{i,j}) \\ v'_{i,j}, if\left|f\left(v'_{i,j}\right) - f(x_{i,j})\right| < \varepsilon \cdot f(x_{i,j}) and \left\|\beta_{v'_{i,j}}\right\| < \left\|\beta_{x_{i,j}}\right\| \\ x_{i,j}, \end{cases} \tag{7}$$

In the formula, $f\left(v'_{i,j}\right)$ and $f(x_{i,j})$ represent the location $v'_{i,j}$ and the original location $x_{i,j}$ of the disease gene detection big data source respectively. $\beta_{v'_{i,j}}$. $\beta_{x_{i,j}}$ represents the norm of the network output weights of the i disease gene detection big data source at the new position $v'_{i,j}$ and the original position $x_{i,j}$ respectively.

Step 5: Number of iterations plus one;

Step 6: The hirer USES the formula to update each solution;

Step 7: According to the input weights, hidden layer bias and training sample weights obtained by Step 4, the fitness function value corresponding to each solution and the corresponding network output weight norm were calculated according to formula (6) and (7). Update the solution with greedy choice strategy;

Step 8: Using roulette selection method, the probability of each solution selected is calculated;

Step 9: The following bees update the solutions in Step 6 according to different probability values;

Step 10: The solution of Step 9 is decoded to obtain the input weight, hidden layer bias and training sample weight of the complex network. The complex network is trained on the training set by these parameters, and the fitness function value corresponding to each solution is calculated on the verification set according to formula (6) and (7) after the output weight is obtained;

Step 11: Update the solution according to Eq. (7);

Step 12: To judge whether the algorithm meets the termination condition, if so, go to Step 9; otherwise, go to Step 8;

Step 13: Bee detection stage: If the solution corresponding to the fitness value of continuous limit generation did not change, then according to the type (8) to generate data processing, calculation of fitness value;

$$x_{i,G} = \left[x_{i,G}^1, x_{i,G}^2, \cdots x_{i,G}^D \right] \tag{8}$$

Step 14: Check whether the algorithm reaches the maximum number of iterations, if so, go to step 15; otherwise, return to step 5 to continue running;

Step 15: The optimal solution is decoded to obtain the optimal input weight, hidden layer bias and training sample weight

Step 16: The data classification model is tested on the test set with the optimal input weight, hidden layer bias and training sample weight, and the final classification result is obtained.

To sum up, the complex network technology is applied to the membership function of the big data of disease gene detection, and the membership function of the complex network of the big data of disease gene detection is established. Combined with the specific implementation steps of the data classification algorithm, the big data classification algorithm of disease gene detection was designed, and the classification of the big data of disease gene detection was realized [12, 13].

3   Experimental Contrastive Analysis

3.1   Data Set and Parameter Setting

The liver cancer gene sequence data of ubiquitination sites were obtained from the prokaryotic ubiquitin-like liver cancer gene database (GenBank). GenBank collected by both experiment and obtain the ubiquitin site location of ubiquitin gene substrates, and the substrate gene structure, functional annotation information can be done (original training sample set and test set), are collected by the experimental determination of the unknown exact ubiquitin locus of ubiquitin gene (can be done without class mark sample set). Follow these steps to build a non-redundant original training sample set, an independent test set and an uncategorized labeled sample set, respectively:

Step1:The original training sample set was constructed by extracting ubiquitin gene substrates with ubiquitin labeling sites from GenBank database. The remaining 180 ubiquitin gene sequences containing 213 ubiquitin sites were redundant with the cd-hit tool, and the similarity threshold of sequence alignment was set to 30% when the CD-HIT tool was used to eliminate redundancy. After de redundancy, the original training samples were composed of 162 gene substrates containing 183 ubiquitination sites and 2258 non ubiquitination sites;

Step2:Twenty liver cancer genes were randomly selected from the de redundant data set as independent test set, which contains 29 ubiquitination sites and 408 non ubiquitination sites;

Step3:From the GenBank database, 1116 gene substrates without specific ubiquitin sites annotated (each liver cancer gene has a known lysine residue) were extracted for the unlabeled sample set. The 1,116 gene substrates encode 14,955 peptides;

From the above data processing process, it is obvious that the positive and negative sample sets of the training set and the independent test set have the characteristics of unbalanced class distribution. In order to simulate the real experimental environment, the following prediction model is built on the original unbalanced sample set, rather than on the artificially balanced sample set. To ensure that the optimal model is obtained after training, the number of hidden layer nodes $\tilde{N}$ and the control parameter C when the AUC is maximized are selected by using a reserved cross validation. The value range of $\tilde{N}$ for the number of nodes in the equilibrium normal and hidden layers of WELM is $\{2 - 12,..., 212\}$ and $\{10,..., 2000\}$, respectively. One peptide in the training set was used in turn to test all other peptides until all of them had been tested. The optimal training model was tested on an independent test set and compared with PUPs, IPUP and imp-pup prediction models. The prediction performance of each model was measured by the classification accuracy of positive samples (SN), negative samples (SP), Matthews coefficient (MCC), the area under the characteristic curve of the subject (AUC) and the total classification accuracy (ACC).

$$MCC = \frac{TP * TN - FN * FP}{\sqrt{(TP + FN) * (TN + FP) * (TP + FP) * (TN + FN)}} \tag{9}$$

3.2 Set up Experimental Data

The disease gene expression data studied in this paper were mainly from the liver cancer data set. Table 1 shows the relevant information of the liver cancer data set.

Table 1. Data sets of experimental liver cancer

The data set	Hepatocellular carcinoma data set	Breast cancer data set	Colon cancer data set
Number of categories	2	2	2
The number of genes	7129	10	2000
Category label	ALL AML	/	Tumor Normal
Number of samples	47/25	683	40/22

In 1999, Golub et al. published the experimental results of the acute leukemia data set in Science. The dataset included a total of 72 acute leukemia samples, each containing 7,129 gene expression data. Of the 7,129 gene expression data, 47 samples were diagnosed with acute lymphoblastic leukemia, including 9 all-t samples and 38 all-b samples. Of these, 25 were diagnosed with acute myelogenous leukemia.

In the experiment of this paper, 38 cases were selected from the training set and 34 from the test set. In ALL test sets, 2 all-t samples, 14 AML samples and 18 all-b samples were set. The adjusted samples included 4 all-t samples, 14 AML samples and 16 all-b samples. In the training set, 8 all-t samples, 11 AML samples and 21 all-b samples were set.

3.3   Determine the Optimal Parameter

For the proposed integrated WELM prediction model, the optimal AUC with the variation of the hidden layer node parameter $\tilde{N}$ was tested on the expanded training set and the independent test set by using the residual cross-validation method, and the $\tilde{N}$ corresponding to the optimal model was found. The curve diagram of the optimal AUC with the change of $\tilde{N}$ on the training set and the independent test set is shown in Fig. 1. The optimal hidden layer node number $\tilde{N}$ is 1800–1960 (the interval is 10). As can be seen from Fig. 1, when the integrated WELM obtained the optimal AUC value of 0.7312 on the independent test set, the equilibrium parameter C was e^{-2}, and the number of network hidden layer nodes $\tilde{N}$ was 1840.

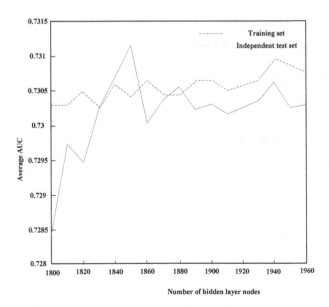

Fig. 1. Curve of UC with the number of hidden nodes in the network

3.4   Analysis of Experimental Results

In order to verify the classification performance of the big data classification algorithm for disease gene detection based on complex network technology. The data sets of liver cancer, breast cancer and colon cancer were used as the data sets of gene expression. And compared with the big data classification algorithm of disease gene detection based on support vector machine and the big data classification algorithm of disease gene detection based on rough set. In order to obtain better classification results, characteristics of three experimental data sets, namely liver cancer data set, breast cancer data set and colon cancer data set, were firstly selected to eliminate the influence of redundant data on experimental results. The input weights and the threshold values of the hidden nodes were randomly selected, corresponding to the data sets of liver cancer, breast cancer and colon cancer. The number of nodes in the hidden layer of complex network technology in these three data sets is set to 30, 3, 3 and 5, respectively. In order to avoid the serious error caused by the instability of complex network technology, the experiment was repeated 30 times during the integration of different Numbers of disease genes, and the average value was calculated. The experiment result is shown in Fig. 2, 3 and 4.

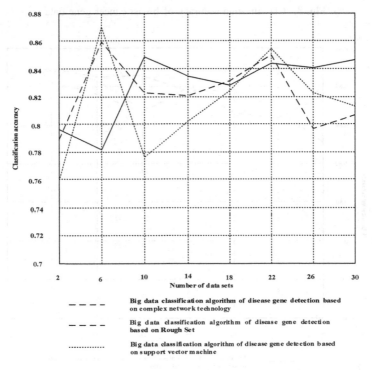

Fig. 2. Classification accuracy of liver cancer gene data set

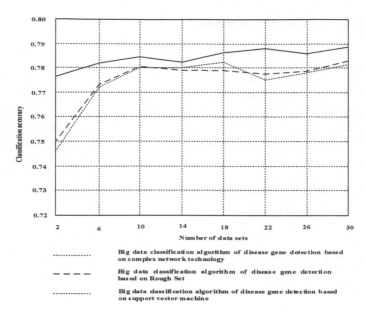

Fig. 3. Classification accuracy of breast cancer gene data

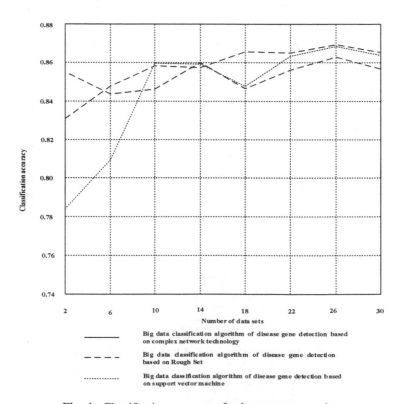

Fig. 4. Classification accuracy of colon cancer gene data

Analysis of Fig. 4 shows that when the number of data sets reaches 30, the average classification accuracy of liver cancer gene data based on the support vector machine-based disease gene detection big data classification algorithm is 0.8, and the liver cancer gene data based on the rough set disease gene detection big data classification algorithm the average classification accuracy is 0.82, while the average classification accuracy of liver cancer gene data based on the complex network technology-based disease gene detection big data classification algorithm is 0.84; The average classification accuracy of breast cancer gene data of the disease gene detection big data classification algorithm based on support vector machine is 0.77, and the average classification accuracy of breast cancer gene data of the disease gene detection big data classification algorithm based on rough set is 0.775, and based on complex network technology the average classification accuracy of breast cancer gene data of the disease gene detection big data classification algorithm is 0.785; The average classification accuracy of colon cancer gene data of the disease gene detection big data classification algorithm based on support vector machine is 0.84, the average classification accuracy of colon cancer gene data of the disease gene detection big data classification algorithm based on rough set is 0.85, and it is based on complex network technology the average classification accuracy of colon cancer gene data of the disease gene detection big data classification algorithm is 0.86; it can be seen that the disease gene detection big data classification algorithm based on complex network technology has a higher classification accuracy of gene data.

In order to verify the effectiveness of the big data classification algorithm for disease gene detection, the variance significance F of various classification algorithms was tested, which was subject to $F = \frac{S_1^2}{S_2^2} \sim F(n-1, m-1)$. Where, S_1 and S_2 respectively represent the sample variance of the classification algorithm, and n and m respectively represent the number of samples. In this paper, the values are 30. For each data set, there are 1, 2,..., 30 disease gene detection big data classification algorithm has been trained, so the degree of freedom is 29. All the data were statistically analyzed and the results are shown in Table 2.

Table 2. Statistical tests of classification algorithms in different gene data sets

The data set	Classification method	The average	The variance	The F value
Liver cancer gene data set	DE-ELM	0.7852	1.5363E-05	
	Bagging	0.7783	8.1556E-05	5.3086
	Boosting	0.7780	6.8460E-05	4.4561
Breast cancer gene data set	DE-ELM	0.8276	4.2114E-04	
	Bagging	0.8070	9.8549E-04	2.3401
	Boosting	0.8053	9.8172E-04	2.3311
Colon cancer gene data set	DE-ELM	0.8546	2.5550E-04	
	Bagging	0.8423	6.6964E-04	2.6209
	Boosting	0.8404	6.0172E-04	2.3550

When the significance level is 0.05, it can be obtained by looking up the F distribution table $F_{0.05}(29, 30) = 1.85, F_{0.05}(29, 24) = 1.9$. Moreover, F values obtained by calculation are all greater than 1.9, so the big data classification algorithm of disease gene detection based on complex network technology proposed in this paper has significant advantages. Therefore, the following conclusions are drawn: Compared with the other two classification algorithms, the big data classification algorithm of disease gene detection based on complex network technology can achieve the same classification accuracy through fewer big data number of disease gene detection, and the heterogeneous classification can tend to be stable in a shorter time.

4 Conclusion

In this paper, a big data classification algorithm for disease gene detection based on complex network technology is proposed. On the basis of complex network technology, the membership function of complex network is constructed, and the data classification algorithm is designed in combination with the process design of the classification algorithm for big data of disease gene detection, so as to realize the classification of big data of disease gene detection. The results show that the big data classification algorithm of disease gene detection based on complex network technology has higher classification accuracy.

5 Fund Projects

Scientific Research Project of Hunan Provincial Department of Education (18C1152)

References

1. Lu, H., Yang, L., Yan, K., et al.: A cost-sensitive rotation forest algorithm for gene expression data classification. Neurocomputing **228**, 270–276 (2017)
2. Suarez-Cetrulo, A.L., Cervantes, A.: An online classification algorithm for large scale data streams: iGNGSVM. Neurocomputing **262**, 67–76 (2017)
3. Xing, W., Bei, Y.: Medical health big data classification based on KNN classification algorithm. IEEE Access **PP(99)**, 1 (2019)
4. Ahsen, M.E., Ayvaci, M.U.S., Raghunathan, S.: When algorithmic predictions use human-generated data: a bias-aware classification algorithm for breast cancer diagnosis. Inf. Syst. Res. **30**(1), 97–116 (2019)
5. Pourpanan, F., Lim, C.P., Wang, X., et al.: A hybrid model of fuzzy min-max and brain storm optimization for feature selection and data classification. Neurocomputing **333**, 440–451 (2019)
6. Shuai, L., Mengye, L., Hanshuang, L., Yongchun, Z.: Prediction of gene expression patterns with generalized linear regression model. Front. Genet. **10**, 120 (2019)
7. Jiang, C., Li, Y.: Health big data classification using improved radial basis function neural network and nearest neighbor propagation algorithm. IEEE Access **99**, 1 (2019)

8. Jian, S., Yunlong, D., Bo, L., et al.: Optimization algorithm of redundant data classification in distributed database scenarios. Boletin Tecnico/Tech. Bull. **55**(16), 54–61 (2017)
9. Liu, S., Lu, M., Li, H., et al.: Prediction of gene expression patterns with generalized linear regression model. Front. Genet. **10**, 120 (2019)
10. Padillo, F., Luna, J.M., Ventura, S.: A grammar-guided genetic programing algorithm for associative classification in big data. Cogn. Comput. **11**(3), 331–346 (2019)
11. Liu, S., Yang, G. (eds.): ADHIP 2018. LNICST, vol. 279, pp. 1–594. Springer, Cham (2019). https://doi.org/10.1007/978-3-030-19086-6
12. Owsiski, J.W., Kacprzyk, J., Shyrai, S., et al.: A heuristic algorithm of possibilistic clustering with partial supervision for classification of the intuitionistic fuzzy Data. J. Multiple-Valued Logic Soft Comput. **31**(4), 399–423 (2018)
13. Liu, S., Liu, D., Srivastava, G., et al.: Overview and methods of correlation filter algorithms in object tracking. Complex Intell. Syst. (2020). https://doi.org/10.1007/s40747-020-00161-4

Text Classification Feature Extraction Method Based on Deep Learning for Unbalanced Data Sets

Li Lin[1(✉)] and Shu-xin Guo[2]

[1] School of Computer Engineering, Jimei University, Xiamen 361021, China
xd220210@163.com
[2] Jilin University of Finance and Economics, Changchun 130117, China

Abstract. In order to fully realize the classified search of text data information, a text classification feature extraction method for imbalanced data sets based on deep learning is proposed. With the help of trestle automatic encoder and depth confidence network, the preliminary definition of text semantic category conditions is completed, and the text semantic classification processing based on depth learning algorithm is realized. On this basis, pre-processing and debugging of text parameters are implemented, and the dimensionality reduction standards related to the text features of the data set to be extracted are established through the expression of the characteristic behavior. The experimental results show that with the application of the new classification feature extraction method, the number of correctly classified documents starts to increase substantially, which meets the practical application requirements for the classification and search of text data information.

Keywords: Deep learning · Unbalanced data sets · Text features · Classification and extraction

1 Introduction

Deep learning is a new research direction in the field of machine learning. It is introduced into machine learning to make it closer to the original goal artificial intelligence. Deep learning is to learn the inherent laws and representation levels of sample data. The information obtained during these learning processes is of great help to the interpretation of data such as text, images, and sound. Its ultimate goal is to enable the machine to be able to analyze and learn like human beings, and to recognize data such as words, images and sounds. Deep learning is a complex machine learning algorithm, which has achieved much better results in speech and image recognition than previous related technologies. Deep learning has made many achievements in search technology, data mining, machine learning, machine translation, natural language processing, multimedia learning, voice, recommendation and personalized technology [1, 2]. Deep learning enables machines to imitate human activities such as audio-visual and thinking, solves many complex pattern recognition problems, and makes great progress in AI related technologies. Deep learning is a kind of machine learning, and machine learning is the only way to realize artificial intelligence. The concept of deep learning stems from the study of artificial neural networks. A multi-layer perceptron with multiple hidden layers is a deep learning structure. Deep learning combines low-level features to form more abstract high-level representation attribute categories or features to discover the distributed feature representation of data. The motivation for studying deep learning is to build a neural network that simulates the human brain for analysis and learning. It mimics the mechanism of the human brain to interpret data, such as images, sounds, and text.

Text classification uses computers to automatically classify and mark text sets (or other entities or objects) according to a certain classification system or standard. According to a set of labeled training documents, it finds the relationship model between document features and document categories, and then uses the relationship model to judge the new document categories. Text classification has gradually changed from knowledge-based method to statistical and machine learning based method [3]. Text classification generally includes text expression, classifier selection and training, classification result evaluation and feedback, etc. The text expression can be subdivided into text preprocessing, indexing and statistics, feature extraction and other steps. The overall function module of the text classification system is: preprocessing: formatting the original corpus into the same format for subsequent unified processing; indexing: decomposing the document into basic processing units, while reducing the cost of subsequent processing; statistics: word frequency statistics, the correlation probability of items (words, concepts) and classification; feature extraction: extracting the characteristics reflecting the document theme from the document Feature; classifier: training of classifier; evaluation: analysis of test results of classifier.

Therefore, this paper proposes a deep learning based text classification feature extraction method for imbalanced data sets. Firstly, this paper introduces the research status and significance of text classification, and expounds the definition, method and process of text classification. Secondly, the text classification algorithm is described, in which KNN and SVM classification algorithm are introduced in detail. The common feature selection methods are introduced and their advantages and disadvantages are analyzed. Thirdly, on the basis of the above, one of the feature selection methods CHL statistical method is improved. Finally, the improved method is verified on the basis of experiments, which shows the feasibility of the improved method.

2 Text Semantic Classification Based on Deep Learning

Text semantic classification is the basic processing link of the application of text classification feature extraction method in unbalanced data set. Under the support of deep learning algorithm, the specific operation process is as follows.

2.1 Stacked Automatic Encoder

The trestle type automatic encoder is a neural network model with multiple hidden layer neurons. A typical trestle type automatic encoder structure is shown in Fig. 1.

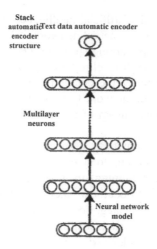

Fig. 1. Structure diagram of the stack type automatic encoder

The trestle type automatic encoder can be regarded as a combination of multiple text data automatic encoders. With the support of the deep learning algorithm, in order to achieve the learning speed of the text data in the unbalanced data set, and to initialize the weight value to a better eigenvalue state, in order to obtain the local optimal value close to the global optimal value, the text classification features to be encoded must be pre trained. In the process of pre training, starting from the input layer, the adjacent two layers are regarded as a separate set of restricted sampling samples for training, and the output of the lower layer is regarded as the input of the higher layer, so as to complete the initialization of the weight [4]. In the process of training, since the deep learning algorithm still has the problem of text gradient dissipation, the unbalanced data set is expanded. At this time, the process of bottom-up propagation can be regarded as encoding the input data, while the process of downward propagation can be regarded as the process of decoding. At the same time, in order to prevent the classification features from overfitting the data and affecting the promotion ability, a certain amount of noise will be added to the weights during fine-tuning.

2.2   Deep Confidence Network

The deep belief network is a directed graph model. With the support of a stack autoencoder, you can specify the classification feature structure of the text information in the query imbalanced data set. The weights between variables derive the state of the hidden variables as shown in Fig. 2.

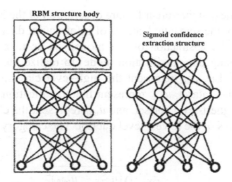

Fig. 2. Structure of depth confidence network

According to Fig. 2, the top two layers of deep learning mechanism can be regarded as an RBM structure subject, while the network below the top layer can be regarded as a directed sigmoid confidence extraction structure. One of the problems of deep confidence networks is how to initialize the feature weights of the text of the data set. It is usually very difficult to optimize the nonlinear deep network with multiple hidden layers, and only the initialization can get better weights, and the network can avoid Fall into a local minimum. The feature extraction method of information initialization based on deep learning mechanism regards two adjacent layers as RBM structure, and trains the network from bottom to top [5, 6]. If the classification characteristics of text data do not change, after unsupervised pre training, the whole network uses supervised learning to fine tune, and finally get the trained DBN data set information. The process of DBN information fine-tuning is similar to that of depth auto encoder, which is also fine-tuning after expansion.

2.3   Text Semantic Category Definition

Text semantic classification, as an important means of data set feature analysis, is widely used in information search, information annotation and other occasions. In some professional fields, unbalanced data set classification has achieved high accuracy, such as big data information recognition, handwritten digit recognition. However, the classification of text features is still a challenging issue due to the reasons for changes in transmission traffic, network transportation rate, data set flexibility, and the complexity of the data set content itself. The semantic-based text classification method starts from the imbalanced data set itself, uses a specific feature extraction method to extract the semantic features in the data, and gradually builds a semantic hierarchy of information on this basis, and finally uses the learned high-level information semantic features sort. The semantic features of data sets are hierarchical, among which the low-level information features are of low abstraction, high correlation with the information data content itself, high-level information features are of high abstraction, and low correlation with the information data content. Therefore, it may be considered to establish a hierarchical learning model, and use unsupervised learning to learn information data to obtain the characteristics of the data itself. As the level increases, the learned features become a higher-order feature description of the input data. Using deep learning model to learn the semantic features of information data, through the increase of depth to improve the abstraction of the model, so as to establish the semantic hierarchy, and then using the classification model to realize the semantic based classification of data sets. Let $s_{\max}$ represent the high-level classification features of the input information of the dataset, $u_{\max}$ represent the classification hierarchy features of the semantics of the highest-level dataset, and $u_{\min}$ represent the classification hierarchy features of the semantics of the lowest-level dataset. The category definition expression of text semantics is:

$$R = \frac{s_{\max}}{2} \sqrt{\frac{y(u_{\max} - u_{\min})^{\varepsilon}}{p^2 - q^2}} \tag{1}$$

Among them, y represents the average transport rate of the unbalanced text data information in the deep learning network, p represents the text feature parameter of the data set information itself, and q represents the text feature parameter in the given semantic model.

3 Text Classification Feature Extraction from Unbalanced Dataset

With the support of the deep learning principle of text semantic classification, according to the application process of text preprocessing, text feature representation, and dimensionality reduction of features to be extracted, the smooth application of text classification feature extraction method for imbalanced data sets is completed.

3.1 Text Preprocessing

From the perspective of the text classification process, regardless of whether Chinese text classification or English text classification is performed, the text used must be preprocessed to remove some useless information. This is to reduce the complexity and complexity of the next steps. The burden of calculation. First, we need to segment words (also known as segmentation). If the text to be classified is Chinese, word segmentation is an essential step, which is to divide the consecutive sentences into individual words (because there is no obvious segmentation mark between a word and a word in Chinese, and the word is the smallest unit in the text that can be used independently). For example, the sentence "I am a student" should be divided into "I/Yes/One/Student". However, if the text to be classified is English, there is no need for word segmentation because the space and punctuation in English have already played a role in word segmentation. The quality of word segmentation has a great impact on the performance of the whole text classification system, which is mainly because the text information used in the later process is all preprocessed text. If the word segmentation is not done well, the whole training text set will be inaccurate, resulting in the accuracy rate of the classification is reduced.

In terms of current development, word segmentation methods include dictionary-based methods, statistical-based methods, and hybrid methods. Among them, the dictionary-based method generally needs to meet the three conditions of the word segmentation dictionary, the order of scanned text, and the matching principle. The order of scanning text generally has three scanning methods: forward, reverse, and bidirectional. The matching principles generally include three methods: forward matching, reverse matching, and bidirectional matching. The word segmentation method based on statistics uses word and word occurrence probability as the basis of word segmentation [7, 8]. The common methods include hidden Markov model, n-ary grammar model and maximum entropy model. The mixed method is to combine two or more methods to segment words. There are two standards of word segmentation: segmentation speed and segmentation precision. Obviously, the accuracy of segmentation is the most important, because it directly determines the quality of classification. In terms of current development, the main limitation of segmentation is that segmentation efficiency is low and the effect of disambiguation is not good.

3.2 Text Feature Representation

At present, there are vector space model, Boolean model and probability retrieval model.

(1) Vector space model

The vector space model defines the correlation between documents as a similarity between them. It believes that the more similar a document is to a user query, the more relevant it is to a user query. Deep belief networks represent documents as a vector in a high-dimensional word space. Each dimension in the vector represents the weight of the corresponding feature in the document. The measure of similarity is the angle cosine.

(2) Boolean model

The Boolean model is a special case of the vector space model. The classic Boolean model can only be used to calculate the correlation between user queries and documents in information retrieval, but it cannot use the model to calculate the deeper similarity of two documents and cannot be used in more text processing. At present, scholars have proposed a variety of extended Boolean models, so that the correlation is no longer simply 0 and 1, but becomes a number between [0, 1].

(3) Probability retrieval model

Both the Boolean model and the vector space model treat the document representation terms as independent items, ignoring the correlation between the representation terms. The probabilistic model takes into account the internal relations between terms and documents, and uses the probability dependence between the terms and between the terms and the documents to retrieve information.

The computer does not have human intelligence. After reading an article, people can have a vague understanding of the content of the article according to their own understanding ability. Fundamentally, the computer can only know 0 and 1. So, like all machine learning problems, if you want the computer to automatically classify text, you need to express the text as a feature, such as words, words, n-grams, phrases, concepts, etc. Obviously, this will lose a lot of information about the content of the article, but this representation can formalize the processing of the text and can achieve better results in text classification.

(a) word

Word is the most widely used feature in text classification. For a document, the most intuitive way is to use words and phrases as the characteristics of the text. For English articles, each word has been separated by a space, and the feature words can be obtained directly. However, due to the changes of word form in English, such as the singular and plural numbers of nouns, the tense changes of verbs, the prefixes and suffixes of words, a process of stem extraction is needed. For Chinese, because there is no pause between words, we need to use dictionaries and special word segmentation techniques.

(b) N-gram items

N-gram is a text representation independent of language. Because the n-gram string method does not consider whether the semantic unit of the text is word, word or phrase at all, but regards the whole text as a string composed of different characters, so it can conveniently represent all kinds of language documents including Chinese and Arabic. For Chinese, n-gram items are generally composed of adjacent words. For English, n-gram can be composed of adjacent words or letters. The N-gram item is a feature of the document, which can avoid huge dictionaries and complicated word segmentation programs. In general, using the same classification method, the effect of word-based text classification is no better than that based on N-gram items. In the case of a small number of features, the classification effect based on N-gram items is better than that based on words. The obvious feature of the N-gram method is the large amount of calculation, so there are not many applications.

(c) Phrase

Phrase notation is widely used in the field of text classification. This notation improves the semantic content of feature vectors and restores some useful information thrown away by word notation. But its expressive power is not obvious. This representation method reduces the statistical quality of feature vectors and makes them more sparse, which makes it difficult for machine learning method to extract statistical characteristics for classification.

(d) Concept

The concept has a higher abstraction. A concept can correspond to a word in a text, or it can correspond to several semantically related words. Using concept space can greatly reduce the dimension of feature space, thus reducing the training time of classifier and the time for similarity comparison. Therefore, concept-based text classification is based on word-based classification in terms of time efficiency; at the same time, because a concept can merge multiple keywords with synonymous relationships, it can avoid an important classification feature due to the dispersion of key times and weaken The weight of the classification; again, mapping one keyword to multiple concepts can avoid the feature ambiguity caused by using only keywords as features. However, because of the complexity of the concept, it will cost unimaginable human and material resources. Secondly, the establishment of concept depends on experts or domain experts, so it has strong subjectivity. Therefore, in the practical application, the concept based application is not ideal.

3.3 Dimensionality Reduction of Features to Be Extracted

Text feature dimensionality reduction is a key step in text classification. An important problem in text mining is the existence of high-dimensional feature space, which is composed of words or phrases in the text. Many traditional methods are difficult to deal with. High-dimensional feature sets are not necessarily all important and beneficial to machine learning, but also increase the burden of machine learning. Without affecting the accuracy of feature classification, it is necessary to reduce the number of high-dimensional features in the text description space. This process is feature dimensionality reduction [9, 10].

The traditional text feature dimensionality reduction method uses a single evaluation function to calculate the feature weight [11–13]. Because the traditional feature evaluation function only pays attention to the single aspect of weight calculation and ignores other important factors, the traditional feature dimension reduction method is not effective. In the process of text mining, it is necessary to consider the influence of multiple factors, effectively combine feature extraction and feature selection, and jointly reduce the feature dimension. The combined feature dimensionality reduction method presented in this paper not only combines the feature items with similar contributions to the classification category into new feature items, but this feature extraction operation has the method of using attribute reduction on the merged feature items. Selected feature selection operation. In this paper, we first use the pattern aggregation theory to fuse the features that have similar contribution to the classification, and then use the method of rough set decision table to connect feature selection with text classification, that is, attribute reduction is carried out according to the importance of features in classification. This method not only reduces the complexity of training, greatly reduces the dimension of feature, but also improves the accuracy of feature dimension reduction.

4 Practical Ability Testing

4.1 Experimental Data and Steps

In order to verify the practical application value of text classification feature extraction method based on deep learning unbalanced data set, the following practical detection experiments are designed. Before text classification, the first step is to prepare training text set and test text set. All text set data used in this experiment are from http://www.nlp.org.cn/ website, from which ten categories of computer, environment, transportation, economy, sports, medicine, education, art, politics and military are extracted for experiment. There are 1887 training document sets and 934 test document sets. The distribution of categories is shown in Tables 1 and 2.

Table 1. Distribution of training concentration categories

Class alias	Computer	Traffic	Economic	Sports	Medicine
Number of training	135	145	215	301	135
Class alias	Surroundings	Education	Art	Political	Military
Number of training	136	150	165	340	165

Table 2. Distribution of test set categories

Category name	Computer	Traffic	Economic	Physical education	Medicine
Number of tests	65	70	108	150	70
Category name	Surroundings	Education	Art	Political	Military
Number of tests	66	73	82	167	83

Taking the text feature classification of unbalanced data sets as an example, the whole experimental process mainly includes the following steps:

First, the selected training text set is segmented.

Secondly, SVM algorithm and KNN algorithm are used to train the segmentation results. Some parameters should be selected during training. For example, feature selection, feature dimension and weight calculation function. These parameters will affect the result of classification. In this process, we mainly test the improved feature selection method.

In the training process of this experiment, the linear kernel function is selected for the kernel function of SVM. The other parameter selections are the same in KNN and SVM, respectively, the feature selection method selects CHI statistical method; the feature dimension selects 1000; the weight calculation selects the established support function.

Third, a classification model (classifier) is formed on the basis of the above.

Fourth, set the parameters of the new test text set. If KNN is used for classification, select 35 for K value, and then classify.

Fifth, after the classification is completed, view the classification results. Take the result of art.

4.2 Number of Documents

This experiment mainly uses the evaluation methods of recall rate (T) and precision rate (P) to evaluate the documents in the test document set, and T and P are expressed by formula 2 and formula 3. In addition to T and P, this experiment also used histogram and confusion matrix to evaluate it.

$$T = \frac{\text{Number of documents classified correctly in a category}}{\text{Total number of documents in this category}} \times 100\% \qquad (2)$$

$$P = \frac{\text{Number of documents classified correctly in a category}}{\text{Number of documents assigned to this category}} \times 100\% \qquad (3)$$

Experimental results 1: the experimental results are analyzed from the number of documents classified by each category. Table 3 compares the number of classified documents before and after improvement.

Table 3. Comparison of the number of documents after KNN classification

Category	The number of documents classified as belonging to this category			The number of correctly classified documents in documents belonging to a certain category			The total number of documents belonging to this category
	Tradition CHI	Improve CHI_1	Improve CHI_2	Tradition CHI	Improve CHI_1	Improve CHI_2	
Traffic	63	65	65	61	63	63	70
Surroundings	57	64	58	52	54	52	66
Computer	58	57	57	57	56	56	65
Education	72	69	68	66	65	65	73
Economic	127	124	121	101	103	102	108
Art	76	80	82	73	77	76	82
Physical education	159	154	154	149	146	147	150
Political	199	199	194	157	161	159	167
Medicine	67	64	69	65	63	65	70
Military	62	58	60	53	53	53	83

It can be seen from Table 3 that the number of documents correctly classified in the improved method is better than that in the traditional method, and the number of some categories is lower than that in the traditional method, for example, in the improved chic method, the number of documents correctly classified in the documents belonging to the categories of computer, education, sports and medicine; in the improved chic method, the number of documents belonging to the counting the number of documents correctly classified in computer, education and sports documents. This may happen because the selected test document appears too little in the training document set. But from the overall classification results, the number of correctly classified documents has increased.

4.3   Recovery and Precision Rates

Experimental result 2: the experimental results are analyzed from the recall and precision of each category. From the number of document classification in Table 3, the recall and precision of each category can be calculated. Take the traffic categories in the traditional Chi method as an example, $T = \frac{61}{70} \times 100\% = 87.143\%$, $P = \frac{61}{63} \times 100\% = 96.825\%$, Table 4 compares the recall rate and precision rate before and after improvement.

Table 4. Comparison of recall rate and recall rate after KNN classification

Category traffic	Tradition CHI		Improvement CHI_1		Improvement CHI_2	
	T	P	T	P	T	P
Environmental science	87.143%	96.825%	90.000%	96.923%	90.000%	96.923%
Computer	78.788%	91.228%	81.818%	84.375%	78.788%	89.655%
Education	87.692%	98.276%	86.154%	98.246%	86.154%	98.246%
Economics	90.411%	91.667%	89.041%	94.203%	89.041%	95.588%
Art	93.519%	79.5285%	95.370%	83.065%	94.444%	84.298%
Sports	89.024%	96.053%	93.902%	96.250%	92.683%	92.683%
Politics	97.333%	93.711%	97.333%	94.805%	98.000%	95.455%
Medicine	94.021%	78.894%	96.407%	80.905%	95.210%	81.959%
Military	92.857%	97.015%	90.000%	98.438%	92.857%	94.203%
Category	63.855%	85.484%	63.855%	91.379%	63.855%	88.333%

It can be seen from Table 4 that the improved classification effect is better than the traditional method as a whole, and the recall rate and precision rate of some categories are lower than the traditional method. For example, in the improved CHIC method, the computer, the recall rate of education and medicine category, the precision rate of environment and computer category; in the improvement CHI: method, the recall rate of computer and education category, the precision rate of environment, computer, art and medicine category. But from the overall classification results, the recall rate and precision rate have been improved.

5   Conclusion

With the development of the network, a lot of information appears on the network. How to find the information we need quickly from the network has become more and more important [14, 15]. Text classification is one of the methods to solve this problem. Therefore, this paper proposes a deep learning based text classification feature extraction method for imbalanced data sets, and defines the text semantic categories through the trestle automatic encoder and deep confidence network. According to the application process of text preprocessing, text feature representation and feature dimension reduction, the smooth application of text classification feature extraction method is completed. The experimental results show that the number of documents correctly classified by this method is more, and the recall rate and recall rate of KNN are improved.

References

1. Chen, W., Liu, X., Lu, M.: Feature extraction of deep topic model for multi-label text classification. Pattern Recogn. Artif. Intell. **32**(9), 785–792 (2019)
2. Wang, Y., He, Y., Zou, H., et al.: WordNG-Vec: a word vector model applied to CNN text classification. J. Chin. Comput. Syst. **40**(03), 37–40 (2019)
3. Song, C., Chen, X., Niu, Q.: Improved feature selection method based on CHI for text categorization. Microelectron. Comput. **35**(09), 80–84 (2018)
4. Han, D., Wang, C., Xiao, M.: Multi-label text classification method based on rotating forest and AdaBoost classifier. Appl. Res. Comput. **35**(12), 141–144 (2018)
5. Yin, Y., Yang, W., Yang, H., et al.: KNN text classification algorithm based on search improvement. Comput. Eng. Des. **39**(09), 231–236 (2018)
6. Tong, X., Guo, P., Xu, P., et al.: Fusing hyperspectral features and image deep features for classification and retrieval of meat. Sci. Technol. Food Ind. **39**(23), 261–266+272 (2018)
7. Xuan, Q., Fang, B., Wang, J., et al.: Pearl multi-feature classification method based on support vector machine. J. Zhejiang Univ. Technol. **46**(05), 5–12 (2018)
8. Hua, S., Hu, S., Gao, L., et al.: Research on fish counting and species recognition system of fishway based on image feature extraction. Water Power **44**(12), 90–9 +128 (2018)
9. Lv, W., Deng, W., Chu, J., et al.: Arrhythmia classification based on feature selection method of S-transform. J. Data Acquis. Process. **33**(2), 306–316 (2018)
10. Liu, S., Bai, W., Liu, G., et al.: Parallel fractal compression method for big video data. Complexity **2018**, 2016976 (2018). https://doi.org/10.1155/2018/2016976
11. Liu, S., Yang, G. (eds.): ADHIP 2018. LNICST, vol. 279. Springer, Cham (2019). https://doi.org/10.1007/978-3-030-19086-6
12. Sridharan, K., Sivakumar, P.: A systematic review on techniques of feature selection and classification for text mining. Int. J. Bus. Inf. Syst. **28**(4), 504–518 (2018)
13. Ferreira, C.H.P., De Franca, F.O., Medeiros, D.R.: Combining multiple views from a distance based feature extraction for text classification. In: IEEE Congress on Evolutionary Computation, pp. 1–8. IEEE (2018)
14. Liu, S., Li, Z., Zhang, Y., et al.: Introduction of key problems in long-distance learning and training. Mobile Netw. Appl. **24**(1), 1–4 (2019)
15. Saikia, L.P., Singh, S.: Feature extraction and performance measure of requirement engineering (RE) document using text classification technique, pp. 1–6 (2018)

Detection Method of Abnormal Behavior of Network Public Opinion Data Based on Artificial Intelligence

Ying-jian Kang$^{(\boxtimes)}$, Lei Ma, and Yan-ning Zhang

Beijing Polytechnic, Beijing 100176, China
kangyingjian343@163.com

Abstract. In order to improve the effect of network public opinion data abnormal behavior detection, an artificial intelligence-based network public opinion data abnormal behavior detection method is proposed. By constructing the network public opinion data model, recognizing the evolution rule of network public opinion data, locating the abnormal data area according to the behavior detection algorithm, and using the probability neural network under artificial intelligence to detect the abnormal data behavior. The experimental results show that the detection method proposed this time is 28.12% and 84.37% higher than the two traditional methods when detecting large-scale public opinion abnormal behavior data. It can be seen that the detection method based on artificial intelligence is not restricted by the volume of network data, and the detection effect is better.

Keywords: Artificial intelligence · Network public opinion · Data abnormal behavior · Detection method

1 Introduction

Internet public opinion refers to the popular Internet public opinion with different views on social issues, which is a form of social public opinion. Through the Internet, it can spread the public opinions and opinions with strong influence and tendentiousness on some hot spots and focus issues in real life. The network public opinion takes the network as the carrier and the event as the core. It gathers the expression, dissemination and interaction of the emotions, attitudes, opinions and opinions of the majority of netizens. However, as a public use platform, the network has abnormal data behavior. Due to the single detection technology, the traditional detection methods will lose some abnormal data when facing massive network information [1]. Therefore, based on artificial intelligence technology, a new detection method for abnormal behavior of network public opinion data is proposed. By constructing the network public opinion data model, the evolution of network public opinion is divided into three stages: public opinion generation period, public opinion diffusion period and public opinion reduction period. The evolution law of network public opinion data in different stages is identified. The abnormal data area is located by using behavior detection algorithm, and abnormal data behavior is detected by artificial intelligence technology. The

S. Liu and L. Xia (Eds.): ADHIP 2020, LNICST 347, pp. 332–344, 2021.
https://doi.org/10.1007/978-3-030-67871-5_30

experimental results show that the method proposed in this paper has a good effect on abnormal behavior detection of network public opinion data.

2 Artificial Intelligence-Based Network Public Opinion Data Abnormal Behavior Detection Method

2.1 Building the Data Model of Network Public Opinion

To construct a network public opinion data model, first assume that the interaction behavior of individual opinions at the micro level is restricted by the trust threshold, that is to say, the interaction behavior can only occur when the distance between the two parties' views is less than the trust threshold, otherwise the two parties will avoid contact and insist on their own Original point of view. Although the interaction rule of this view is simple, it vividly reveals the phenomenon of "different ways do not conspire" in the process of interpersonal communication. Therefore, through the continuous iteration of this interaction mode, we can explore some rules of the aggregation process of group views from a macro perspective. The model breaks through the limitation of the individual point of view binary or can only take values within a limited range of values. It is believed that the individual point of view can be any real number within a given range of values. Among the categories of models. In addition, the model assumes that any two individual groups randomly selected from the group may have viewpoint interaction behavior, as long as their viewpoint distance is less than the trust threshold [2].

Let the group size be W, i, and j are two random individuals in the group, and the viewpoint values at time t are expressed as $u_i(t)$ and $u_j(t)$, respectively, and $u_i(t), u_j(t) \in [0, 1]$, given a trust threshold φ, is a constant between the range $[0, 1]$, If $|u_i(t) - u_j(t)| \leq \varphi$, then:

$$\begin{cases} u_i(t+1) = u_i(t) + \lambda(u_j(t) - u_i(t)) \\ u_j(t+1) = u_j(t) + \lambda(u_i(t) - u_j(t)) \end{cases} \tag{1}$$

Otherwise:

$$\begin{cases} u_i(t+1) = u_i(t) \\ u_j(t+1) = u_j(t) \end{cases} \tag{2}$$

In formula (1): Parameter λ is the convergence coefficient of the model, which will affect the speed of system convergence. Adjust the value of the convergence coefficient λ to define the nature of the group. Let λ take the value in the interval $[0, 0.5]$. When $\lambda = 0$, all individuals always adhere to their own views without any change; when $\lambda = 0.5$, both sides of the view interaction will get the average of the two views. The above two cases correspond to the interaction groups in two extreme cases respectively. Generally speaking, When the value of table λ is small, it is corresponding to the individuals with stronger strategies. They are not easy to change their own views, and when they are large, they tend to adopt a compromise view interaction strategy. When $\lambda > 0.5$, it means that after two individuals interact with each other, they respectively choose the views that are more inclined to each other, that is, their views have been transposed. At this time, the model constructed believes that this kind of situation rarely occurs in real life, so it is necessary to make $\lambda \in [0, 0.5]$ [3]. In order to simplify the model, the value of convergence coefficient is usually fixed, that is, the rules of opinion interaction are as follows:

$$
\begin{aligned}
&\left|u_i(t) - u_j(t)\right| \le \varphi, \begin{cases} u_i(t+1) = u_i(t) + 0.5\left(u_j(t) - u_i(t)\right) \\ u_j(t+1) = u_j(t) + 0.5\left(u_i(t) - u_j(t)\right) \end{cases} \\
&\left|u_i(t) - u_j(t)\right| > \varphi, \begin{cases} u_i(t+1) = u_i(t) \\ u_j(t+1) = u_j(t) \end{cases}
\end{aligned}
\tag{3}
$$

In the network public opinion data model, trust threshold φ has an important influence on the aggregation process of group opinion. Through the model simulation, it can be found that when $\varphi \ge 0.5$, the group tends to form a consensus, that is, all individuals in the group ultimately hold the same views on a given issue. With the decrease of φ value, the group gradually divides into two or more opinion groups, and members of each opinion group share the same views.

2.2 Identify the Evolution of Internet Public Opinion Data

From its generation to its final demise, Internet public opinion is in a complete dynamic change process, and presents or follows some internal rules to run. Therefore, on the basis of the constructed public opinion data model, the evolution of network public opinion is divided into three stages: public opinion generation period, public opinion diffusion stage and public opinion restoration stage, and the evolution law of network public opinion data in different stages is found out. Figure 1 is the life cycle curve of network public opinion obtained from the model analysis.

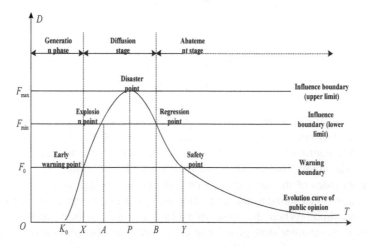

Fig. 1. Life cycle curve of online public opinion

In the figure, OT represents the time axis; OP represents the evolution degree of network public opinion. Based on the analysis of the existing public opinion life cycle of the emergency network, combined with the key characteristics of "five points", the public opinion of the emergency network can be divided into three stages: public opinion generation stage, public opinion diffusion stage and public opinion reduction stage.

2.2.1   Generating Law of Network Public Opinion

The generation of network public opinion is a special risk factor and risk influence formed in the network due to the interference of a "nature economy society" system. Previous studies have shown that analyzing and grasping the generation law of public opinion is the basic premise and key to effectively guide and manage public opinion. The generation rule of online public opinion believes that during the generation of network public opinion, it mainly undergoes four triggers: "trigger-agglomeration-hot discussion-burst", and each of them presents corresponding evolutionary rules: shape mutation rule, superposition focus law, resonance convergence law, group polarization law. The state generation of network public opinion can also be regarded as "state break" or "state break", and the control factor of this "state break" belongs to a dependent variable, and the "state break" condition can be regarded as a fixed critical limit value. Therefore, based on this value, we can build a catastrophe simple function generated by public opinion of emergency network [4].

Assuming that the damage status and social impact of a network event are a risk and crisis independent variable function x, the event stakeholders are a risk dependent variable function y, and the calculation formula of public opinion risk value is:

$$\mu = f(x, y) \tag{4}$$

Where: μ is the public opinion risk index; $f(x, y)$ is the risk function of x and y. If the risk value of public opinion does not exceed the risk critical limit value F_0, i.e. $\mu \leq F_0$, it is a sub stable balance, i.e. it belongs to the incubation period of network public opinion. During this period, Internet public opinion was basically stable or under control, and Internet public opinion was relatively calm. If the control factor energy continues to increase due to some internal or external forces, the risk value of public opinion will reach or exceed the risk critical value Limit value F_0, i.e. $\mu \geq F_0$, Then it will release potential energy and seriously destroy the inherent state of public opinion. The state of public opinion balance will suddenly "interrupt" or "break", forming a real public opinion risk. Figure 2 below is a schematic diagram of the generation of Internet public opinion.

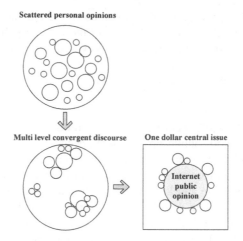

Fig. 2. Schematic diagram of network public opinion generation rule

2.2.2   The Law of Diffusion of Internet Public Opinion

Due to the characteristics of freedom, interaction and complexity of Internet public opinion, its diffusion and development is not a simple linear rise or decline, but a complex process. But generally speaking, the diffusion of network public opinion mainly refers to the process of transmission and change from small to large, from weak to strong, from recessive to dominant after the generation of network public opinion. In essence, it is a process of "strengthening and magnifying" after the generation of online public opinion. Generally speaking, in the stage of network public opinion diffusion, along the logical structure of "strengthening and amplifying", its diffusion is going through four stages of "popularity—reinforcement—repetition—sublimation". It presents certain evolutionary laws: linear asymptotic law, ripple divergence law, interference interaction law and spiral ascent law.

The dissemination effect of public opinion and the amount of information transmission show a linear positive relationship. The linear diffusion evolution of public opinion is a structural description of the process of public opinion communication. That is to say, every link and process of public opinion dissemination are closely linked after the emergence of public opinion in emergency network. Suppose that in the online public opinion, communicator a may spread to communicator b, while communicator b develops to communicator c, etc. The propagation process can be expressed as $W_a \rightarrow W_b \rightarrow W_c \rightarrow W_d \rightarrow \cdots$, and the public opinion is amplified in a linear mode to form large-scale public opinion, and the scale Public opinion is the development and value of communicator a to communicator n. At this time, the scale public opinion function can be expressed as:

$$W = \sum_{i=1}^{n} \sigma_i \tag{5}$$

Where: σ_i represents the scale index of i data. When the scale of public opinion is consistent with the analysis scale in the first section, that is, when W reaches a certain extreme value, the network public opinion will then spread, forming a gradual mode [5].

2.2.3 Law of Reducing Public Opinion on the Internet

Every material movement in the world is a process of alternate development of prosperity and decline. Prosperity is a kind of development of material movement, and decay is also a kind of development. Similarly, the network public opinion also develops along a rise and fall. The reduction of network public opinion mainly refers to the process of network public opinion gradually decreasing and declining from large to small, strong to weak, and changing from hot-spot events to common events after the generation and spread of network public opinion. With the proper handling and resolution of the event, the social resources driven by the network public opinion are gradually exhausted, the public's attention to the event shows fatigue, the network media and traditional media pay less attention to and report, and the development of the network public opinion lacks a new power mechanism, so the network public opinion begins to gradually enter a slowly subsided reduction stage. In the process of reducing public opinion on the Internet, some have disappeared in a broken manner, and some have repeatedly weakened. But generally speaking, it is the process of progressing from prosperity to decline along the evolution path of "conflict-order change-fading-dying". In this process, each public opinion reduction node accordingly exhibits certain laws: conflict blocking law, substitution transfer law, defocusing and fragmentation law and natural dissipation law [6]. Figure 3 is the conflict resistance function diagram of network public opinion reduction.

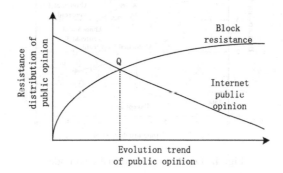

Fig. 3. Network public opinion reduces conflict blocking function

In the figure, point Q represents the intersection point between public opinion reduction and conflict resistance. From this point on, public opinion drops rapidly. At this point, it realizes the recognition of evolution rule of network public opinion data.

2.3   Behavior Detection Algorithm Locates Abnormal Data Area

In the actual environment, the data is not immutable, it changes with the time environment, and new abnormal behavior public opinion data appears, so the behavior detection algorithm is used to divide the information categories. Most of the time, the network is in a normal state, and the network data is normal or abnormal data. With the change of time and environment, the progress of network technology will be accompanied by the continuous change of network abnormal behavior. At this time, new network data will appear, whose characteristics are completely different from the previous data characteristics, that is, the abnormal behavior data will also be different [7].

In the behavior detection algorithm, comparing the distance $d(z, h_i)$ between the test sample and the center point of the nearest anomaly class H_i and the maximum distance D_i, $d(z, h_i) > D_i$ between the center point of the anomaly class and other points in the class, it is determined that a new anomaly class appears, The point is defined as a new anomaly point. When there are other test points that determine the new anomaly class, D_i is the distance between the two, and then the anomaly classification model is updated; otherwise, the test sample is determined as an i type anomaly. Suppose that in the test sample, the object is z, H_i is the center of mass of cluster i, h is the center of mass of all points, and the abnormal node is located. Figure 4 is the positioning result [8].

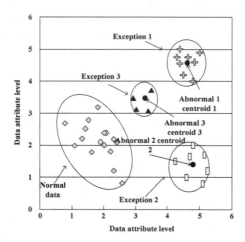

Fig. 4. Location of abnormal data nodes

According to the selected behavior detection algorithm, we can locate the abnormal behavior data in the network public opinion data.

2.4   Artificial Intelligence Technology Detects Abnormal Data Behavior

Known probability neural network is one of the artificial intelligence indexes, which is suitable for data separation. The intelligent network originates from radial basis function network, which can make use of simple linear structure to expand the nonlinear learning algorithm. The main principle of probabilistic neural network is to combine density function estimation and Bayesian decision theory. Under certain conditions, the network can detect abnormal behavior of public opinion data.

The behavior area of abnormal data is E, the prior probability is $V_1 = p(E_1)$, $V_2 = p(E_2)$, and $V_1 + V_2 = 1$. Given the input vector $\gamma = [\gamma_1, \gamma_2, \ldots, \gamma_n]$, in order to get a set of observation results, the classification basis is as follows:

$$E = \begin{cases} E_1, p(E_1|\gamma) > p(E_2|\gamma) \\ E_2, otherwise \end{cases} \qquad (6)$$

Where: $p(E_1|\gamma)$ is the probability of occurrence of γ, the posterior probability of category E_1 [9, 10]. According to the Bayesian formula, the posterior probability is equal to:

$$p(E_1|\gamma) = \frac{p(E_1)p(\gamma|E_1)}{p(\gamma)} \qquad (7)$$

In classification decision-making, input vectors should be classified into categories with higher posterior probability. In practical application, risk and loss should be considered. For example, the loss caused by the wrong classification of the E_1 sample into the E_2 category, or the wrong classification of the E_2 sample into the E_1 category is often very different, so the classification rules need to be adjusted. Define behavior β_i as the behavior of assigning the input vector to h_i, and ε_{ij} as the loss caused by taking behavior β_i when the input vector belongs to h_j, then the expected risk of behavior β_i is:

$$R(E_1|\gamma) = \sum_{j=1}^{n} \varepsilon_{ij} p(E_1)\delta_i \qquad (8)$$

In the formula: δ_i represents the probability density function of E_1. So far, the probability network in artificial intelligence technology is used to detect the abnormal behavior of network public opinion data [11, 12].

3   Experiment and Analysis

In order to verify the reliability of the proposed detection method, two traditional detection methods are selected and applied to the detection of abnormal behavior of network data. The proposed method is used as experimental group A, and the two traditional methods are used as experimental group B and experimental group C respectively. Choose a website as the experimental test background condition. It is known that there is abnormal behavior data on this website. The three detection methods are used to locate the abnormal behavior data in the network public opinion data database, and the positioning differences of the three detection methods are compared.

3.1   Experiment Preparation

Three sets of network public opinion data sequences with different data volumes are selected, and the abnormal data are hidden in the three data sequences respectively. Figure 5 below is a statistical diagram of the data volume of different data sequences.

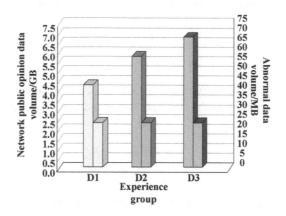

Fig. 5. Statistical chart of data series

In the figure, D1 is a public opinion data sequence with a small volume; D2 is a public opinion data sequence with a moderate volume; D3 represents a public opinion data sequence with a large volume. The two statistical results in each group represent the total data volume and abnormal data volume of the sequence. It is known that under the premise of different data sequence volume, the data volume of abnormal behavior of network public opinion is the same, both of which are 22.5 MB. D1 group and D3 group were used as test variables to detect three groups of public opinion data series. Get and analyze the experimental test results.

3.2   The First Set of Experimental Results

Group D1 with relatively small data volume is taken as the first group of experimental test objects, and Fig. 6 below is the experimental test results.

According to the test results shown in the above figure, in the face of small data series, the three detection methods can detect the abnormal behavior data, without missing any abnormal network public opinion data. According to the statistics, the detection rate of three groups of methods is 100%, and the difference between them is also 0. It can be seen that three detection methods are applicable to the detection of abnormal behavior of network public opinion data with small volume.

3.3   The Second Set of experimental results

In order to ensure the authenticity and reliability of the experimental test results, group D3 with relatively large data volume is taken as the second group of experimental test objects, as shown in Fig. 7 below.

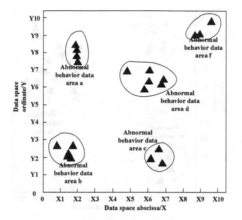

(a) Experiment A test results

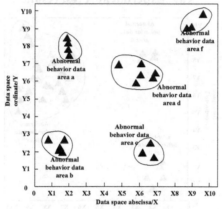

(b) Test results of group B

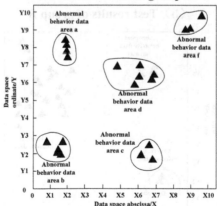

(c) Experiment C test results

Fig. 6. The first group of experimental test results

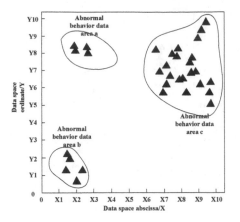

(a) **Experiment A test results**

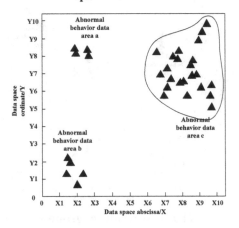

(b) **Test results of group B**

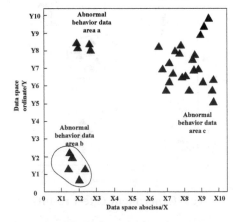

(c) **Experiment C test results**

Fig. 7. The second set of experimental test results

According to Fig. 7, in the face of large-scale data series, experimental group a can also fully detect abnormal behavior data, while experimental group B and experimental group c lose a lot of abnormal behavior information. The detection rate of the three methods is as shown in Table 1 below.

Table 1. Statistical results of detected bit rate

Group	Arrival rate	Difference from the rate of full arrival
Group A	100%	0
Group B	71.88%	28.12%
Group C	15.63%	84.37%

According to Table 1, the detection of group A is completely in place; the detection of group B exists 28.12%. The detection rate of group C was only 15.63%. It can be seen that the traditional two detection methods are not suitable for the detection of abnormal behavior of large-scale network public opinion data. It can be seen from the results of the two sets of experimental tests that the detection method proposed this time can perform data abnormal behavior detection in large-scale network public opinion data.

4 Conclusion

Artificial intelligence, also known as intelligent machinery and machine intelligence, is the intelligence shown by machines made by human beings. Usually, artificial intelligence presents human intelligence through ordinary computer programs. Artificial intelligence is applied to abnormal behavior detection to provide more reasonable technical support for detection methods. The proposed detection method fully realizes the detection of abnormal behavior data. However, the total number of tests in this experiment is less. In future research, we should expand the experimental data, increase the number of experiments, and strengthen the persuasion of test results [13].

References

1. Xiwei, W.A.N.G., Ruonan, J.I.A., Duo, W.A.N.G., et al.: Research on the development trend of artificial intelligence research in library and information field. Libr. Inf. Serv. **63** (01), 70–80 (2019)
2. Longjie, S., Kaijun, Y.: Research on big data analysis model of library user behavior based on internet of things. Comput. Eng. Softw. **40**(06), 113–118 (2019)
3. He, H., Haojiang, D., Jun, C.: Network abnormal behavior detection technologies based on traffic-feature modeling. J. Netw. New Media **8**(04), 11–20 (2019)
4. Lianhai, S., Ying, L.: Multivariate autoregression based algorithm for anomaly detection in rating data. Comput. Eng. Des. **39**(06), 1629–1632 1652 (2018)
5. Huanli, P.A.N.G., Hongyan, L.I.: Intelligent detection simulation for crowded pedestrian abnormal behavior. Comput. Simul. **35**(11), 405–408 (2018)

6. Jia, Y., Xiaonan, L., Yang, Z., et al.: Abnormal behavior detection for campus email systems based on big data analysis. J. Commun. **39**(S1), 116–123 (2018)
7. Xiaolin, W.U., Fuyuan, C.A.O.: An unsupervised outlier detection algorithm for categorical matrix-object data. J. Shenzhen Univ. (Sci. Eng.) **36**(01), 33–42 (2019)
8. Baisheng, G.U.A.N., Chunjiang, B.I.A.N., Shuichun, F.E.N.G., et al.: Abnormal action recognition of interaction research based on neural network. Electron. Des. Eng. **26**(20), 1–5 (2018)
9. Xiao, L., Yi, L., Yefeng, J., et al.: Overall design of artificial intelligence application based on monitoring big data platform. Power Syst. Big Data **22**(04), 37–42 (2019)
10. Shuai, L., Weiling, B., Nianyin, Z., et al.: A fast fractal based compression for MRI images. IEEE Access **7**, 62412–62420 (2019)
11. Fu, W., Liu, S., Srivastava, G.: Optimization of big data scheduling in social networks. Entropy **21**(9), 902 (2019)
12. Liu, S., Li, Z., Zhang, Y., et al.: Introduction of key problems in long-distance learning and training. Mob. Netw. Appl. **24**(1), 1–4 (2019)
13. Liu, S., Glowatz, M., Zappatore, M., et al. (eds.).: e-Learning, e-Education, and Online Training, pp. 1–374. Springer International Publishing (2018)

Query Optimization Method for Massive Heterogeneous Data of Internet of Things Based on Machine Learning

Yun-wei Li[1] and Lei Ma[2(✉)]

[1] Beijing Youth Politics College, Beijing 100102, China
liyunwei@bjypc.edu.cn
[2] Beijing Polytechnic, Beijing 100016, China
malei235@tom.com

Abstract. In view of the problem that the traditional query optimization method of massive heterogeneous data of the Internet of things can not describe the data characteristics clearly, which results in the long execution time of data query, a query optimization method of massive heterogeneous data of the Internet of things based on machine learning is designed. It divides the massive heterogeneous data query level of the Internet of things, and extracts the data characteristics according to the hierarchical structure and the Dirichlet smoothing method in machine learning. The feature data is transformed into a query tree, and a dynamic data dictionary is constructed. The data dictionary is referred to the traditional query optimization method of massive heterogeneous data in the Internet of things. At this point, the query optimization method for massive heterogeneous data of the Internet of Things based on machine learning is designed. The test link of the construction method shows that the use effect of this method is better than the original method and the method based on artificial intelligence technology.

Keywords: Machine learning · Dirichlet smoothing method · Heterogeneous data · Internet of things

1 Introduction

With the further development of the Internet of things technology, in the face of growing data, the traditional storage architecture due to poor scalability, in the long run, the storage environment will become increasingly complex, resulting in high energy consumption [1]. Unlike traditional storage systems, distributed cloud storage systems can store massive amounts of information, efficiently manage large-scale files, and provide good query efficiency. But because the cloud storage system is based on Internet technology, it mainly stores small file data, such as small picture streams and small video streams, but in the Internet of Things, it needs to repeatedly access massive picture streams and video stream data. Because the Internet of Things needs to collect various information such as sound, light, heat, electricity, chemistry, location, etc., and the content of information captured by different types of sensors and information receivers is very different.

S. Liu and L. Xia (Eds.): ADHIP 2020, LNICST 347, pp. 345–356, 2021.
https://doi.org/10.1007/978-3-030-67871-5_31

The distributed storage of massive data and the data processing between heterogeneous databases have become an inevitable trend. Through the integration of data information and hardware devices in various databases, a heterogeneous database system is formed which is logically unified and physically independent [2, 3]. This kind of cross-database query and multi-table connection needs to increase data redundancy to ensure the reliability and availability of heterogeneous database systems, but the physical distributed storage of data and the need for such data redundancy make heterogeneity become higher. The query processing of the database adds more content and difficulty. Therefore, query optimization of heterogeneous database system plays an important role.

This article first analyzes the structure of the heterogeneous database of the Internet of Things and the query processing process, and understands that in the query process of the heterogeneous database system, the query request should be converted into a global query tree and converted. Through the principle of equivalence, the query tree can be decomposed into a local query tree to further optimize and determine the final query execution plan. Such query optimization process needs to consider local response time and transmission cost. At present, the existing query optimization algorithms can only achieve the shortest local response time or the lowest transmission cost, and can not take into account the dual optimization of time cost and space cost. Moreover, when using dynamic algorithms to achieve query optimization, the computing power is insufficient, and the algorithm often falls into the local optimization and can not get the global optimal solution that conforms to the characteristics of heterogeneous databases. Therefore, in this study, a query optimization method for massive heterogeneous data in the Internet of Things based on machine learning is designed.

2 Query Optimization Method of Massive Heterogeneous Data in Internet of Things Based on Machine Learning

2.1 Hierarchical Division of Massive Heterogeneous Data Query in Internet of Things

Query processing, as the most important link in the multi data integration system, directly affects the query efficiency of the system. How to ensure the correctness of query processing, and how to answer query efficiently are all the tasks to be completed by query processing [4]. The query processing of the massive heterogeneous database of the Internet of Things is more complicated. A query in a global mode needs to be decomposed into query fragments that can be run on each site. After the corresponding operations of these query fragments are executed at the site, the results must be returned to the global query Mode, and summarize the execution results of all the fragments as the final output of the system. But for users, query fragmentation, local processing, result summary processing and other processes are transparent. Users only care about the execution results of global queries. In the query optimization of a heterogeneous database system, the communication cost and the time complexity of local response must be considered at the same time. Therefore, these two types of query optimization standards are often used at the same time, each of which depends on the execution environment and data size. The standard weights will vary accordingly.

In order to achieve the query effect required by users, the query process of massive heterogeneous database system is generally divided into four levels from the structure: query decomposition, data localization, global optimization and local optimization, as shown in the figure below (Fig. 1).

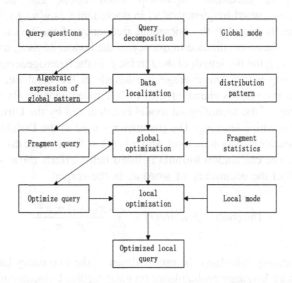

Fig. 1. Structure hierarchy

The most important work in query processing is query hierarchy. Query hierarchy refers to the decomposition of m-sql queries based on mediator mode into sub query sets based on wrapper mode [5]. In a multi-data integration system, the user query writes a query statement for the Mediator mode. There is no actual data in the Mediator mode. The query decomposition needs to convert the M-SQL query according to the specific situation according to the mapping relationship between the Mediator mode and Wrapper mode One or more queries based on Wrapper mode. Based on the above-mentioned query hierarchical division results of the massive heterogeneous data of the Internet of things, the massive heterogeneous data of the Internet of things is classified, and the classification results are used as the data base of the query optimization results.

2.2 Data Feature Extraction

After determining the query hierarchy of the Internet of things massive heterogeneous data, we need to generate a rich set of features for each query vocabulary, which represent the relationship between the candidate extension word and the original query, and the importance of the candidate extension word in the corpus. These features can be generally divided into two categories: the value of one type of feature depends on the specific input query, and the other type of feature is independent of the specific query [6]. Since SVM is used for training, all features need to be normalized. In normalization, given a search topic and a set of candidate expansion words, the characteristics of each candidate expansion word are changed proportionally, and the characteristics of the current expansion word are normalized between 0 and 1 by the maximum value of the same characteristic in the training data.

Set in the physical network heterogeneous database, there are N query information $A : a_1, a_2, \ldots, a_n$; a_n candidate expansion word queue can be expressed as $U : ua_1, ua_2, \ldots, ua_n$; word frequency of u_n in document f_i is $g(u_n, f_i)$; word frequency of u_n in Y corpus Is $g(u_n, Y)$; the document frequency of the lexical item u_n in the corpus C is $g(u_n)$; the co-occurrence frequency of the lexical items u_n and u_m can be set to $co - ocur(u_n, u_m)$; the file length of the i th file f_i in the heterogeneous database is fl_i; $avgfl$ is the heterogeneous The average file length in the database. Use the above settings to obtain the various characteristics of IoT heterogeneous data.

The probability of the monolingual model is calculated by the Dirichlet smoothing method [7, 8] in machine learning. The parameter α used in the Dirichlet smoothing is set according to each evaluation data set, and α is set as the average document length of this test data set. The calculation formula is listed below. Here, do/n is used to model the probability P of the occurrence of word u_n in the corpus.

$$Unigram - fwature(t) = \sum_{f_i \in f} \frac{g(t, u_n) + \alpha \frac{do}{n}}{fl + \alpha} \tag{1}$$

This design mainly calculates the probabilities of the two unary language models, calculates the unary language probabilities on each feedback document, and then adds up these probability values. Consider all the feedback documents as a whole, and calculate the probability of the unitary model on this whole.

The BM25 weight is a well-known formula in the field of information retrieval. It is used to express the importance of this word in pseudo-feedback documents. The higher the value of BM25, the more important this word is in this document. This value indirectly depends on the original query item. For a given candidate extension word, the BM25 weight formula is as follows:

$$BM25 - fwature(t) = \frac{(h_i + 1)g}{h_i(1 - j) + j\frac{fl}{avgfl}} \log \frac{n - do + 0.5}{do + 0.5} \tag{2}$$

Among them, $h_i \in [1.0, 2.0]$ and j are usually set to 0.75. The weight of BM25 is similar to the calculation of the previous unary language model. There are two cases to calculate the characteristics of the weight of BM2s, that is, calculate the BM25 weight of the word on each pseudo-feedback document, and then add these weights together; Or take all the pseudo-feedback documents as a whole and then calculate the BM25 weight.

In this study, the main features of the global corpus are the monolingual model of the candidate extension word in the whole corpus and the gdo value in the whole corpus. The calculation formula is:

$$g * do(a_i) = g(u_n, Y) * \log \frac{Corpus - Size}{g(u_n)} \tag{3}$$

In order to calculate the similarity between the candidate expansion word and the original query, for a given candidate expansion word, extract mutual information and Pearson's chi-square statistics from the pseudo-feedback document [9, 10]. The higher the value of these statistics, the closer the relationship between the extended word and the original query word. The original query submitted by the user is composed of multiple keywords, so the similarity between the extension words and each keyword is accumulated here. As shown below, both statistics need to calculate the co-occurrence between the expansion word w and the original query term a, using the following statistics on the co-occurrence between the two words:

$$MI - fwature(t) = \sum_{w \in a} MI(a, w) \tag{4}$$

$$\beta - fwature(t) = \sum_{w \in a} \frac{n(c_{11}c_{22} - c_{21}c_{12})^2}{(c_{11}c_{12})(c_{11} + c_{21})(c_{12} + c_{22})(c_{21} + c_{22})} \tag{5}$$

Among them, (a, w) represents the query word a and the expansion word w. The joint probability on the pseudo feedback document set. c_{11} represents the co-occurrence of the query term a and the expansion term w; c_{12} represents the frequency where only the query term a appears without the expansion term w; Qiu: c_{21} represents the frequency where the query term a does not appear and only the expansion term w appears; c_{22} represents both the query word a does not appear, nor does the frequency of the expansion word w appear. Obtain the characteristics of the query information and the corresponding generated information through the above calculation, and set the collected characteristics to the database storage format.

2.3   Building a Data Dictionary

Through the research on the original data query method, it can be seen that the user query request is converted into a global query tree, and by accessing the global data dictionary, the relevant database information, site information and other global description information are obtained, so as to perform the subsequent query decomposition operation. In the process of setting up the global data dictionary [11, 12], there are some points to be noted, such as when a field appears in multiple tables, multiple records should be set up, otherwise conflicts will occur. If the table of the same field is distributed in the database of different sites, multiple records need to be established according to the site conditions.

In view of the problems in the use of the original query method, the global data dictionary is converted into a dynamic data dictionary. The dynamic data dictionary is similar to the global data dictionary but different. It can take query commands as input, and can use dynamic data fields to describe the dynamic status of global sites and relational tables of heterogeneous databases. A dynamic data dictionary mainly includes the following contents (Table 1):

Table 1. Dynamic data font information table

Information no. Name content	Information no. Name content	Information no. Name content
1	Serial number	
2	Site number	
3	Table number	Table numbers in order
4	Table name	
5	Number of records in the table	
6	Field name	The fields of the same table are arranged in order; if a field appears in multiple tables at the same time, these fields are matched to different tables and added to the dictionary
7	Whether the field has an index	
8	Field data length	
9	Select field means that the value of a field must meet a certain condition in a query	Select field means that the value of a field must meet a certain condition in a query
10	Is it the result field? In the query, the value of the field is the content displayed after the query	Is it the result field? In the query, the value of the field is the content displayed after the query
11	Whether it is a contact field means that in the query, the value of the field is the connection condition between the table and the table	Whether it is a contact field means that in the query, the value of the field is the connection condition between the table and the table
12	Number of values allowed for the field	
13	Network performance matrix The performance ratio of the site itself to other network connections and network transmission speeds	Network performance matrix The performance ratio of the site itself to other network connections and network transmission speeds
14	Computational Performance Array Computational performance indicators of sites in the network	Computational Performance Array Computational performance indicators of sites in the network

The dynamic data dictionary stores the dynamic state information of each site and the table in the site. When the site works normally, the network performance and computing performance status can be stored normally. If the site is abnormal or loses connection, it will be represented by infinity or a negative number. The data dictionary is introduced into the original data query method, and the rest use the original data query method to design content. At this point, the query optimization method for massive heterogeneous data of the Internet of Things based on machine learning is designed.

3   Method Application Test

3.1   Test Platform Construction

In order to effectively simulate the massive heterogeneous data network of the Internet of things, HBase is used to build the distributed data storage network of the Internet of things.

This article uses a completely distributed operating mode to build an experimental platform. The experimental platform built in this article is based on Hadoop-1.0.4 version, HBase uses a version based on HBase-0.94.8 modified source code and recompiled, the JDK version is jdk-6u45-linux-i586. The specific hardware and software configuration is as follows (Table 2):

Table 2. Database cluster configuration

Cluster distribution Cluster composition Parameters	Cluster distribution Cluster composition Parameters	Cluster distribution Cluster composition Parameters
operating system	UbuntuKylin 14.04	32-bit
Hardware parameters	CPU	Intel E7000(Binuclear 、 2.66 GHz)
	Memory	Master node 2 GB, Slave node 1 GB
	hard disk	250 GB(5400 RPM)
HBaseConfiguration	hbase.hregion max.filesize	10 GB
	hbase.hregion memstore.flush.size	128 MB
	hbase.client write.buffer:	2 MB
MySQL Configuration	Cluster mode	Standalone
	default-storage-engine	INNODB
	innodbes buffer-pool size	1230 MB
	transaction-isolation:	READ-COMITTED

After completing the above configuration, start Hadoop and HBase in turn. Before starting Hadoop, you need to format the HDFS file system, and then start Hadoop and HBase on the cluster through the start all. SH and start HBase. Sh scripts respectively. When viewing the process through JPS after startup, if there are JPS, namenode, secondarynamenode, jobtracker, hmaster, hquorumpeer on the master node, and JPS, datanode, tasktracker, hregionserver, and hquorumpeer on the slave node, it indicates that the startup is successful. You can view the status of HDFS through http: / / masterhost: 50070 /, and HBase through http:! / masterhost: 60010 /. So far, the establishment and start-up of the experimental environment has been completed.

3.2   Test Methods

In order to ensure the validity of the method test, this paper studies the effects of these three methods by comparing the design method with traditional optimization methods and optimization methods designed by artificial intelligence technology. In this test, the test data set will be set as the test data sample. The specific data is shown below (Table 3).

Table 3. Test data set

Data set	Dataset a dataset B	Dataset a dataset B
Data set size	2048	10240
Path expression	1315	6560
Suffix array	14887	74380
Remove redundant data items	9311	44540

Use the design method and the other two methods to query the above data set, and set the query statement as shown below (Table 4).

Table 4. Query statement

Query statement Matching method Category	Query statement Matching method Category	Query statement Matching method Category
X1	Backward	Resources
X2	Backward	Attribute
X3	Backward	Resources
X4	Backward	Resources
X5	Backward	Resources
X6	Backward	Attribute
X7	Forward	Attribute

Set the above query statement as the design method and the other two methods in the article, and query the test data set.

3.3   Test Indicators

In this test, the test index is set as follows: the search times in half and the execution time are used as the test index, and this index reflects the difference between the design method in the article and the other two methods.

Half search times and execution time are the important embodiment of query ability of massive heterogeneous data. Through this index, we can directly understand the use effect of optimization method through data.

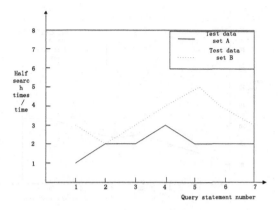

(a) Test results of design method in this paper

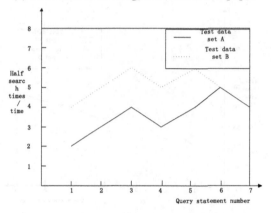

(b) Test results of traditional methods

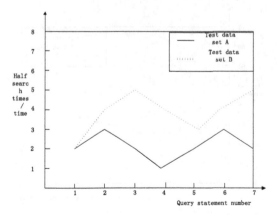

(c) Test results based on artificial intelligence method

Fig. 2. Half-find test results

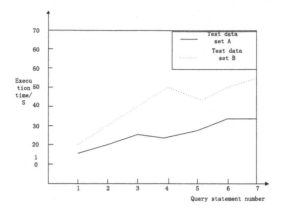

(a) Test results of design methods in the article

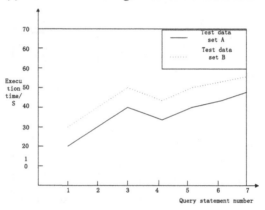

(b) Test results of traditional methods

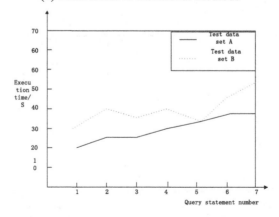

(c) Test results based on artificial intelligence methods

Fig. 3. Execution time test results

3.4 Half-Find Test Results

According to the above test results, among the seven query statements, the method proposed in this paper needs the least number of half search. The reason is that the method proposed in this paper builds a data dictionary, which can reduce the number of search in half by indexing. It can be found from the data in the table that the method proposed in this paper is able to maintain the minimum number of halving searches even when the data set is significantly larger than the other two methods (Fig. 2).

3.5 Execution Time Test Results

In traditional methods and artificial intelligence methods, query processing time is almost the same, but in the design method of this article, it is half of the first two time. It can be seen from the data that under the premise that the query result is equal to the left and right comparison times, machine learning can greatly shorten the query time and improve the query efficiency. It can be seen that the method proposed in this paper has a good balance between query efficiency and accuracy (Fig. 3).

4   Conclusion

This paper studies the massive heterogeneous data processing of the Internet of things from a new perspective. While some achievements have been achieved, there are still some deficiencies and areas to be improved. In the future research, we should improve them. Finally, I hope that the work of this paper can play a reference role for the development of distributed heterogeneous data query processing and other related fields [13].

References

1. Hu, F., Li, C., Wang, M., et al.: SQL injection detection scheme based on machine learning. Comput. Eng. Des. **40**(06), 1554–1558 (2019)
2. Guo, S., Guo, Z., Hu, N., et al.: IoT-oriented heterogeneous data conversion model. Periodical Ocean Univ. China **49**(06), 140–146 (2019)
3. Li, G., An, J.: Internet of Things-oriented heterogeneous entities relation service model. Internet Things-Oriented Heterogen. Entities Relat. Serv. Model **46**(02), 131–140 (2019)
4. Zhang, P., Wang, Y., Jiang, M., et al.: Spatio temporal data retrieval and prediction system based on K-D tree and machine learning. Comput. Eng. Softw. **39**(08), 215–218 (2018)
5. Sang, H., Guo, W.: Research on key model of index semantic query based on massive heterogeneous data. J. Fuzhou Univ. (Nat. Sci. Ed.) **46**(03), 324–329 (2018)
6. Xia, R.: Design of machine learning web service engine based on spark. Command Control Simul. **40**(01), 113–117 (2018)
7. Wang, H., Dai, B., Li, C., et al.: Query optimization model for blockchain applications. Comput. Eng. Appl. **55**(22), 34–39+171 (2019)
8. Liu, S., Li, Z., Zhang, Y., et al.: Introduction of key problems in long-distance learning and training. Mob. Netw. Appl. **24**(1), 1–4 (2019)

9. Le, Y.: Design and research of query optimization algorithm for large scale database. Bull. Sci. Technol. **35**(09), 66–69+74 (2019)

10. Shuai, L., Gelan, Y.: Advanced Hybrid Information Processing, pp. 1–594. Springer International Publishing, Heidelberg (2019)

11. Xing, B., Lv, M., Jin, P., et al.: Energy-efficiency query optimization for green datacenters. J. Comput. Res. Dev. **56**(09), 1821–1831 (2019)

12. Fu, W., Liu, S., Srivastava, G.: Optimization of big data scheduling in social networks. Entropy. **21**(9), 902 (2019)

13. Liu, S., Lu, M., Li, H., et al.: Prediction of gene expression patterns with generalized linear regression model. Front. Genetics **10**, 120 (2019)

Heterogeneous Big Data Intelligent Clustering Algorithm in Complex Attribute Environment

Yue Wang[1] and Jian-li Zhai[2(✉)]

[1] Software College & Nanyang Institute of Technology,
Nanyang 473000, China
wangyue66531@163.com

[2] Huali College Guangdong University of Technology,
Guangzhou 511325, China
Zhaijianli2033@163.com

Abstract. In order to improve the stability of heterogeneous big data mining operations in complex attribute environment, such as data analysis and cleaning, a heterogeneous big data intelligent clustering algorithm is established. The data cleaning classification method is applied to clean the parameter space in complex attribute environment, and the regular term of sparse subspace clustering is introduced to eliminate the irrelevant and redundant information of heterogeneous big data, and the intelligent clustering index of heterogeneous big data is obtained. By measuring the clustering results, the design of heterogeneous big data intelligent clustering algorithm in complex attribute environment is completed. The experimental results show that the heterogeneous big data intelligent clustering algorithm in complex attribute environment has strong stability in the process of data analysis and cleaning.

Keywords: Complex attribute environment · Heterogeneous big data · Clustering algorithm · Cleaning data

1 Introduction

In recent years, with the increasing utilization of network resources, various industries pay more and more attention to heterogeneous big data mining, especially in complex attribute environment, big data has a lot of characteristic parameters. Has affected the user to the big data utilization degree. For this reason, people need to use the database to carry on the reasonable planning and the effective mining to the heterogeneous big data. Scientific research institutions have proposed some heterogeneous big data mining methods in complex attribute environments, but the mining work in complex attribute environments requires a series of operations such as data analysis, cleaning, conversion, and integration. As a result, the method proposed in the past can not have strong accuracy, stability and practicability at the same time in the mining work [1].

Heterogeneous big data intelligent cluster analysis uses data modeling technology to simulate and analyze the internal structure and distribution of data. From the point of view of data mining, heterogeneous big data intelligent clustering is an unsupervised algorithm. In the absence of prior knowledge, clustering algorithm is used to divide data and

S. Liu and L. Xia (Eds.): ADHIP 2020, LNICST 347, pp. 357–366, 2021.
https://doi.org/10.1007/978-3-030-67871-5_32

form marker clusters. The research directions of the theory of cluster analysis include the following aspects: First, the ability to process different types of data. Most of the existing algorithms are applied to the analysis of numerical data, but many kinds of data types need to be faced in practical application. Therefore, the limitation of the algorithm in the data processing ability hinders the popularization and application of the algorithm. Second, the ability to identify clusters of arbitrary data shapes. Most of the existing clustering algorithms use standard Euclidean distance to complete similarity measurement tasks, so this algorithm tends to identify spherical clusters. The cluster shape of the actual medium-high dimensional data is mostly non-spherical, so improving the ability of the algorithm to recognize clusters of arbitrary shapes is the key to improving the clustering effect.

In the complex attribute environment of big data mining method, the choice of heterogeneous database is particularly important. Therefore, the RDBMS big data mining method under the complex attribute environment is proposed. By cleaning the parameter space of complex attribute environment and adopting the distributed idea to improve the practicability of the method, the accuracy and stability of mining heterogeneous big data are effectively improved.

2   Design of Heterogeneous Big Data Intelligent Clustering Algorithm Based on Complex Attribute Environment

2.1   Cleaning Parameter Spaces for Complex Attribute Environments

The purpose of cleaning parameter space is to meet the quality requirements of data analysis, so as to fully guarantee the correctness of data analysis [2]. Data cleaning refers to the discovery and correction of corrupt or erroneous records in a recordset, table, or database, and then the replacement, correction or deletion of identified dirty data that is incomplete, incorrect, inaccurate or irrelevant. The process of achieving data consistency. The parameter space cleaning classification methods are as follows:

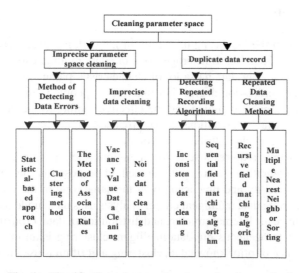

Fig. 1. Classification of parameter space cleaning methods

Figure 1 is a parameter space cleaning method. The first is the cleaning of the imprecise record attributes of the parameter space data sets, and the recognition of the exception attributes in the parameter space data sets [3]. The core idea is to give weight to each attribute first, then count the average value and standard deviation of each attribute field value, and then set a confidence interval for each attribute. Based on whether the attribute value is within the confidence interval to determine whether the attribute is abnormal. The clustering algorithm can judge whether the attribute is abnormal according to the distance between the attribute value and the cluster center, and use pattern recognition knowledge to find the abnormal attribute [4].

The problem of data cleaning is regarded as a statistical inference problem of structured text data in complex attribute environment. It is a classical tool for representation and reasoning of inconsistent knowledge. Before defining Bayesian networks, this paper first gives the corresponding formulas as the theoretical basis. Let Ω be the sample space of experiment E, A is an event of E, Ω is a partition of $p(a) > 0, b_1, b_2, b_3, \ldots \ldots . b_n, p(b_i) > 0, (i = 1, 2, \ldots, n)$. Then,

$$p(b_i|a) = \frac{p(b_i|a)}{\sum j = 1 p(b_i|a)} \tag{1}$$

$D = T_1, T_2, \ldots, T_n$ represents the input of structured data that contains dirty data. $T_I \in D$ represents one or more tuples with dirty data for the value of the m attribute [5]. Given a candidate replacement set C. Tuple T for possible dirty data in D, it can clean up the database by replacing $T_I \in D$ with a candidate cleanup tuple T with $P_{R(T*|T)}$. Using Bayesian rules in complex attribute environments, it is necessary to take into account in multi-source cleaning that each data source may involve different data fields and different forms of data exist, so the reasons for producing inaccurate data are varied. Inexact data problems in multi-source heterogeneous data environments can be summarized as follows: first, error data: errors in data may be caused by improper data collection or irregular data input, resulting in errors of varying degrees in the data [6]. Second, naming conflict: a naming conflict occurs when the same name is used for a different object or when a different name is used for the same object. Third, data heterogeneity: different representations of the same objects from different sources, such as different component structures, different data types, and different integrity constraints. Fourth, data redundancy: different representations of data from different sources have different version errors. Fifth, in a multi-source heterogeneous environment, even if the same attribute name and data type exist, there may be different value representations or different interpretations across the data source.

2.2   Introducing Regular Terms for Sparse Subspace Clustering in Complex Attribute Environments

When dealing with low-dimensional datasets, traditional clustering algorithms try to find clusters in all dimensions of datasets. But in complex attribute environments, there are usually many independent dimensions. These independent dimensions will hide the existing clusters in the noise data and interfere with the results of the traditional clustering algorithm, while the real correlation data will be distributed on the low-dimensional structure which can represent the characteristics of the clustering algorithm [7]. In addition, in a very high-dimensional dataset, the distribution of data objects is sparse, and all the data objects are almost equal to each other. This makes a single measure of distance meaningless and can lead to dimensional disaster. Therefore, in this design, the regular term of sparse subspace clustering in complex attribute environment is introduced to eliminate the irrelevant and redundant information in the data set, and clustering is only carried out on the related dimensions. Because the observed data dimension is usually higher than its essential correlation dimension, it is theoretically possible to reduce the dimension of the original space without losing any information. In practice, the dimensionality reduction method is often used to reduce the dimension of high-dimensional data before clustering [8]. There are two commonly used methods of data dimensionality reduction: feature extraction and feature selection.

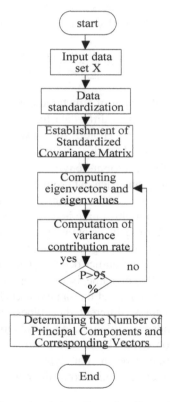

Fig. 2. High-dimensional data dimensionality reduction flowchart

Figure 2 is a concrete operation flow. In the first step, feature extraction is a preprocessing method in projection space, which makes the clustering algorithm only use a small number of newly selected features to cluster. Feature extraction by introducing the regular term of sparse subspace clustering in complex attribute environment, creating linear combination induction data set of data attributes, discovering the potential structure, generating and selecting new feature vectors, so as to achieve dimension reduction [9]. The second step is to keep the relative distance between the original data objects without deleting any original data attributes when introducing the regular term of sparse subspace clustering in the complex attribute environment, which makes the influence from independent dimensions still exist. Therefore, when there are a large number of independent attributes masking clusters in the dataset, feature extraction will not be able to get the desired effect. In the third step, feature selection is a method to eliminate redundant information by analyzing the entire dataset. It selects the optimal subset from the original data set by searching for various feature subsets and using some criteria to evaluate these subsets. Common search strategies include random search, sampling search and greedy sequential search. Step 4, the evaluation criteria follow two basic models: the wrapper model and the filter model. The fifth step, according to most of the work of supervised learning, finally, select the accuracy measure and classification label to complete the introduction of sparse subspace clustering rules in the complex attribute environment.

2.3 Setting Heterogeneous Big Data Intelligent Clustering Index

Heterogeneous big data clustering validity refers to whether a given fuzzy partition is suitable for all data. The validity index of heterogeneous big data cluster can be used to directly measure the quality of the given clustering results. The good clustering results should be as compact as possible and as far as possible between the clusters [10]. Different literatures put forward different scalar validity measures, but none of them is completely applicable to the evaluation of all clustering results. In this chapter, six validity indicators are selected to evaluate the clustering results.

The first indicator coefficient used to measure the number of "overlaps" between clusters and define it as a formula (2):

$$PC = \frac{1}{N} \sum_{I=1}^{C} \sum_{I=1}^{N} U_{IJ} \tag{2}$$

In formula (2), U_{IJ} indicates the extent to which data point j belongs to category i, C and N indicate the number of clusters and the total number of data samples, respectively.

The second index coefficient, classification entropy (CE): measure the ambiguity of cluster partition and define it as formula (3):

$$CE = \frac{1}{N} \sum_{I=1}^{C} \sum_{I=1}^{N} U_{IJ} \log(U_{IJ}) \tag{3}$$

In formula (3), indicators PC and CE measure whether the clustering results are clear. The higher the value of PC, the more compact the class is, the lower the value of CE, the better the clustering effect [11, 12].

The third indicator coefficient, Partition index (SC): it is the ratio of the sum of the compactness within the cluster and the separation between the clusters. It is the sum of the individual cluster validity measures normalized by dividing by the fuzzy cardinality of each cluster, which is defined as a formula (4):

$$SC = \sum_{i=1}^{e} \frac{\sum_{i=1}^{n} U_{IJ}}{N_I \sum_{K=1}^{C} |V_K - V_I|} \tag{4}$$

In formula (4), N_I represents sample j of the dataset, V_K and V_I are the i and k cluster centers, respectively. SC can be used to measure the quality of different partitions with the same number of clusters. The lower the value of SC is, the better the clustering results are.

The fourth indicator coefficient, separation index (S): in contrast to partition index SC, the separation index uses the minimum distance separation to achieve the effectiveness of the partition. The smaller the value of S is, the farther the separation between classes is, and the better the clustering results are.

The fifth indicator coefficient, the purpose of XB is to quantify the ratio of total changes within a cluster to the separation of clusters. The size of XB can measure the degree of compactness and separation between clusters. The smaller the corresponding value, the more compact the cluster is and the farther the separation between clusters is, the better the clustering result is.

2.4   Realization of Heterogeneous Big Data Intelligent Clustering Computation

In order to detect and eliminate duplicate records in data sets, it is necessary to solve the problem of how to determine whether the two records are duplicated or not, and to evaluate the similarity of data, that is, the problem of data matching. The simplest attribute set can be obtained by reducing the above attributes. According to the simplest attribute set, the data of the related attributes are extracted, a data table is synthesized, and then the data table is cleaned with similar duplicate data. Therefore, the corresponding records of the records are compared, and the similarity is calculated.

The idea of basic sorting neighborhood law can be summarized in three steps: the first step is to create a sort key: calculate the sort key for each record in the dataset by extracting the relevant field or part of the field. Step 2, sort data: sort the entire data set or part of the data set according to the keys created in the step. Third, data duplication identification: sliding a fixed-size window according to the order of the records, comparing each record with the other records in the window. If the window size window, each new record enters the window compared to the previous record, A record was found to be a "match" for $W \sim 1$. In fact, the accuracy of duplicate record detection depends largely on the sort keyword created, which directly affects the matching efficiency and accuracy. If you do not select keywords correctly, you may miss a large number of duplicate records. First, because two duplicate records may be far away from the physical location after sorting, they may never be simultaneously located in the same sliding window and cannot be identified as duplicate records.

Secondly, it is difficult to determine the sliding window size W. If the W is too large, the comparison time will increase, resulting in some comparison unnecessary; if W is too small, some duplicate records cannot be detected. When the size of all the duplicated clusters in the dataset varies greatly, no matter how the size of W is selected, it is not appropriate [13, 14]. In addition, for the whole matching process, the time complexity of the algorithm is O, where n is the total record of the data set, and the flow of heterogeneous big data intelligent clustering algorithm is shown below (Fig. 3).

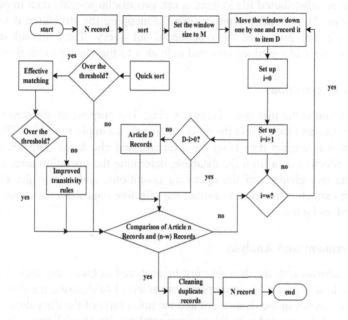

Fig. 3. Heterogeneous big data intelligent clustering proc

Firstly, considering that the window size W is difficult to determine in the SNM algorithm, we analyze the attributes and sort the data sets many times, which makes the repeated records more aggregated, thus entering the same sliding window at the same time. Second, when matching fields, the algorithm assigns a special weight to each attribute, and introduces the concept of effective weight, multiplies the weight by the similarity of the corresponding non-empty attribute, and then combines them to obtain the similarity of the entire record. And it is used to determine whether the two records duplicate the value. Thirdly, in the process of attribute selection, the rationality of the selection is proved by checking the similarity among m specific windows. So far, the design of heterogeneous big data intelligent clustering algorithm in complex attribute environment has been completed.

3   Experimental Conclusion

3.1   Experimental Environment

In order to verify the effectiveness of the algorithm in this paper, simulation experiments are carried out under Matlab 7.0, VS2010 + opencv2.4.13, windows 10, Intel (R) Xeon (R) CPU e5-2603v4 @ 2.20 GHz operating system and 32 GB memory.

Prototype experiment based on hadoop cluster, hive cluster, sqoop cluster and so on. Hadoop is a distributed file system. It can process large-scale data in parallel with hadoop cluster. Hive is mainly responsible for mapping the structured data file into a database table and providing the function of sql query. Then the sql statement is transformed into a MapReduce task and uploaded to the cluster to implement.

3.2   Data Preparation

Data preparation is the first step of data cleaning. The purpose of this step is to outline the process data and then select the most suitable data sample to model. The main task of this step is to extract data from the database, and check the data set. In order to extract the effective data from the database, determine the operation area, requirement analysis and any changes of the operating conditions, and ensure the efficiency of information extraction in order to extract the effective data from the database through samples and variables.

3.3   Experiment and Analysis

Before the contrast test, the data set must be analyzed to know the data characteristics of the data. In this paper, two parameters are selected to describe the data source: the number of attributes in the data record and the noise ratio of the dirty data contained in the data record. Compared with the experimental results, the following is shown:

In Fig. 4, we can see that the number of tuples increases as the number of records entered increases. Figure 4 shows that the proportion of similar duplicates and inconsistencies in the data set is still very high, some have reached more than 25%, the lowest is also more than 10%. It is shown that in the complex attribute environment, not only the data record scale is huge, the number of data attributes also reaches the level of one million, and the data contains the data of different data quality problems of considerable scale, and the number of properties eliminated exceeds the number of verification set. It is shown that the algorithm designed in this paper is relatively stable and the verification experiment is consistent with the real data experiment. It is shown that the experiment design is reasonable and the results are in line with the reality.

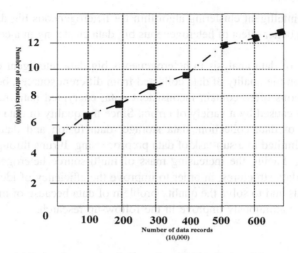

(a)Experimental results of traditional algorithms

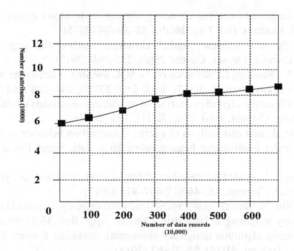

(b)Experimental results of this algorithm

Fig. 4. Experimental results

4 Conclusion

Due to the large number of heterogeneous big data clustering attributes in traditional heterogeneous big data intelligent clustering algorithms in complex attribute environments, this paper proposes a new heterogeneous big data intelligent clustering algorithm. In a complex attribute environment, by cleaning up the parameter space of the complex attribute environment, introducing sparse subspace clustering rule items, setting heterogeneous big data intelligent clustering indicators, so as to realize heterogeneous big data intelligent clustering calculation. Experimental verification shows that the intelligent clustering algorithm for heterogeneous big data proposed in this paper has a better effect of heterogeneous big data clustering in a complex attribute environment.

The quality of data analysis in heterogeneous big data intelligent clustering algorithm depends on the quality of data collected from different sources, because data sets in real applications often contain inconsistent data, encrypted data, noise values and data integration caused by a variety of errors. Since the quality of data often fluctuates in the process of data collection, data storage, data fusion and data analysis, data cleaning is not limited to a sub-task of data preprocessing. It runs through every link of data processing. Facing the increasing mass of multi-source heterogeneous data and more complex data structures, in order to improve the efficiency of identifying similar duplicate records and to solve the quality problem of data because of imprecise data, it is necessary to continuously improve in the follow-up research.

References

1. Anonymous: Large data optimal clustering algorithms in cloud computing environment based on PSO. Electron. Des. Eng. **26**(19), 86–89+94 (2018)
2. Qujie: Research on intelligent parallel clustering method for large data in virtual environment. Comput. Measur. Control **25**(6), 257–260 (2017)
3. Yi, M., Ting, X., Shaobin, L.: Research on NoSQL distributed large data mining method in complex attribute environment. Sci. Technol. Eng. **17**(09), 244–248 (2017)
4. Chunhua, H.: Clustering algorithm analysis of multidimensional data de-duplication in large data environment. Comput. Prod. Circ. **32**(11), 151 (2017)
5. Anonymous: Prediction and analysis of energy consumption behavior of integrated energy system users under multi-source heterogeneous large data. Smart Power **46**(10), 92–101 (2018)
6. Li, B.H., Junhua, C., et al.: Distributed clustering algorithms of attribute graph under multi-agent architecture. Comput. Sci. **44**(S1), 407–413 (2017)
7. Linjing, W., Lulu, N., Bin, G., et al.: Sparse fractional feature selection clustering algorithms based on entropy weighting in large data. Comput. Appl. Res. **35**(8), 59–60 + 69 (2018)
8. Houlisa: Clustering algorithm design for eliminating redundant features in large data sets. Mod. Electron. Technol. **41**(14), 56–58+62 (2018)
9. Xiaoyu, C., Xiaojing, L., Haiying, M.: A fast automatic clustering algorithm for large data. Comput. Appl. Res. **34**(9), 2651–2654 (2017)
10. Xiaoyan, T.: A large data text clustering algorithm based on word embedding and density peak strategy. Innov. Appl. Sci. Technol. **6**, 90–90 (2017)
11. Fu, W., Liu, S., Srivastava, G.: Optimization of big data scheduling in social networks. Entropy **21**(9), 902 (2019)
12. Sun, G., Liu, S. (eds.): ADHIP 2017. LNICST, vol. 219. Springer, Cham (2018). https://doi.org/10.1007/978-3-319-73317-3
13. Shuai, L., Weiling, B., Nianyin, Z., et al.: A fast fractal based compression for MRI images. IEEE Access **7**, 62412–62420 (2019)
14. Liu, S., Li, Z., Zhang, Y., et al.: Introduction of key problems in long-distance learning and training. Mob. Netw. Appl. **24**(1), 1–4 (2019)

Research on Clustering Algorithm of Heterogeneous Network Privacy Big Data Set Based on Cloud Computing

Ming-hao Ding$^{(\boxtimes)}$

Department of Computer and Software Technology, Tianjin Electronic
Information College, Tianjin 300350, China
dingrui8562@126.com

Abstract. With the rapid development and application of global information technology, big data era has come. China's information security strategy needs to consider the complexity and timeliness of large-scale and heterogeneous network security behavior in big data's time. In order to solve the problem of inaccurate and randomness of single clustering algorithm, a clustering algorithm based on cloud computing for heterogeneous network privacy big data set was proposed. The algorithm utilized the advantages of cloud computing to collect and extract features of big data sets. Then the similarity method was used to carry out the mining process of big data sets, so as to realize the clustering calculation process of big data sets. The algorithm was verified on the UCI dataset. The results showed that the efficiency and accuracy of the cloud computing-based big dataset clustering algorithm were better than the existing ones, indicating that the algorithm design and update strategy were effective.

Keywords: Cloud computing · Heterogeneous network · Big data set · Clustering algorithm · Similarity

1 Introduction

Clustering analysis has been studied for many years and formed a systematic method system [1]. Clustering is an unsupervised machine learning method that takes a group of physical or abstract objects. According to the degree of similarity between them, they are divided into several groups, so that the similarity of data objects in the same group is as large as possible, and the similarity of data objects in different groups is as small as possible [2]. However, the single clustering algorithm has the problems of unstable results and large randomness. Existing research tends to combine the results of clustering large data sets to overcome the shortcomings of clustering.

Research on the clustering of private data sets in heterogeneous networks has appeared in recent years [3], and it has attracted wide attention from all walks of life. However, how to generate the optimal clustering dataset and select the best merging strategy, especially the clustering fusion algorithm for large datasets of classification attributes, is still an unsolved problem. Therefore, it is necessary to conduct research on the generation and mining of cluster members to get the best clustering results.

S. Liu and L. Xia (Eds.): ADHIP 2020, LNICST 347, pp. 367–376, 2021.
https://doi.org/10.1007/978-3-030-67871-5_33

A cloud computing based heterogeneous network privacy big data set clustering algorithm research is proposed in this paper, and gives the fusion method and strategy of big data set. Firstly, the attributes of each large data set are divided according to the value, and the features are collected and extracted to obtain the initial cluster members. Then, the optimal fusion clustering results are obtained through continuous adjustment and mining. In order to verify the validity of the cloud computing-based heterogeneous network privacy big data set clustering algorithm designed in this paper, the experimental demonstration is carried out. The experimental results show that the cloud computing-based big data set clustering algorithm can improve the data clustering effect and ensure the accuracy of the clustering results, which is extremely effective.

2   Design of Large Data Set Clustering Algorithm

The dispersed large data set matrix is used as an input set of the cloud computing clustering algorithm, and the feature coefficients of each column in the matrix are respectively calculated by the pair of data features. And comparing the matrix characteristics of each big data set within a given threshold, and determining whether the number of points around the point that are greater than the threshold is greater than the data feature. If the characteristic coefficient of a point and any other point is greater than the matrix feature, and the number of features around it and its characteristic coefficient are greater than or equal to the matrix feature, then the point is the core data point, and all the points with the same density as the core point are of one type, and the other type is noise point.

By mining all common feature data in the big data set, and collecting and extracting the characteristics of the common data, once again, using cloud computing technology, mining the clustering features of big data sets to achieve clustering calculation of big data sets, the cloud computing based clustering algorithm flow is shown in Fig. 1.

2.1   Big Data Set Feature Collection and Extraction

Suppose there are n data points $X = \{x_1, x_2, x_3, \cdots x_n\}$, which contain m attributes, the i-th attribute has k_i different values, and the i-th attribute has a weight of ω_i. This paper adopts the simplest and most easy to understand method of generating cluster members [4], that is, by attribute value division, the function expression of the attribute division rule R_i of the i-th big data set is:

$$R_i = |C_{i,1}, C_{i,2}, C_{i,3} \cdots C_{i,j}| 1 \leq i \leq m \tag{1}$$

Where, $C_{i,j}$ c represents the j-th data feature of the segmentation result, $\sum_{i=1}^{m} \omega_i = 1$.

Inspired by the literature [5], the data on each attribute is divided into a cluster member, and a unified method is used to divide the different data subsets to obtain the characteristic relationship among the cluster members. Thus, clustering results $R = \{R_1, R_2, R_3, \cdots, R_m\}$ of m cluster members can be obtained, and each cluster member R_i has k_i matrix features.

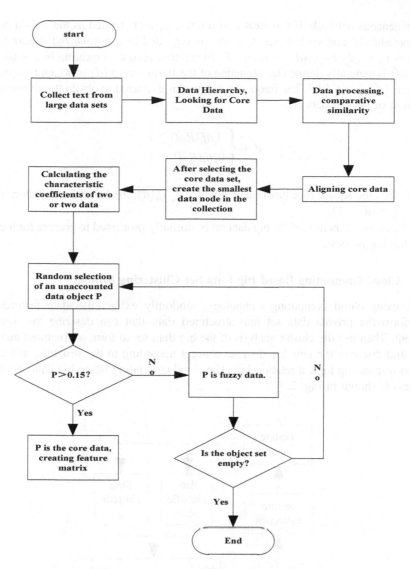

Fig. 1. Cloud computing based clustering algorithm flow chart

The feature collection method for similarity [6] refers to the collection process of metadata in the big data set representing the privacy of heterogeneous networks. By constructing the feature matrix, the clustering combination partition method of multiple large data sets is found, and the similarity between any two data points is used to describe and define the clustering features of large data sets. First of all, we must use cloud computing technology to extract features from the privacy big data sets in heterogeneous networks, and classify them according to their characteristics [7]. Secondly, the content characteristics of the metadata are represented, and the metadata is regarded as a vector space generated by a set of privacy orthogonal terms in a heterogeneous network. If t_i is treated as a term, $w_i(d)$ is treated as the weight of t_i in the metadata d, and each metadata d can be regarded as a normalized feature vector $V(d) = t_{1i}, w_i(d), t_{2i}, w_i(d), \cdots t_{ni}, w_i(d)$. In general, all data appearing in d is taken as t_i; $w_i(d)$ is generally defined as a function of the frequency $tf_i(d)$ in which t_i appears in d, i.e. $w_i(d) = \vartheta(tf_i(d))$. The frequency function is extracted to obtain the characteristic function of the big data set:

$$\vartheta = \begin{cases} 1, tf_i(R_id) \geq 1 \\ 0, tf_i(R_id) = 1 \end{cases} \tag{2}$$

Where, the square root function of ϑ is $\vartheta = \sqrt{tf_i(d)}$; the logarithm function of ϑ is $\vartheta = \log(tf_i(d) + 1)$.

The feature function of the big data set is similarly processed to prepare for the next data mining process.

2.2 Cloud Computing Based Big Data Set Clustering Mining

First, using cloud computing technology, randomly extract metadata features and transform the private data set into structured data that can describe the metadata content. Then use the cluster analysis of the big data set to form a structured metadata tree, and discover the new big data set concept according to the structure, and obtain the corresponding logical relationship. The cloud computing-based big data set mining process is shown in Fig. 2.

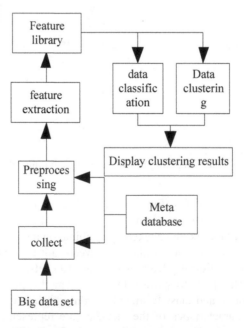

Fig. 2. Big data set mining process diagram

Since the amount of data in a private big data set in a heterogeneous network is very large, the dimension used to represent the metadata feature vector is also very large, and may even reach tens of thousands of dimensions. Therefore, we need to extract the network term with higher weight as the feature item of the metadata to achieve the purpose of dimension reduction of the feature vector. Then, the feature clustering mining process of big data sets is carried out. The big data set cluster mining process is as follows:

(1) Select some of the most representative data features from the original features.
(2) According to the similarity method principle [8], select the most influential feature data set.
(3) Transforming original features into fewer new features by means of mapping or transformation in cloud computing technology [9, 10].
(4) Using the evaluation function method [11], each feature in the feature set is independently evaluated and given an evaluation score, and a predetermined number of best features are selected as feature subsets of the big data set.

Let there be a sample set $X = \{X_1, X_2, X_3, \cdots, X_n\}$ to be classified, and n is the number of elements in the sample, and c represents the number of target clusters. Then there is the following data mining matrix for n elements corresponding to class c:

$$\mu_c = \begin{bmatrix} \mu_1, \mu_2 \cdots \mu_n(n \leq \vartheta) \\ \vdots \\ \mu_{c1}, \mu_{c2} \cdots \mu_{cn}(c \leq \vartheta) \end{bmatrix} \tag{3}$$

Where, μ_{cn} represents the mining matrix feature of the n-th element to the c-type data ($1 \leq i \leq c, 1 \leq j \leq n$), and meeting $\min J(X, \mu, v) = \sum_{i=1}^{c} \sum_{j=1}^{n} \mu_{cn} d_{cn}^2$, then the problem of clustering this multivariate data set is converted into a simple problem of finding the minimum value of the objective function.

The cloud computing-based big data set clustering mining process is based on the original big data set mining, adding cloud computing technology, adding constraints to the objective function to enforce the cluster search that satisfies the condition, and making the monitoring information constrained clustering search process.

The above assumes an unmarked big data set $X = \{X_1, X_2, X_3, \cdots, X_n\}$, $X_n \in R_n$, divide it into class K, which is $C_1, C_2, C_3, \cdots, C_k$, and the mean of none class is $M_1, M_2, M_3, \cdots, M_k$. Suppose the number of samples in the K-th class is N_K, then

$$m_K = \frac{1}{N_K} \sum_{i=0}^{k} X_i, \quad K = 1 \cdots K.$$

According to the European distance and intra-class error squares and criteria, the objective function of cloud-based big data set clustering is $J = \sum_{i=1}^{k} K = 1 \sum_{i=1}^{Nk} |X_i - m_K|^2$. When the algorithm is initialized, the center of each class is randomly selected, so the selection of the initial center determines the quality of the clustering results. After the introduction of cloud computing technology, a large data set formed by a small number of labeled samples, the large data set contains all K clusters, and each class contains at least one sample to implement a cloud computing-based big data set mining process.

2.3　Implementation of Large Data Set Clustering Algorithm

The cloud computing-based heterogeneous network privacy big data set clustering algorithm is implemented as follows:

The clustering algorithm requires two parameters ε and μ when executed, known as

$$\mu_c = \begin{bmatrix} \mu_1, \mu_2 \cdots \mu_n (n \le \vartheta) \\ \vdots \\ \mu_{c1}, \mu_{c2 \cdots \mu_{cn}} (c \le \vartheta) \end{bmatrix}, \text{ and } \varepsilon \text{ represents the spatial dimension of the heteroge-}$$

neous network, up to the dimension [12–14], so no orientation analysis is done.

Search for the number of core data points by checking the ε-domain dimension of the arriving data point in the current time. If the ε field of any data point P contains at least μ data points, create a data matrix with data point P as the core point. Then, by means of breadth search, the data points that can be directly clustered from these core data points are aggregated, and all the obtained density from the data point P is assigned to one class.

If P is the core data point, the cluster data points starting from point P are marked as the current class, and the next step is extended from the center of the matrix. If P is not a core data point, then when the algorithm clusters, the next data point will continue to be processed, in order, until a complete cluster core data point is found. Then select an unprocessed core data point to start expansion, and get the next clustering process, in sequence, until all data points are marked [15, 16].

For data points that are not added to the clustering matrix, they are noise points, and temporarily store them in the invalid area. If the number of data in the invalid area exceeds the maximum range of the preset threshold, the calling algorithm clusters the data in the temporary storage area, and deletes the already clustered data points from the temporary storage area. The dynamic data of the quadratic clustering is recorded as Q, and the clustering calculation process for Q is as follows:

(1) A large data set of mixed attribute features is processed by using different distance calculation methods, and new data point features are calculated by using Eq. (2).
(2) Perform online maintenance on the characteristics of large data sets, and perform mining processing after maintenance.
(3) The clustering algorithm is executed, and if there is data that is not clustered, it is placed in the temporary storage area.
(4) The data feature matrix is again mined and the clustering algorithm is executed until the core data points are found.

The cloud computing technology is used to guide the clustering implementation process of big data sets [17, 18], which solves the problem that the single algorithm clustering quality is not high. First enter the data point $x \in X = \{d_1, d_2, \cdots, d_n\}$ in memory, d_1 represents the data point in memory. The implementation of the big data set is replaced by a triple, which is equivalent to the center point with the weight to participate in the clustering, the number of data points is the weight, and the clustering result is the mark set, then there is $labels = U^k$. Outputting K sets of disjoint big data matrices $\{X_1\}^{k=1}$ of x, and the objective function is $J \in \sum_{i=1}^{k} J = ik$, then the local optimal clustering process of the big data set is obtained.

So far, the cloud data-based heterogeneous network privacy big data set clustering algorithm design is completed.

3 Simulation Experiment Demonstration and Analysis

In order to ensure the effectiveness of the cloud computing based clustering algorithm for heterogeneous network privacy big data sets, the simulation experiments are carried out.

Set the experimental object to the privacy UCI data set of a heterogeneous network, and perform clustering calculation on it.

In order to ensure the effectiveness of the experiment, the traditional algorithm and cloud based clustering algorithm are compared, and the accuracy of the two algorithms is statistically analyzed. The experimental results are shown in Table 1, Table 2 and Fig. 3.

Table 1. Traditional clustering algorithm results

Category	Network dimension	Flow	Privacy agencies	Accuracy	Data matrix
Error rate (%)	0.59	0.85	0.25	0.36	0.48
Cluster velocity measurement (v/ms)	23.6	15.4	53.6	41.2	25.9

Table 2. Cloud computing based clustering algorithm results

Category	Network dimension	Flow	Privacy agencies	Accuracy	Data matrix
Error rate (%)	0.12	0.05	0.02	0.11	0.03
Cluster velocity measurement (v/ms)	68.3	72.5	95.6	82.6	75.6

According to the data in Tables 1 and 2, the error rate of traditional clustering algorithm is higher than that of this clustering algorithm, and its clustering speed is

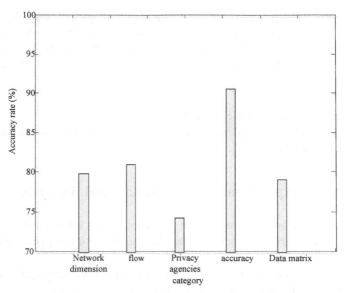

(a) Accuracy of traditional clustering methods

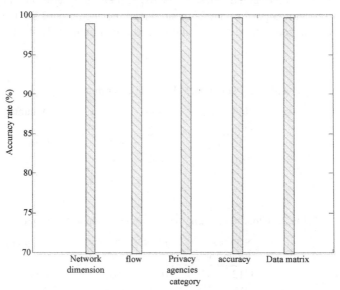

(b)The accuracy of clustering method in this paper

Fig. 3. Accuracy Analysis of two clustering algorithms

slower than that of this clustering algorithm. It shows that the heterogeneous network privacy big data clustering algorithm has better effect than the traditional algorithm.

According to Fig. 3, the clustering accuracy of this algorithm can reach 99%, while that of the traditional clustering algorithm is only 91%, which indicates that the clustering accuracy of this algorithm is higher than that of the traditional algorithm.

To sum up, cloud computing based clustering algorithm in the heterogeneous network privacy big data set clustering process, regardless of the clustering speed, or deal with each heterogeneous network privacy structure, traffic and dimension, it is better than the traditional algorithm processing results. It can be seen that the cloud computing based heterogeneous network privacy big data agglomeration algorithm not only improves the clustering accuracy of private data sets in heterogeneous networks, but also improves the stability of the calculation process. The clustering error increases to zero gradually.

4   Conclusion

This paper analyzes and designs the clustering algorithm of heterogeneous network privacy big data sets based on cloud computing, and uses the advantages of cloud computing technology to collect and extract the matrix features of large data sets. Combining the similarity method, mining large data sets and realizing the clustering algorithm design of big data sets. The experimental results show that the cloud computing-based clustering algorithm designed in this paper is extremely efficient. When performing clustering calculation of the privacy big data set in the heterogeneous network, it greatly improves the accuracy of the clustering calculation, and can effectively reduce the clustering error, save the calculation time, and improve the working efficiency of the clustering algorithm. It is hoped that the research in this paper can provide theoretical basis and reference for China's heterogeneous network privacy big data set clustering algorithm.

References

1. Xinchun, Y.C., et al.: Application of multi-relational data clustering algorithm in internet public opinion pre-warning on emergent. Int. Engl. Educ. Res. 1, 16–19 (2019)
2. Dongmei, C.: Discussions on big data security and privacy protection based on cloud computing. Comput. Knowl. Technol. 15(15), 101–103 (2018)
3. Ma, R., Angryk, R.A., Riley, P., et al.: Coronal mass ejection data clustering and visualization of decision trees. Astrophys. J. Suppl. Ser. 236(1), 14–17 (2018)
4. Mingbo, Pan: Research on privacy protection algorithms for network data in large data environment. Microelectron. Comput. 34(7), 101–104 (2017)
5. Wenzheng, Z., Zaiyun, W., Afang, L.: Relevant analysis of big data security and privacy protection based on cloud computing. Netw. Secur. Technol. Appl. 15(4), 59–63 (2018)
6. Hodi: Analysis of big data security and privacy protection in cloud computing. Electron. World 25(16), 98–101 (2017)
7. Yang, H.: Privacy protection of large data security based on cloud computing. Netw. Secur. Technol. Appl. 25(11), 86–87 (2017)

8. Arman, O., Rahmat, K., Mehrdad, T.H., et al.: Direct probabilistic load flow in radial distribution systems including wind farms: an approach based on data clustering. Energies **11**(2), 310–311 (2018)
9. Jinbo, X., Jun, R., Lei, C., et al.: Enhancing privacy and availability for data clustering in intelligent electrical service of IoT. IEEE Internet Things J. **6**(2), 1530–1540 (2018)
10. Yating, W.: Exploration of big data security privacy and protection based on cloud computing. J. Heihe Univ. **15**(6), 85–87 (2018)
11. Liu, S., Li, Z., Zhang, Y., et al.: Introduction of key problems in long-distance learning and training. Mob. Netw. Appl. **24**(1), 1–4 (2019)
12. Shudong, H., Yazhou, R., Zenglin, X.: Robust multi-view data clustering with multi-view capped-norm K-means. Neurocomputing **311**, 197–208 (2018)
13. Sun, G., Liu, S. (eds.): ADHIP 2017. LNICST, vol. 219. Springer, Cham (2018). https://doi.org/10.1007/978-3-319-73317-3
14. Allab, K., Labiod, L., Nadif, M.: A semi-NMF-PCA unified framework for data clustering. IEEE Trans. Knowl. Data Eng. **29**(1), 2–16 (2017)
15. Li, T., Pintado, F.D.L.P., Corchado, J.M., et al.: Multi-source homogeneous data clustering for multi-target detection from cluttered background with misdetection. Appl. Soft Comput. **60**, 436–446 (2017)
16. Fu, W., Liu, S., Srivastava, G.: Optimization of big data scheduling in social networks. Entropy **21**(9), 902 (2019)
17. Chang, X., Wang, Q., Liu, Y., et al.: Sparse regularization in fuzzy -means for high-dimensional data clustering. Cybern. IEEE Trans. **47**(9), 2616–2627 (2017)
18. Liu, S., Lu, M., Li, H., et al.: Prediction of gene expression patterns with generalized linear regression model. Front. Genet. **10**, 120 (2019)

Research on Collaborative Classification of E-Commerce Multi-attribute Data Based on Weighted Association Rule Model

Yi-huo Jiang[✉]

Fuzhou University of International Studies and Trade, Fuzhou 350202, China
hbgv96012@126.com

Abstract. Because the association between multi-attribute data of e-commerce is not obvious, the traditional collaborative classification method of e-commerce multi-attribute data has the problem of low classification accuracy. Therefore, the weighted association rule model is introduced to realize the optimal design of collaborative classification method of e-commerce multi-attribute data. Firstly, the weighted association rule model is built, and the multi-attribute data is mined and cleaned under the e-commerce platform. Taking the processed e-commerce data as the sample, the multi-attribute data classification index of e-commerce is determined. Through setting project weight, e-commerce data attributes and calculating multi-attribute relevance, multi-attribute data collaborative classifier is obtained. In the weighted association rule model, the collaborative classifier is used to get the multi-attribute data collaborative classification results of e-commerce. Compared with the traditional collaborative data classification methods, it is concluded that the accuracy of collaborative data classification is improved under the e-commerce platform of clothing and food 24.22%.

Keywords: Weighted association rule model · E-commerce · Multi-attribute data · Data collaborative classification

1 Introduction

E-commerce usually refers to a new type of business which is carried out by the buyer and the seller without meeting each other in a wide range of business activities all over the world, under the open Internet environment, to realize online shopping of consumers, online transactions and online e-payment among merchants, as well as various business activities, transaction activities, financial activities and related comprehensive service activities Business operation mode [1–3]. In the new business operation mode of e-commerce, one of the most important problems faced by merchants is how to obtain the market demand information of commodities in time and actively, develop the purchasing potential of customers, and find the corresponding superior commodities as soon as possible, so as to adjust the business plan as quickly as possible [4–6]. E-commerce, as the business model of information society, is developing at a faster speed than people expected. The U.S. government decided in the morning that e-commerce is a major information construction project, and formulated the "global

© ICST Institute for Computer Sciences, Social Informatics and Telecommunications Engineering 2021
Published by Springer Nature Switzerland AG 2021. All Rights Reserved
S. Liu and L. Xia (Eds.): ADHIP 2020, LNICST 347, pp. 377–388, 2021.
https://doi.org/10.1007/978-3-030-67871-5_34

e-commerce framework". From the perspective of China's domestic situation, the overall situation of Internet development determines the success or failure of e-commerce marketing. According to the report of China Internet Network Information Center, as of June 30, 2018, the number of Chinese Internet users reached 80200. However, with the increasing scale of e-commerce websites, the types of goods stored in websites and the relationship between these types are becoming more and more complex. In order to facilitate the users' query and respond to the requirements of the market, e-commerce websites must first classify a large amount of commodity data, and then achieve the goal of information extraction and intelligent market decision-making [7–10].

E-commerce multi-attribute classification is one of the most important technologies in the application field of e-commerce voucher, and many algorithms have been proposed so far. E-commerce multi-attribute classification is a technology which constructs a classifier according to the characteristics of data set and assigns a class to the unknown class samples. The process of constructing classifier is generally divided into two steps: training and testing. In the training stage, the characteristics of the training data set are analyzed to produce an accurate description or model of the corresponding data set for each category. In the test phase, we use the description or model of category to classify the test and test its classification accuracy. E-commerce website can use the classification algorithm in data mining, through training and testing, construct the classifier of commodity information and category, and realize the automatic classification of commodities [11–13].

At present, multi-attribute data classification methods in e-commerce platform include decision tree classification method, support vector machine classification method and K nearest neighbor classification method, but there are some problems in current classification methods, such as low classification accuracy and long classification time. In order to solve the above problems, a weighted association rule model is proposed. General association rule mining assumes that all items in the database have the same importance. When calculating frequent item sets, only the frequency of items is considered. But in some application areas, users pay different attention to different projects, that is, the importance of projects is different. In order to reflect the importance of the project, project weighting is introduced. After project weighting, when mining the weighted association rules, the frequency of project set appearing in transaction database and the weight of project should be considered comprehensively. At the same time, the weighted process can not only distinguish the importance of the project, make the mining results more reasonable, but also greatly improve the efficiency of the algorithm. Because in the association rule algorithm, the main machine operation time will be consumed in the stage of generating frequent item sets. If the irrelevant items with small weight are cut off in the early stage of generating frequent sets, the time complexity of the algorithm can be effectively reduced. Through the introduction of weighted association rule model, classification collaboration can be realized, so as to improve the ability of e-commerce multi-attribute data classification.

2   Design of Collaborative Classification Method for Multi-attribute Data in E-Commerce

From a mathematical point of view, e-commerce multi-attribute data classification is a process of mapping. It maps the data of unspecified categories to the original categories. The mapping can be one-to-one or one to many, because in some cases, a product can be associated with multiple categories, which can be expressed as follows:

$$f : A -> B \tag{1}$$

Where a is the set of commodities to be classified and B is the set of categories in the classification system. The mapping rule of e-commerce multi-attribute data classification is that the system summarizes the classification rules according to the information of each classification in the analyzed samples, and then establishes the discrimination formula and rules. When the new data comes, according to the discrimination rules, determine the commodity related categories.

2.1   Building Weighted Association Rule Model

The essence of association rule mining is to find hidden patterns or causal relationships between projects in a large number of complex information data carriers. Because the theoretical basis of association rule technology is easy to understand and accepted by people, and the extensive and far-reaching application space in the future, a large number of researchers have conducted extensive research and improvement on it, and algorithms with representative ideas of the times have emerged. The weighted association rule model can be described as: set D as transaction database, transaction number as N, I as all item sets in the database, and the weight set corresponding to I as W, where w_j indicates the importance of project i_j. $Support(X)$, $Confidence(X)$ support and confidence of data sample x, respectively. w min sup is the weighted support threshold. The form of the weighted association rules discussed is:

$$X \Rightarrow Y \tag{2}$$

If there are association rules in formula 2 in e-commerce multi-attribute data set D, the support degree is the percentage of the number of data containing $(X \cup Y)$ in the total number of transaction database d, that is, the probability of occurrence of event $(X \cup Y)$ is:

$$Support(X \Rightarrow Y) = P(X \cup Y) = \frac{N_{X \cup Y}}{N_{total}} \tag{3}$$

Where and respectively represent the number of tuples and the total number of tuples containing X and y. Then the weighted support degree of weighted association rules is:

$$W \ \text{sup}(X \Rightarrow Y) = \left(\sum_{i_j \in X \cup Y} w_j \right) \times Support(X \Rightarrow Y) \tag{4}$$

If the weighted support of itemset x satisfies the condition in formula 5, it is called weighted frequent itemset.

$$W \ \text{sup}(X) \geq w\text{min sup} \tag{5}$$

In addition, external confidence is a determinacy measure used to express the validity of association rules. If rule $X \Rightarrow Y$ of transaction data set D is used, confidence *Confidence* is defined as the ratio of the number of transactions X and Y occur at the same time in D to the number of transactions x only occur, that is, conditional probability $P(Y|X)$ is:

$$Confidence(X \Rightarrow Y) = P(Y|X) = \frac{N_{X \cup Y}}{N_{total}} \tag{6}$$

The principle and steps of the proposed model algorithm for weighted association rules are as follows: Taking the user's one-time login as the transaction division, the user resource download transaction table is generated according to the recent download record table. Recently, the storage structure of the download record table is user ID, login time, user questions, query time, downloaded documents, download time. Then, the resource weight table is generated according to the recently downloaded resource scale and article description table, and the coverage of frequent itemsets set by users is accepted to generate the minimum support min*Support*. According to resource weight table and user resource download transaction table, weight association rules are generated by frequent item set discovery based on Apriori algorithm. In the first step, scan the database, search the maximum transaction length size in the database and return the value. In the second step, access resource weight table and generate resource weight sequence in descending order. In the third step, according to the returned result size and Weight set w calculation:

$$\Delta = \frac{1}{\sum_{j=1}^{Size} w_j} \tag{7}$$

Minimum weighted support is generated from min*Support* and Δ. In the fourth step, Apriori algorithm is used to generate frequent sets with min*Support* as the minimum support degree, and the weighted support degree of item set is filled. w*minSupport* is used as the minimum weighted support to filter the weighted frequent item set, and weighted association rules are generated based on the weighted frequent set.

2.2 Mining E-Commerce Data

According to the user's behavior, such as user registration information, user rating data, and user browsing behavior, the collaborative classification method of e-commerce multi-attribute data establishes the user's behavior interest model. Therefore, it is necessary to mine the corresponding e-commerce user information data. The data mining process is shown in Fig. 1.

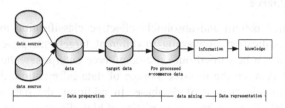

Fig. 1. Data mining flow chart of e-commerce

The user data used mainly includes the following three categories: user profile, user browsing record and user behavior characteristics. A real-time acquisition program is installed on the e-commerce platform, which stores all the running data in the platform into the memory, and takes it as the data sample of e-commerce multi-attribute data collaborative classification.

In order to ensure the effectiveness of the collaborative classification results of e-commerce multi-attribute data, and reduce the time consumed by classification work, the preliminary e-commerce data is cleaned. According to the different data quality problems, the cleaning of e-commerce data includes three parts: the cleaning of removing advertising words and commodity titles. The rule-based method is used to find out the advertising words in the product information and remove them. This step is an offline processing step. We design a rule-based method to clean the advertising words. Therefore, we establish a rule base of advertising words and clean the advertising words in the product information based on the rule base. All the rules in the rule base are described in the form of regular expressions. According to the different categories of advertising words, we divide the rules in the rule base into three categories, which are characteristic word rules, specific part of speech combination and rules of relations between goods. For goods in different categories, except for advertising words, other similar rules are relatively small, among which the rules in the first and second categories are added to the rule base by artificial settings, The third kind of rules are obtained by machine statistical learning. There are still some data quality problems in the product title after removing the advertisement words, such as the emergence of special punctuation, the abnormal segmentation and combination of useful information of the product, the repetition and contradiction of the product information, etc., After removing the advertisement words, the result is "10 times light changing 920000 screen Nikon digital camera s8100v/s9100 is better than s8000s8200", and the symbol "V" should replace s8000s8200 with a space, which is an abnormal combination phenomenon and contradicts the description of s8100s9100. Regular expression is used to eliminate the problems of special punctuation and abnormal segmentation and combination of commodity information. At the same time, the relative position of words in commodity title is recorded, so that the attribute tuples matching in part of speech tagging have different weights, so as to reduce the interference of commodity information conflict on entity recognition.

2.3   Determine the Multi-attribute Data Classification Index of E-Commerce

There is no unified pattern and absolutely effective classification method for multi-attribute data classification. According to different target enterprises, multi-attribute data classification can be carried out. In e-commerce enterprises, the most prominent characteristics of data are the massive storage of data information, the fast update of data, and the dynamic change of database. In addition, the data behavior data in e-commerce model often has the characteristics of high dimension, many variables and incompleteness. Based on the operation and storage characteristics of e-commerce multi-attribute data, the corresponding classification index is determined, and the classification index system is shown in Fig. 2.

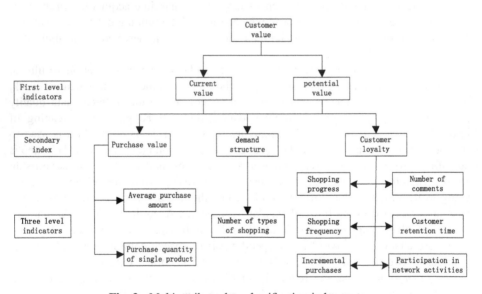

Fig. 2. Multi attribute data classification index system

In the multi-attribute data classification index system of e-commerce, network activity participation is a qualitative index, which is measured by two dimensions of high and low network activity participation, and other indexes are quantitative indexes.

Project refers to the document resources in e-commerce platform. The novelty of resources is the most important parameter to attract users. That is to say, the description time of resources is selected as the parameter to generate the weight of resources. The specific project weight is set as follows: set the resource sequence in the resource recent download scale as $(d_1, d_2, \cdots, d_n)$, the corresponding description time series is $(t_1, t_2, \cdots, t_n)$. Suppose T_{max} is the latest time in the description time series, T_{min} is the oldest time, Then the calculation formula of the weight value of resource d_n is:

$$\eta_j = a + (1+a) * \frac{(t_j - T_{min})}{(T_{max} - T_{min})} \tag{8}$$

Among them, α is a constant, and the specific value can be determined according to the requirements of e-commerce platform.

E-commerce data sets usually have different kinds of attributes, including character attributes and numerical attributes. The numerical attributes can be divided into sequential attributes, discrete value attributes and continuous value attributes. Character class attributes are usually external categories of key values. Numerical attribute is the quantitative record of variables, in which the order attribute is to arrange the key values in order and express the order with numbers, and the discrete value attribute is the discrete value key value without operational significance, while the continuous value attribute is the most common numerical attribute. Table 1 shows the list of attribute words of e-commerce data.

Table 1. List of attribute words of e-commerce data

E-commerce data	Attribute word			
	Acer	Adapter	......	Zoom
X_1	valueser [1]	valueser [2]	......	valueser[n]
X_1	valueser [1]	valueser [2]	......	valueser[n]
......	valueser [1]	valueser [2]	......	valueser[n]
X_1	valueser [1]	valueser [2]	......	valueser[n]

Improve the relevance between projects in order to predict users' rating of projects more accurately. The revised formula for calculating the project forecast score of users is:

$$P_{a,p} = \bar{R}_p + \frac{\sum_{n \in MAI} AC(p,n)(R_{a,n} - \bar{R}_n)}{\sum_{n \in MAI} AC(p,n)} \tag{9}$$

$AC(p, n)$ is the confidence level of association rules between item P and item n. Mai is the most recently associated set of items for item P.

2.4   Install Data Collaborative Filtering Classifier

Because there are some differences in the number of positive and negative samples in the small classifier, the classifier chooses the class weighted association rule model classifier. According to the classification principle of weighted association rule model, a set of training samples is given l, Training sample (x_i, y_i) with space dimension D, according to the data attribute association relationship of e-commerce represented by formula 10, two kinds of data are classified based on the mining data samples.

$$H : \omega \cdot x + b = 0 \tag{10}$$

The specific data classification process can be expressed as follows:

$$\begin{cases} H_1 : y = \omega \cdot x + b = +1 \\ H_2 : y = \omega \cdot x + b = -1 \end{cases} \tag{11}$$

Where ω is the reciprocal of the distance from H_1 to h, when the value meets condition 12, the sample can be separated accurately, that is to say, it meets the following requirements:

$$y_i[(\omega \cdot x_i) + b] - 1 + \xi_i \geq 0 \tag{12}$$

ξ_i is the relaxation factor. Penalty factor C is introduced into the mathematical model of the classifier, and the classification function is as follows:

$$f(x) = sign\left(\sum y_i C(x_i, x) + b\right) \tag{13}$$

In the data collaborative filtering classifier, the penalty parameters are divided into C_+ and C_-, the corresponding penalty factors are positive and negative.

2.5   Implement Collaborative Classification of E-Commerce Multi-attribute Data

The mining and processing of e-commerce data is input into the weighted association rule collaborative classifier, and the similarity between the electronic data and each attribute is calculated respectively, so as to get the multi-attribute data collaborative classification results of e-commerce. The similarity calculation formula is:

$$sim(u, v) = \frac{\sum_{i \in I_{uv}} (R_{ui} - \bar{R}_u)(R_{vi} - \bar{R}_v)}{\sqrt{\sum_{i \in I_{uv}} (R_{ui} - \bar{R}_u)^2} \sqrt{\sum_{i \in I_{uv}} (R_{vi} - \bar{R}_v)^2}} \tag{14}$$

Where $sim(u, v)$ represents the similarity between u and V of e-commerce data, R_{ui} and R_{vi} represent the attributes of data u and V to e-commerce data I respectively, $\bar{R}_u$ and $\bar{R}_v$ set of items for data u and V. Set the threshold value of e-commerce multi-attribute data collaborative classification as χ, if the calculation result in formula 14 is greater than the threshold value, then the e-commerce data u and V have the same attribute and can be divided into the same category, otherwise, calculate the similarity of the next group of data until all the e-commerce data mined are classified.

3 Comparative Experimental Analysis

The purpose of the experiment is to test the performance of the algorithm and algorithm under a single minimum support degree, the performance of the classification method under the multi weighted association rule model, and compare the test results under different environments, and analyze the data classification results before and after the application of the multi-attribute data collaborative classification method in e-commerce.

3.1 Experimental Environment and Preliminary Preparation

The operating environment of the experiment is window xp operating system, inter (R) core (TM) 2 Duo T6500 (2.10 GHz) CPU, 2G memory, written in C++ language. In the experiment, IBM data generator was used to generate different data sets with different transaction number, different project number and different average transaction width under XP system. The parameters of each group of experiments are the same except that the contrast parameters are variable. IBM is a classic data set synthesis tool, which is used to generate standard experimental data in association rule mining research. Due to the real-time change of e-commerce data, in order to reflect the collaborative design of classification methods, the implementation environment of e-commerce multi-attribute data collaborative classification method based on weighted association rule model is different e-commerce platforms.

Because the weighted association rule model is applied in the design of e-commerce multi-attribute data collaborative classification method, the relevant operation parameters of the model need to be set, and the model setting interface is shown in Fig. 3.

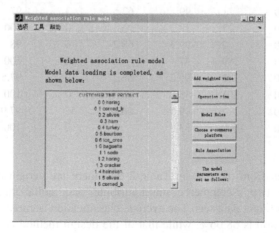

Fig. 3. Parameter setting interface of weighted association rule model

3.2   Experimental Process

Set the accuracy rate as the measurement index of the classification method, and the solution method of the index is the coincidence rate of the set e-commerce classification data and the classification method output data. In order to form the experimental comparison, the traditional e-commerce multi-attribute data collaborative classification method is set as the experimental comparison method and applied to the same e-commerce platform. Through the real-time collection and classification of e-commerce multi-attribute data, the classification results are output, and the experimental results about the accuracy rate are calculated.

3.3   Analysis of Comparative Experimental Results

Under the environment of clothing e-commerce, the experimental results about the accuracy rate of classification obtained through the statistics and statistics of data are shown in Table 2.

Table 2. Experimental results of clothing e-commerce environment classification

Data set	Class	Dataset size/MB	Traditional e-commerce multi-attribute data collaborative classification method			Design Collaborative classification method of multi-attribute data in E-commerce		
			Wrong score/MB	Accuracy %	Total accuracy %	Wrong score/MB	Accuracy %	Total accuracy %
1	A	25	0	100	76	0	100	100
	B	25	12	52		0	100	
2	A	100	1	99	63.5	3	97	98.5
	B	100	72	28		0	100	
3	A	300	2	99.3	68.2	6	98	99.0
	B	300	189	37		0	100	
4	A	750	6	99.2	66.2	19	97.5	98.7
	B	750	501	33.2		0	100	
5	A	1500	7	99.5	66.4	32	97.9	98.9
	B	1500	1000	33.3		0	100	

It can be seen from the data in Table 2 that there are certain differences in the accuracy of the two e-commerce multi-attribute data collaborative classification methods under different data sets. The average classification accuracy of the traditional classification method is 68.06%, while that of the design method is 99.02%, which is 30.96% higher than that of the design method.

In the same way, the collaborative classification results of e-commerce multi-attribute data are obtained under the food e-commerce platform, as shown in Table 3.

Table 3. Classification experiment results of clothing e-commerce environment

Data set	Class	Dataset size/MB	Traditional e-commerce multi-attribute data collaborative classification method			Design Collaborative classification method of multi-attribute data in E-commerce		
			Wrong score/MB	Accuracy %	Total accuracy %	Wrong score/MB	Accuracy %	Total accuracy %
1	A	25	8	68	84	0	100	100
	B	25	0	100		0	100	
2	A	100	36	64	82	3	97	98.5
	B	100	0	100		0	100	
3	A	300	126	58	79	6	98	99.0
	B	300	0	100		0	100	
4	A	750	274	63.5	81.7	19	97.5	98.7
	B	750	0	100		0	100	
5	A	1500	571	62	81	32	97.9	98.9
	B	1500	0	100		0	100	

Through the calculation of the data in Table 3, the average classification accuracy of the two multi-attribute data collaborative classification results is 81.54% and 99.02% respectively, compared with the classification accuracy of the design method increased by 17.48%. By synthesizing the classification results of multi-attribute collaborative data in different e-commerce environments, it is found that the design method can stabilize the classification accuracy of data above 99%, so it has high application performance.

4 Conclusion

It can be seen from the above that with the growing e-commerce market, the increasing number of commodities and the increasingly diversified levels of participants, it is increasingly difficult to provide valuable information for users. At this time, it is necessary to classify all kinds of goods in different levels to extract information intelligently from e-commerce market. It can quickly query the corresponding commodities for both sides of the market transaction, determine the purchase scheme of commodities, and then complete the placement strategy of commodities and recommend the commodities that may be of interest to users. The extraction of these information is completed on the premise of classification. The above classification methods will effectively improve the classification efficiency and accuracy of e-commerce market commodity data, and better serve businesses and customers.

References

1. Cheng, C.H., Chen, C.H.: Fuzzy time series model based on weighted association rule for financial market forecasting. Expert Syst. **4**(35), 110–115 (2018)
2. Cagliero, L., Garza, P., Kavoosifar, M.R., et al.: Discovering cross-topic collaborations among researchers by exploiting weighted association rules. Scientometrics **116**(2), 1273–1301 (2018)
3. Subbulakshmi, B., Deisy, C.: An improved incremental algorithm for mining weighted class-association rules. Int. J. Bus. Intell. Data Min. **13**(3), 291–308 (2018)
4. Murugan, I., Nabhan, A.R., Subramanian, A.: A weighted association rule mining method for predicting HCV-human protein interactions. Curr. Bioinform. **13**(1), 73–84 (2018)
5. Liu, S., Yang, G. (eds.): ADHIP 2018. LNICST, vol. 279. Springer, Cham (2019). https://doi.org/10.1007/978-3-030-19086-6
6. Fernandes, D.S.F., Domingues, M.A., Vaccari, S.C., et al.: Latent association rule cluster based model to extract topics for classification and recommendation applications. Expert Syst. Appl. **112**(6), 34–60 (2018)
7. Liu, H., Yang, S., Gou, S., et al.: Terrain classification based on spatial multi-attribute graph using polarimetric SAR data. Appl. Soft Comput. **68**(24–38), S1568494618301510 (2018)
8. Fu, W., Liu, S., Srivastava, G.: Optimization of big data scheduling in social networks. Entropy **21**(9), 902 (2019)
9. Gou, J., Hou, B., Ou, W., et al.: Several robust extensions of collaborative representation for image classification. Neurocomputing **348**(5), 120–133 (2019)
10. Liu, S., Lu, M., Li, H., et al.: Prediction of gene expression patterns with generalized linear regression model. Front. Genet. **10**, 120 (2019)
11. Xiao, H.-g., Deng, G.-q., Wen, T.A.N., et al.: A weighted association rules mining algorithm based on matrix compression. Meas. Control Technol. **37**(3), 10–13 (2018)
12. Gupta, K.O., Chatur, P.N.: Gradient self-weighting linear collaborative discriminant regression classification for human cognitive states classification. Mach. Vis. Appl. **31**(3), 1–16 (2020)
13. Liu, S., Liu, D., Srivastava, G., Połap, D., Woźniak, M.: Overview and methods of correlation filter algorithms in object tracking. Complex Intell. Syst. 1–23 (2020). http://doi.org/10.1007/s40747-020-00161-4

Design of Distributed Multidimensional Big Data Classification System Based on Differential Equation

Pei-ying Wang[✉]

Tianhe College of Guangdong Polytechnical Normal University,
Guangzhou 510540, China
wangpeiying258@sina.com

Abstract. In today's more distributed and disorderly network environment, how to organize this information simply and effectively, so that users can quickly obtain potentially valuable data is a common problem in all fields. The commonly used classification systems are based on genetic algorithms and orthogonal decomposition. These two types of systems have high memory usage and low classification accuracy. Aiming at the above problems, a distributed multidimensional big data classification system based on differential equations is designed. The system design is mainly divided into three parts: the first design system overall framework; the second design system hardware, including multidimensional data integration module, central processing module, storage module, result output and display module; third, designing multidimensional big data according to differential equation Classification software main program. The results show that compared with the big data classification system based on genetic algorithm and the big data classification system based on orthogonal decomposition, the classification accuracy of distributed multidimensional big data classification system based on differential equation is improved by 8.75% and 6.75%, and the system memory occupancy rate is improved. Reduce by 35% and 12%.

Keywords: Differential equation · Distributed multidimensional big data · Classification system

1 Introduction

In recent years, with the gradual development and widespread application of computers and the Internet, the amount of data in the Internet has gradually increased, but the rich data resources have made users face greater challenges. The large amount of data scattered and disorder has greatly increased people's The difficulty of using network information. Therefore, it is necessary to design a big data classification system to help users quickly and efficiently obtain the required information in a large amount of network data [1]. Big data is the concept of demand driven. With the popularization of database system and the expansion of Internet services, the data available to enterprises or individuals is expanding, and the existing technology is difficult to meet the data analysis needs in the era of big data. Therefore, we need to explore new theories and

S. Liu and L. Xia (Eds.): ADHIP 2020, LNICST 347, pp. 389–398, 2021.
https://doi.org/10.1007/978-3-030-67871-5_35

methods to support the application of big data. Although 4 V attributes of big data have been widely discussed, most of them describe the representation of big data, so it is difficult to abstract a unified data format. Therefore, it is necessary to find out the technical features that can be used for data formatting.

At present, there are many network big data classification systems, and relevant scholars have achieved good results. For the application requirements of big data with the main technical characteristics of distribution and mobility, reference [2] takes distributed data flow as the data expression carrier, and then designs the corresponding big data classification model and mining operator. At the same time, to solve the key problems of big data classification mining, the algorithm corresponding to the key steps is constructed, which proves the rationality of the micro cluster merging technology and the sample data reconstruction method in theory. Experiments show that the proposed classification model and algorithm of big data based on distributed data flow can not only greatly reduce the communication cost between network nodes, but also improve the global mining accuracy by about 10% on average (compared with the existing typical algorithm DS means). Although the time cost is slightly higher than DS means, the difference between them is very small under different data capacity tests, and the time climbing trend is similar. However, the storage module of the system does not use hierarchical structure, and the system memory occupation rate is high. KNN algorithm is a kind of big data classification algorithm with simple idea and easy implementation, but when the training set is large and there are many characteristic attributes, its efficiency is low and its time cost is large. To solve this problem, reference [2] proposes an improved KNN classification algorithm based on fuzzy C-means, which introduces the fuzzy c-means theory on the basis of the traditional KNN classification algorithm. In order to reduce the number of training sets, the sub cluster is used to replace all the sample sets of the sub cluster. Thus, the workload of KNN classification process is reduced and the classification efficiency is improved. KNN algorithm is better applied to data mining. The theoretical analysis and experimental results show that the algorithm can effectively improve the classification efficiency of the algorithm in the face of large data, and meet the needs of data processing, but the system's big data classification accuracy is low.

Aiming at the shortcomings of the above systems, a distributed multidimensional big data classification system based on differential equations is designed. The system design is mainly divided into three parts: system framework design, system hardware design and system software design. Finally, compared with the big data classification system based on genetic algorithm and the big data classification system based on orthogonal decomposition, the classification accuracy of distributed multidimensional big data classification system based on differential equation is improved, and the system memory occupancy is reduced. It can be seen that the performance of this system is better.

2 Big Data Classification System Based on Differential Equation

Classification systems have always played a very important role in the field of life and engineering. Speech recognition, handwriting recognition, identity recognition, etc. are all areas of classification system discussion. Because of its wide application value, the design and application of classification systems have always been valued [2].

In the past, organizing and organizing a large collection of original documents by manual means is not only time-consuming and laborious, but the effect may not be ideal. By directly filtering and classifying the data through the computer and submitting the parts that the user really needs to the user, the user can be freed from the cumbersome data processing work. Differentiating different types of data more quickly, systemizing a large amount of disordered data, greatly improving the utilization of information. Through the automatic data classification system, it can help users to organize and obtain information well, which is of great significance in improving the speed and accuracy of information retrieval, and has important research value [3].

A differential equation is a mathematical equation used to describe the relationship between a class of functions and their derivatives. It is widely used and can solve many derivative-related problems. Many kinesiology and dynamics problems involving variability, such as the resistance of air to the falling motion of the velocity function, can be solved by differential equations. Differential equations are different from linear equations, quadratic equations, higher-order equations, exponential equations, logarithmic equations, trigonometric equations, and equations. It is not the process of finding the relationship between the known number and the unknown, the equation of the column, and the process of finding the solution of the equation. Rather, the process of finding one or several unknown functions that satisfy certain conditions is the most viable mathematical branch equation. This time, the differential equations are combined with the big data classification system to design a distributed multidimensional big data classification system based on differential equations.

2.1 Overall Framework of Big Data Classification System

Big data classification is not simply to find one or several fixed values. In the past, the staff needed to find the relationship between each data and establish a relationship function to complete the classification process. However, based on the Internet cloud computing, a distributed multidimensional big data classification model is established, and different data can be effectively classified by differential equation algorithm. The classified distributed multidimensional big data includes unstructured environmental data and semi-structured environmental data [4]. The structure of the big data classification system is shown in Fig. 1.

From the perspective of the data classification system, the system includes five modules: multidimensional data integration module, central processing module, storage module, result output and display module.

The distributed multidimensional big data classification system described above shows the basic laws followed by data changes. As long as the corresponding differential

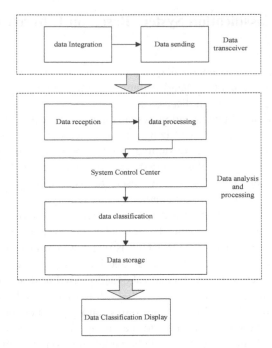

Fig. 1. Structure of big data classification system based on differential equation

equations are listed, the connections and differences between different data can be found and then classified.

2.2   Hardware Design of Distributed Multidimensional Big Data System

(1) Multidimensional data integration module

Distributed multidimensional big data in various forms, such as papers, bibliographies, conference records, journals, etc. These diverse data resources are often heterogeneous (structured, semi-structured, unstructured), so how to automatically migrate these massive, scattered, and heterogeneous data to a central station according to statistical criteria is data classification. The basis [5].

The multi-dimensional data integration mainly completes the operation through the CP2210 integrated chip CP2210, realizes big data acquisition, and then transmits the collected data to the central processor through the network interface. The specific process is as follows: The power supply sends 6 V voltage, which is transmitted to the voltage regulator of the MCU through the REGIN pin of the microcontroller, adjusts the voltage to the 4 V voltage required for the operation of the MCU, and sends the remaining 2 V voltage to the remaining components through the VDD pin. The MCU exchanges information with P3, P4 and other I/O pins. The signal obtained from the network passes through the signal adjuster, and the P25 pin of the single chip reaches the A/D converter, and the A/D converter converts the signal into corresponding data, thereby completing the collection of the network data.

(2) Central processing module

The role of the master chip is to control the operation of the entire system, all the programs of the system need to be written on the master chip. There are currently four main control chips on the market, such as microcontroller, FPGA, ARM and DSP. Among the above four main control chips, the data processing capability and operation speed of the DSP are optimal, which is in line with the design goal of the system. Here, the TMS320DM642 chip in the C6000 series specially designed for audio processing is selected from TI. The main hardware included in the TMS320DM642 is program memory FLASH, power supply circuit, clock circuit, reset circuit and JTAG port [6].

(3) Storage module

The data storage part includes three parts: FLASH, SDRAM and CF card. FLASH memory has the function of electric erasing and writing in the system, and the information is not lost after power-off. It is used to save the system self-starting code and system program code. This system uses ATMEL's AT29LV020 FLASH chip, which is a NOR type FLASH chip. The total capacity is 256 KB and the data bus is 8 bits. When the EMIFA boot mode is selected by the DSP, the program is automatically loaded from the CE1 space after power-on, so the FLASH must be connected to the CE1 space of the EMIF. The SDRAM memory has a high access speed. It is used to store the system running code and temporary image data. The system uses four Samsung SDRAM K4S561632E, each of which is 16 bits, 32 MB, and CE0 connected to the DSP's EMIF interface. space. The CF card is connected to the CE2 space of the EMIF to store the original image data and the recognition result [7].

(4) Result output and display module

A display is a window in which a person interacts with a robot. In the design of the system, a touch panel, that is, a touch screen, is selected as a display device for sorting results. The system design uses the TPC1063H touch panel produced by Shenzhen Kunlun Tongzhou Technology Co., Ltd., and its composition is shown in Fig. 2.

This touch screen has high performance: it is equipped with Cortex-A8/1G Hz main frequency CPU, which has fast response speed and fast communication speed, which can bring more extreme and smooth operation experience; More serial ports: There is a 232 + 2 485 communication serial port, and the integration of multiple serial ports makes the product easier to use; Complete compatibility: adapt to the market's mainstream touch screen manufacturers opening size [8].

2.3   System Software Design Based on Differential Equation

Differential equation is a kind of mathematical equation which describes the relationship between function and its derivative. Its solution is usually function, while the solution of equation in elementary algebra is usually numerical. The differential equation must first reduce the dimensionality of environmental data that can undergo

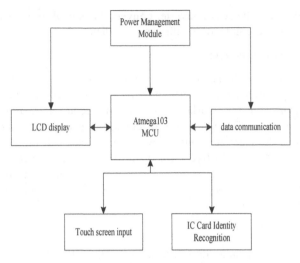

Fig. 2. TPC1063H touch panel

continuous changes, and minimize the possibility of data variation. After the data is reduced from high dimensionality to low dimensionality, each data is orthogonalized between the same dimensions, and the relationship between the data is judged by the cross result, and finally the same type of environmental protection data is grouped together and counted [9, 10].

In the case where the selected environmental data has different dimensions and the difference is relatively large, it is necessary to find the principal components between the different coefficients for analysis. Principal component analysis (PCA) is a statistical method. Through orthogonal transformation, a group of variables that may have correlation are transformed into a group of linear uncorrelated variables, and the transformed group of variables is called principal component. In practical projects, in order to analyze the problems comprehensively, many variables (or factors) related to this are often put forward, because each variable reflects some information of this project in varying degrees. Principal component analysis is first introduced by K. Pearson for non random variables, and then H. hotellin extended this method to the case of random vectors. The size of information is usually measured by the sum of squares or variance of deviations. The principal component analysis method makes the variables appear ellipsoidal distribution, and the distribution area is three-dimensional, and the linear relationship is extremely strong, and the analysis has significant significance. The formula for calculating the differential equation is as follows:

$$Y_i = k_i X = k_{i1} X_1 + k_{i2} X_2 + , \ldots, k_{in} X_n (i = 1, 2, \ldots, n) \tag{1}$$

In formula (1), Y represents the principal component data in i data, k_i represents the corresponding feature vector in the variable covariance matrix.

The data is put into the covariance matrix or the correlation coefficient matrix to find the eigenvector corresponding to the largest eigenvalue [11, 12]. The direction of the eigenvector corresponds to the covariance matrix variability, and the matrix direction represents the direction of the main environmental data [13]. The direction corresponding to the second largest eigenvalue is the direction of data mutation, which is orthogonal to the first eigenvector, and the degree of orthogonality can reflect the relationship between environmental data. The eigenvectors are used to measure the proportion of data in different directions, and the largest eigenvector is used to form a reference frame for dimensionality reduction. It should be pointed out that the number of feature vectors selected is lower than the dimension of the original data.

The general solution has been the main goal of differential equation in history. Once the expression of the general solution is found, it is easy to get the special solution needed by the problem. The expression of the general solution can also be used to understand the dependence on some parameters, so that the parameter value is appropriate, the corresponding solution has the required performance, and it is also helpful for other research on the solution. Later development shows that there are not many cases in which the general solution can be obtained. In practical application, it is necessary to find the special solution satisfying certain specified conditions. Of course, general solutions are helpful to study the properties of solutions, but people have shifted the focus of research to the problem of definite solutions. Differential equations can find and classify environmental data, and environmental data in the same category should have as high a homogeneity as possible, while categories should have as high a heterogeneity as possible [14]. The grouping is done by finding the distance and similarity between the environmental data. The specific steps are as follows:

(1) Create N observation points and seek K familiar data;
(2) Record the distance between two pairs of different observation points;
(3) The near observation points are unified into one category, and the distant observation points are counted into another category. Finally, the distance between groups is maximized and the distance within the group is minimized.

The data classification first uses the K-means classification method to obtain the ideal classification model, and classifies a large number of multidimensional data samples, and then uses hierarchical classification to find out the number of interaction classifications and analyze to provide accurate similarity information [15]. According to the similarity of the formation of big data, the hierarchical map is drawn to make the class division more intuitive and accurate. Different samples are self-contained. The differential equation is used to calculate the distance between the classes and the distance between samples.

$$M(x, y, z) = N(x, y, z) \cdot Z(x, y, z) \tag{2}$$

In formula (2), M represents the distance between classes, x, y, z represent different vector directions, $G(x, y, z)$ represents the vector direction of the first type of data, $Z(x, y, z)$ represents the vector direction of the second data. Inter-sample distance calculation process:

$$T(x, y) = \frac{e}{2g} \qquad (3)$$

In formula (3), $T(x, y)$ represents the distance between samples in the horizontal and vertical directions, e represents a regular vector constant, g indicates the number of types of data. After the above calculation is completed, the two types of data with the smallest distance are unified and the two types of data are the largest, until all the big data classification ends.

3 Experiment Analysis

3.1 Lab Environment

The data used in the experiment comes from the network information database. The system needs two computers. The system hardware configuration is: Intel Rean-core 3 GHz processor, 32 GB memory.

3.2 Parameter Settings

The data types used in the experiment are: meteorological data, geological data, economic data, transportation data, etc., and they are numbered as S 1, S 2, S 3, S4, etc.; the data size is 2000, respectively. 1500, 1700, 1800, etc. [12].

3.3 Result Analysis

The classification performance of distributed multidimensional big data classification system based on differential equation, big data classification system based on genetic algorithm and big data classification system based on orthogonal decomposition is compared. The experiment uses three systems to classify experimental big data.

(1) Classification accuracy

It can be seen from Table 1 that the average classification accuracy of big data in this system is 86.25%. Compared with the big data classification system based on genetic algorithm and the big data classification system based on orthogonal decomposition, the accuracy is improved by 8.75% and 6.75%. Because the system designed in this paper uses differential equation to design software content, and then improves the accuracy of data classification.

(2) System resource occupancy

As can be seen from Fig. 3, the CPU usage of the system is 15%, the memory usage is 20%, and the total occupancy is 35%. The total occupancy rate of the big data classification system based on genetic algorithm is 58%. The total occupancy rate of big data classification systems based on orthogonal decomposition is 70%. It can be seen that the resource occupancy rate of the system is significantly lower than the other two

Table 1. Big data classification accuracy results obtained by the three systems

Data type	Differential equation	Genetic algorithm	Orthogonal decomposition
S1 (%)	85	87	95
S2 (%)	88	90	95
S3 (%)	85	88	94
S4 (%)	87	88	96
Average value (%)	86.25	88.25	95

systems, which proves that the system has better performance. The reason for this result is that in the big data classification system designed in this paper, the content of hardware design specially designs data integration module, which reduces the occupancy rate of system resources.

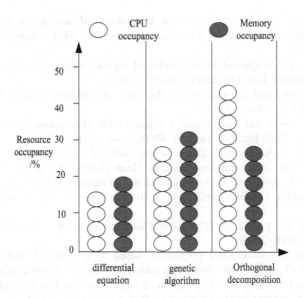

Fig. 3. System resource occupancy rate

4   Conclusion

In summary, for the big data classification system based on genetic algorithm and the big data classification system based on orthogonal decomposition, the classification accuracy is low and the system occupancy rate is high. The distributed multidimensional big data classification system based on differential equation is designed. The biggest feature of this system design is the application of differential equations. It has been verified that the accuracy of system classification is improved and the system resource occupancy rate is reduced. It can be seen that the performance of the system is improved. For the current distributed multi-dimensional data classification accuracy can not be significantly improved, the possible reason is that the data structure is different, resulting in the data structure can not meet the requirements of the current classification system. In the future, we can use the integrated heterogeneous classifier to meet the data samples of different data structures, so that the classification accuracy can be better improved.

References

1. Yin, S.: Research on the classification technology of large environmental data based on differential equation. Environ. Sci. Manag. **43**(247(6)), 126–129 (2018)
2. Mao, G., Hu, D., Xie, S.: Large data classification model and algorithms based on distributed data flow. J. Comput. Sci. **1**, 161–175 (2017)
3. Huang, S., Lyu, Y., Peng, Y., et al.: Analysis of factors influencing rockfall runout distance and prediction model based on an improved KNN algorithm. IEEE Access **7**, 66739–66752 (2019)
4. Xiaofeng, Z., Yingtao, C.: Improved technology of association mining based on mathematical model of partial differential classification. Mod. Electron. Technol. **40**(8), 36–38 (2017)
5. Min, F., Jun, L.: Design and implementation of big data classification system based on web network. Electron. Des. Eng. **26**(8), 106–109 (2018)
6. Zhe, X.: Design and research of web big data classification system. Comput. Knowl. Technol. **13**(17), 216–217 (2017)
7. Kexing, Z.: Design and implementation of feature data classification system in network big data platform. Mod. Electron. Technol. **40**(8), 25–28 (2017)
8. Luo, X., Cha, Z., Xu, H., et al.: Design of large data automatic classification and processing system based on cloud computing. Comput. Meas. Control **25**(10), 278–280 (2017)
9. Liu, S., Liu, D., Srivastava, G., et al.: Overview and methods of correlation filter algorithms in object tracking. Complex Intell. Syst. (2020). https://doi.org/10.1007/s40747-020-00161-4
10. Liu, S., Bai, W., Liu, G., et al.: Parallel fractal compression method for big video data. Complexity **2018** (2018)
11. Lu, M., Liu, S.: Nucleosome positioning based on generalized relative entropy. Soft Comput. **23**(19), 9175–9188 (2018). https://doi.org/10.1007/s00500-018-3602-2
12. Liu, B., Liu, C.: Automatic classification of large data stored in cloud data management system using content text categorization method. Electron. Technol. Softw. Eng. (20), 179–180 (2017)
13. Tang, Z., Srivastava, G., Liu, S.: Swarm intelligence and ant colony optimization in accounting model choices. J. Intell. Fuzzy Syst. **38**, 2415–2423 (2020)
14. Weihs, C., Ickstadt, K.: Data science: the impact of statistics. Int. J. Data Sci. Anal. **6**(3), 189–194 (2018)
15. Thanigaivasan, V., Narayanan, S.J., Iyengar, S.N., et al.: Analysis of parallel SVM based classification technique on healthcare using big data management in cloud storage. Recent Pat. Comput. Sci. **11**(3), 169–178 (2018)

Efficient Feature Selection Algorithm for High-Dimensional Non-equilibrium Big Data Set

Shuang-cheng Jia[(✉)] and Feng-ping Yang

Alibaba Network Technology Co., Ltd., Beijing 100102, China
xindine30@163.com

Abstract. When the traditional algorithm is used to calculate the feature classification of high-dimensional non-equilibrium and large data set, it is easy to appear the problem of low accuracy and recall rate of feature selection. Therefore, a feature selection algorithm based on granular fusion is designed. By using the regularization feature of the data, the original big data aggregate is transformed into a small-scale data subset. On the basis of this, the feature selection function of the data particle is obtained. Finally, the weight fusion calculation of each feature subset is carried out. The feature classification of high-dimensional non-equilibrium big data set is realized. The experimental results show that the feature selection algorithm based on granular fusion can realize the feature selection and recall of high dimensional unbalanced data sets. The accuracy of the method is higher than that of the traditional method, which shows that the method is feasible and effective.

Keywords: High dimensional data · Non-equilibrium feature · Granulation fusion · Feature selection

1 Introduction

Feature selection, as one of the preprocessing steps of data analysis and mining, is widely used in the fields of machine learning and pattern recognition [1]. With the rapid development of network and data acquisition technology, data sets with ultra-high dimension and unbalanced data are emerging constantly. Ultra-high-dimensional unbalanced data usually has a large number of redundant, independent features, which makes feature selection more difficult. The massive size of the data greatly affects the computing efficiency of feature selection, and sometimes ordinary microcomputers cannot even load all of the data. Therefore, the exploration is more efficient. The feasible feature selection algorithm for massive high-dimensional non-equilibrium data has important theoretical and practical significance. The granular fusion computing theory is an imprecise solution method to study big data's analysis. On the premise of guaranteeing the value of the data, the data scale is reduced, and the input of the problem is converted into multiple information grains from the original big data set. Can significantly reduce the size of the amount of data [2], the application of granular fusion computing theory to large-scale data processing has attracted the attention of many scholars. Liang et al. used clustering technology to granulate big data on cloud platform to reduce the loss of data information and improve the efficiency of time and

S. Liu and L. Xia (Eds.): ADHIP 2020, LNICST 347, pp. 399–408, 2021.
https://doi.org/10.1007/978-3-030-67871-5_36

resource utilization [3]. Yuan et al. aimed at large-scale time series data, the fuzzy information granulation method is used for granulation, and support vector machine is used for regression analysis and prediction on the particle, so as to improve the speed of time series data analysis [4].

Inspired by the above research, this paper proposes and designs a feature selection algorithm based on granular fusion. Making use of the advantages of granulation fusion theory, the selection process of high-dimensional non-equilibrium data sets is transformed into the feature selection process of small data scattered points. In order to ensure the effectiveness of the data feature selection algorithm designed in this paper, the experimental results show that the feature selection algorithm based on particle fusion has the advantages of traditional algorithm, and its accuracy and recall rate are higher than the traditional algorithm. It shows that the proposed algorithm is effective and practical.

2   Design of Feature Selection Algorithm for High-Dimensional Data Based on Granular Fusion

Facing the severe challenge to the traditional feature selection algorithm caused by massive high-dimensional non-equilibrium data, based on the granulation fusion perspective, this paper proposes a feature selection algorithm based on data set based on the granulation fusion theory. The algorithm is divided into three steps: granulation, feature selection on granulated data grains, fusion of granulated features to select [5], as follows:

2.1   High Dimensional Data Granulation Processing

In feature selection of high-dimensional non-equilibrium data set based on granulation fusion theory, statistical random sampling theory is used to split the original large-scale data set in the process of granulation [6]. In determining the size of the data set, the variance of the whole data set must be calculated first, and then, according to the attribute characteristics of the massive data, the hierarchical sampling theory is used to granulate the data set [7]. After granulation, the high-dimensional non-equilibrium large data sets are represented as shown in Fig. 1.

In this paper, the granulation of massive data is realized based on the theory of granulation fusion. The particle size can be calculated directly according to the size of the original data set N and given parameters r, which will greatly improve the efficiency of granulation of massive data, and meanwhile, the data volume contained in each particle is greatly reduced, and the feasibility of data particle selection in a subsequent single-machine environment is ensured [8].

When calculating, first enter the dataset X of the sample size N, and then output the eigenvalues of P data grains:

$$p = \frac{(X - X_P)}{N^r} \tag{1}$$

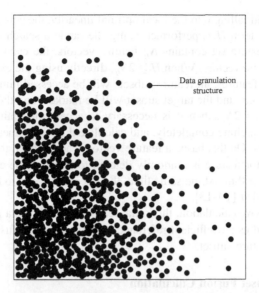

Fig. 1. Representation of granulated large datasets

Of which, p represents the eigenvalues of data grains; According to the theory of granulation fusion, the range of parameter r is as follows: $0.5 \leq r \leq 0.9$.

The above calculation realizes the granulation process of high-dimensional non-equilibrium large data sets, in which the eigenvalue p of the data grains involved can be calculated as a parameter of feature selection. The analysis shows that the p value is determined by X_P and N^r, and the exact p value is obtained, then the backup output is carried out, which provides the basis for the calculation of the characteristics of the data grains in the next step.

2.2 Data Particle Feature Selection

According to the original feature space, the candidate feature subset of the data particle is generated and used as the input of the feature selection algorithm [9]. For the search of feature items, the starting point of the search determines the search direction, so to start the search from the empty set H_i, the subsequent search process is the process of adding the selected features to the candidate feature subset in turn, then the function expression of the forward search is as follows:

$$H = \frac{x|pw_i|}{q_1} + \frac{x|pw_i|}{q_2} + \cdots + \frac{x|pw_i|}{q_n} \tag{2}$$

Of which, H represents forward search term of empty set H_i; $\frac{x|pw_i|}{q_1}$ represents degree of granulation of unbalanced coefficient x when the probability of scatter point is 1. In the same way, $\frac{x|pw_i|}{q_n}$ represents the degree of granulation of the disequilibrium coefficient x when the probability of scatter point is n.

In order to avoid falling into the local optimal linearity, the constraint calculation of the forward search term H is performed using the random search strategy: Assuming that the original feature set contains N_q feature vectors, the candidate feature subset may have $2N_q$ feature vectors. When $H \geq 2N_q$, directly using the exhaustive method to search spatial data features, all feature subsets will be accessed immediately according to the search direction, and the target subset will be marked, and the evaluation criteria will be given. If $H < 2N_q$, then it is necessary to access the optimal results of each feature subset by searching completely, and calculate the complementary balance factor of the data using N_q. On this basis, a feature vector is added from the candidate feature set, and a non vector feature is randomly deleted to improve the efficiency of feature selection in the algorithm. At the same time, the uncertainty due to high computational complexity is avoided [10–13].

Through the above calculation, the initial feature N_q of the data particle is obtained, and N_q is only used as a coefficient reference to prepare for the fusion calculation and selection of the feature subset.

2.3   Feature Subset Fusion Calculation

Feature subset evaluation is one of the important steps in the feature selection process of big data set. Every candidate feature subset needs to be evaluated by evaluation criteria. By introducing the relevant content of granular fusion theory, the fusion calculation process of feature selection of large data sets is transformed into the evaluation process of feature subset. The fusion evaluation form of the subset is shown in Fig. 2.

Figure 2 shows that the fusion of data particle subset mainly includes unidirectional fusion, pattern fusion, euclidean distance fusion, equal distance fusion and independent fusion. Introduce the above model into this calculation and write it down as u_i, and $u_i = \{u_1, u_2, u_3, \cdots, u_i\}$, the independent standard and the fusion standard are combined to evaluate the characteristics by the intrinsic attributes of the data.

At the same time, according to the specific learning algorithm, the distance criterion is used to measure the similarity between sample data in order to represent the contribution or effectiveness of features to classification and recognition. Feature selection using distance measures is generally based on the following assumptions:

The samples belonging to different classes of feature subsets are taken as the distance criterion f_1, and the absolute value coefficients of f_1 is calculated and measured in the form of $f_1 \rightarrow f_n$. If that measure set is found to be a null set, that $f_1 \rightarrow f_n = 0$, then we stop the transmission of high-dimensional data and continue to wait for the mining function in the data buffer until $f_1 \rightarrow f_n \neq 0$, then get the characteristic selection coefficient of the high-dimensional non-equilibrium data grains:

$$\sigma = \frac{1}{f_1 \rightarrow f_n \neq 0} \sum_{j=1}^{i} |u_{ij}| \tag{3}$$

Of which, σ represents characteristic selection coefficients of high-dimensional non-equilibrium data grains; u_{ij} represents Information entropy characteristics of data.

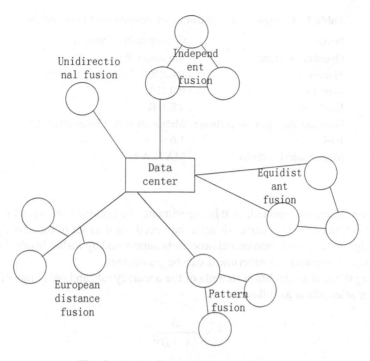

Fig. 2. Fusion form of data granular subset

If the selected big data feature can be constrained by σ, it shows that the feature of the data has the metric and can be recalled effectively. The correlation criterion of σ is used to select the feature subset of the data. The feature of big data can be constrained by σ. This indicates that the feature of the data has the metrology and can be recalled effectively. If the large data characteristic to be selected does not meet the constraint criterion of σ, the measurement performance of the data feature is not high, the effective recall rate is low, the feature selection is abandoned, the selection range is redefined until the large data characteristic to be selected can be restricted by σ, and the characteristic selection of the high-dimensional non-balanced large data set is finished according to the correlation coefficient between the features.

3 Experiment

In order to ensure the validity of the feature selection algorithm based on granulation fusion designed in this paper, the experimental analysis is carried out. The experimental environment is shown in Table 1:

The object of the experiment is a high-dimensional unbalanced and large data set, and the testability detection is carried out. The testability detection results accord with the standard, which shows that the experimental data have practical significance. At the same time, in order to ensure the preciseness of the experiment, the traditional data feature selection algorithm is used for comparison, and the accuracy and recall rate of

Table 1. Configuration information of development environment

Name	Configuration situation
Operating system	Microsoft Windows XP
Processor	Intel(R)Celeron(R) 2.6 GHz
Memory	6.0 GB
Hard disk	4.0 GB
Database management software	Microsoft SQL Server 2010 R2
JDK	1.6
Mathematical software	MATLAB

the two algorithms are counted. In this experiment, the accuracy and recall rate of the feature selection of the high-dimensional unbalanced big data set of the two algorithms are investigated, and the experimental process is supervised by Fisher Score. Therefore, the illustrative result of the experiment can be guaranteed.

Among them, the calculation formula of the accuracy rate and recall rate of big data set feature selection is as follows:

$$\gamma = \frac{A}{(A+B)^2} \tag{4}$$

$$\lambda = \frac{C}{(C+B)^n} \tag{5}$$

Of which, γ represents the accuracy rate of the characteristic selection of the large data set, λ represents the recall rate selected on behalf of the large data set feature, γ and λ are considered in percentage; A represents number of constraints representing data; B represents clustering coefficients with negative data constraints; C represents data set parameter mean; n represents constant, represents the number of experiments.

The experimental process is as follows: firstly, two kinds of feature selection algorithms are used to select the most valuable feature subset, then the original data is projected into the low-dimensional feature subspace to cluster, and the clustering algorithm adopts the simple mean algorithm. Finally, the feature selection results of large data sets are modified and backed up.

The traditional algorithm and this algorithm compare the feature selection recall rate of high-dimensional non-equilibrium large data set as shown in Table 2.

From the Table 1, we can see that the performance of the proposed algorithm is obviously better than that of the traditional feature selection algorithm. Although the value of data is lower than the characteristic value of constraint information, the clustering performance of data is greatly improved after global or local monitoring. The results show that the data feature selection algorithm based on granular fusion can effectively use the theory of granular fusion and unsupervised information for feature selection, improve the data recall rate, and verify the effectiveness of the algorithm.

Table 2. Comparison of maximum recall rates of dataset feature (%)

Data set	Sample	Features	Category	Traditional algorithm	Algorithm in this paper
Heart	270	16	2	66.7	74.6
Sphere	362	36	2	72.1	92.3
Sonar	246	49	3	59.6	86.4
Digits	189	16	4	78.3	91.6
Wine	4563	42	8	92.1	99.6
Image	961	26	9	59.3	86.4
Zoo	143	15	10	89.3	98.9

At the same time, in the data set Wine, the data recall rate of the two feature selection algorithms is ideal, which can make use of more features to achieve the highest clustering performance. However, the proposed algorithm still has some advantages, and its maximum data recall rate is 99.6%. It greatly increases the efficiency of feature selection for large data sets.

The accuracy of feature selection for high-dimensional unbalanced large data sets is compared by the two algorithms, as shown in Fig. 3.

In Fig. 3(a) shows the accuracy of feature selection of large data sets under the traditional method; (b) shows the accuracy rate of feature selection of large data sets in this paper. The analysis shows that, no matter how many pairs of constrained data are used, the accuracy of the proposed method is higher than that of the traditional method. At the same time, for the large data set in the form of global variance, the accuracy of feature selection between the two algorithms is not much different. But we can still see the advantages of this algorithm.

In addition, when using pairwise constraint data for feature selection, the accuracy of this algorithm for Sphere data selection is low, which is due to the linear characteristics of Sphere data. However, the accuracy of feature analysis of linear data is not high, which leads to the low accuracy of feature selection. In addition, all of the algorithms in this paper have overall advantages.

To sum up, the accuracy and recall rate of the proposed algorithm are higher than those of the traditional method, so it can be said that the high-dimensional non-equilibrium big data feature selection algorithm designed in this paper is effective and feasible.

In order to further verify the effectiveness of the algorithm in this paper, the efficient selection time of the high-dimensional unbalanced large data set features of the algorithm in this paper and the traditional algorithm is compared and analyzed. The comparison result is shown in Fig. 4.

According to Fig. 4, as the number of experiments increases, the efficient selection time of the high-dimensional unbalanced large data set features of the algorithm in this paper and the traditional algorithm is gradually increasing, but the efficient selection time of the high-dimensional unbalanced large data set features of the algorithm in this paper is steadily increasing. The time is within 20 s, while the traditional algorithm's high-dimensional unbalanced large data set feature efficient selection time is unstable,

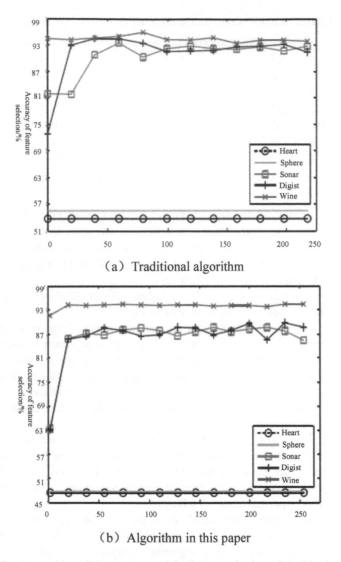

（a）Traditional algorithm

（b）Algorithm in this paper

Fig. 3. Comparison of the accuracy of the feature selection of the big data set

and the time is within 60 s, indicating that the high-dimensional unbalanced large data set feature efficient selection time of this algorithm is short.

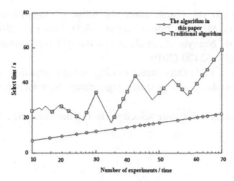

Fig. 4. Comparison and analysis of efficient selection time

4 Conclusion

A feature selection algorithm based on granularity fusion is designed for high-dimensional unbalanced large data sets. The original big data is aggregated into a small-scale data subset by using the regularization features of data. On this basis, the feature selection function of data particles is obtained. Finally, the weight of each feature subset is calculated. F realizes the natural classification of high-dimensional unbalanced large data sets. The effectiveness of the algorithm is verified by experiments. However, there are still a series of deficiencies in the research process of this paper. I hope that the next research can make theoretical analysis and practical test again to improve the applicability of the algorithm.

References

1. Zhe, J., Jianjun, H.: Prediction of lysine glutarylation sites by maximum relevance minimum redundancy feature selection. Anal. Biochem. **550**, 1–7 (2018)
2. Xiaofeng, N.: Research on locally discrete text data mining in high-dimensional data sets. Mod. Electron. Technol. **40**(19), 138–141 (2017)
3. Dan, Z., Chunming, W.: Feature selection based on improved quantum evolutionary algorithm. Comput. Eng. Appl. **54**(1), 146–152 (2018)
4. Hongxiang, D., Qiuyu, Z., Moyi, Z.: FCBF feature selection algorithm based on normalized mutual information. J. Huazhong Univ. Sci. Technol. (Nat. Sci. Ed.) **45**(1), 52–56 (2017)
5. Xin, Z., Haitao, W., Xuehong, C.: Feature selection algorithm based on random forest in Hadoop environment. Comput. Technol. Dev. **28**(255(07)), 94–98+104 (2018)
6. Zhao, X., Zhang, L.: High-dimensional unbalanced data set classification algorithm based on SVM. J. Nanjing Univ. (Nat. Sci.) **54**(2) (2018)
7. Liping, Y., Yunfei, L.: Zhu World Bank: anomaly detection algorithm based on high-dimensional data stream. Comput. Eng. **44**(1), 51–55 (2018)
8. Liu, S., Liu, D., Srivastava, G., et al.: Overview and methods of correlation filter algorithms in object tracking. Complex Intell. Syst. (2020). https://doi.org/10.1007/s40747-020-00161-4
9. Liu, S., Bai, W., Liu, G., et al.: Parallel fractal compression method for big video data. Complexity **2018**, 2016976 (2018). http://doi.org/10.1155/2018/2016976

10. Hongjun, Z.: Research on visualization algorithm of high-dimensional data in multi-dimensional data sets. Microelectron. Comput. **34**(5), 110–113 (2017)
11. Shuai, L., Weiling, B., Nianyin, Z., et al.: A fast fractal based compression for MRI images. IEEE Access **7**, 62412–62420 (2019)
12. Gaber, M.M., Philip, S.Y.U.: Data stream mining in fog computing environment with feature selection using ensemble of swarm search algorithms. New Gener. Comput. **25**(1), 95–115 (2018)
13. Li, J., Liu, H.: Challenges of feature selection for big data analytics. IEEE Intell. Syst. **32**(2), 9–15 (2017)

Industrial Automation and Intelligent Control

Risk Prediction Pattern Matching Method of Construction Project Management System in Big Data Era

Qiu-yi Li[✉]

Fuzhou University of International Studies and Trade, Fuzhou 350202, China
lqyl3799475957@163.com

Abstract. In view of the influence of the scale of pattern information set on the pattern matching of risk prediction, aiming at improving the performance of pattern matching of risk prediction, this paper puts forward the method of pattern matching of risk prediction of construction project management system in the era of big data. In the era of big data, the central idea of risk prediction pattern matching algorithm is analyzed. Based on the description of risk prediction pattern matching algorithm, the specific implementation steps of risk prediction pattern matching algorithm are designed, the design of risk prediction pattern matching algorithm is completed, and the risk prediction pattern matching process of construction project management system is combined to achieve the risk prediction pattern matching. The experimental results show that, under different experimental platforms, compared with other risk prediction pattern matching methods, the accuracy of risk prediction pattern matching method of construction project management system in the era of big data is higher.

Keywords: Big data era · Construction engineering · Project management system · Risk prediction model

1 Introduction

With the rapid development of computer technology, construction project management system has become an important guarantee of social development. There are many important construction project management information stored, transmitted and processed in the construction project management system, such as macro-control decision-making, commercial economic information, bank funds transfer, stock and securities, energy and resource data, scientific research data and other important information, which will inevitably attract all kinds of human attacks from all over the world (such as information leakage Information stealing, data tampering, data deletion, computer virus, etc. [1, 2]. At the same time, the construction project management system will also be tested by flood, fire, earthquake, electromagnetic radiation and other aspects. Construction project management system has become one of the serious social problems. Construction project management system is a comprehensive discipline involving computer science, network technology, communication technology, cryptography technology, information security technology, applied mathematics, number theory,

S. Liu and L. Xia (Eds.): ADHIP 2020, LNICST 347, pp. 411–424, 2021.
https://doi.org/10.1007/978-3-030-67871-5_37

information theory and other disciplines [3]. It mainly refers to the protection of the hardware, software and the data in the construction project management system from accidental or malicious reasons, damage, change and leakage, continuous and reliable operation of the system, and uninterrupted network services. Obviously, with the further development of the network, the threat brought by the construction project management system will increase day by day.

At present, there are also individuals and research organizations, mainly represented by Santonatos and Mike Fisk, who actively carry out the research on risk prediction pattern matching methods. China has also stepped up research in this area, such as Wang Yongcheng of Shanghai Jiaotong University, Song Hua of Tsinghua University, and other universities such as Hebei University, northwest Polytechnic University, Harbin Polytechnic University, Nanjing Normal University, etc. It can be seen from the relevant materials and literatures that at present, the research on the risk prediction pattern matching method for the construction project management system still stays in the single pattern matching method, while the research on the multi pattern matching method mainly focuses on the method overview, testing and some corresponding improvements on the existing methods [4]. Although these improved methods have achieved some results, the overall effect is not very ideal, mainly because the method speed is limited by the number of intrusion rules or the space consumption of the implementation method is too large, so the practicability of the method used in the construction project management system is not strong. At the same time, it can also be seen that it is difficult to propose a new risk prediction pattern matching method. Therefore, since adopting the multi pattern matching method for management, the multi pattern matching method introduced by the construction project management product has not changed much, and the main methods still continue to be used, such as the famous snort construction project management system with open source code AC or WM method is adopted; WM method is adopted for easy guard construction project management system; AC-BM method is adopted for Tiantian construction project management products of Beijing Qiming Xingchen company and management series products of Lenovo online.

As the core technology of construction project management system, risk prediction pattern matching technology has always been the focus of the industry. According to the statistics of relevant researchers, in the filtering and detection of the construction project management system, the pattern matching module takes up 70% of the execution time and 80% of the program instructions of the system; about 30% of the risk problems are caused by the low efficiency of data package detection in the construction project management system [5, 6]. With the increase of network bandwidth and the expansion of matching pattern set, the performance of pattern matching method has become the bottleneck of network security system. Therefore, further study and improvement of pattern matching technology is of great significance to improve network fluency, performance of network security system and security of network security system.

In this context, this paper proposes a risk prediction pattern matching method for construction project management system. This paper analyzes the core idea of the risk prediction pattern matching algorithm, designs the risk prediction pattern matching algorithm, and combines the management system to realize the risk prediction pattern matching. In order to verify the effectiveness of the proposed method, an experimental simulation experiment was conducted. Experimental results show that the proposed method is more accurate.

2 Design of Risk Prediction Pattern Matching Method

2.1 Improved Design Risk Prediction Pattern Matching Algorithm

In the construction project management system, the risk prediction pattern matching algorithm is a widely used pattern matching algorithm. One of the most important characteristics of the risk prediction pattern matching algorithm is that in the process of matching the risk prediction pattern strings, many useless risk prediction patterns can be skipped, that is, no useless risk prediction patterns are matched. Through this kind of skip matching, we get a higher execution efficiency [7]. Some experimental data show that the matching speed of risk prediction pattern matching algorithm is about 3–5 times that of other matching algorithms.

The central idea P(*pattern*) of risk prediction pattern matching algorithm is to assume that the length of risk prediction pattern is m. The left most risk prediction model *pattern* and the left most risk prediction model of the shilling risk prediction model string T(*text*) are aligned, and then the last risk prediction model $t(m)$ is compared with its corresponding risk prediction model in *text*, that is, matching from right to left. When a mismatch is found, the algorithm adopts two heuristic rules, which are heuristic rules It is: bad risk prediction pattern (BC) rule and good suffix (GS) rule to calculate the moving distance of the pattern risk prediction pattern string, and realize the leaping ergodic matching [8].

Description of bad risk prediction model (BC) Rules.

When a risk prediction mode in the mode risk prediction mode string P is different from a risk prediction mode in the text risk prediction mode string t, a bad risk prediction mode appears. The risk prediction mode matching algorithm moves the mode risk prediction mode string to the right, makes the most right corresponding risk prediction mode in the mode relative to the bad risk prediction mode, and then continues to match. The move function is as follows:

$$BadChar(c) = \begin{cases} m \\ m - j \end{cases} \tag{1}$$

After aligning *pattern* with *text* the left, there are two possibilities for comparing the results of the last risk prediction model in *text* and its corresponding risk prediction model in *pattern*.

The first situation is that $t(m)$ does not appear in any of the *pattern*, so we can move m risk prediction modes to the right of *pattern*, and then compare the last risk prediction mode with the corresponding risk prediction mode $(t(2m))$ in *pattern*.

The second case is: if $t(m)$ is the j risk prediction mode in *pattern*, then we can move *pattern* M-J risk prediction mode to the right.

Good suffix GS rule description.

If a good suffix has been matched, and there is another same suffix in the pattern, GS rule considers the matched situation and determines a new moving distance, which is only related to pattern risk prediction pattern string P.

The steps of risk prediction pattern matching algorithm are as follows:

Step 1: preprocessing, the algorithm moves the pattern risk prediction pattern string to the right as far as possible according to the two pre calculated arrays. The calculation of *Skip*[] risk prediction model and risk prediction model *Shift*[], represent BC rule and GS rule respectively.

Step 2: compare the text risk prediction model string and model risk prediction model string one by one from right to left, and continue to compare the single risk prediction model matching. If it reaches the far left of the pattern risk prediction pattern string, the matching occurs successfully and the output is output; if the risk prediction pattern mismatch occurs, the third step is turned.

Step 3: take the array corresponding *Skip*[] and *Shift*[] to the mismatch risk prediction mode and the largest value in the array as the moving distance, and move the mode risk prediction mode string to the right. If the end of the text risk prediction mode string has been reached, the algorithm exits; otherwise, it will go back to the second step.

The first mock exam algorithm is the first mock exam algorithm to search for a single pattern risk prediction pattern in building engineering project management system. In a single pattern risk prediction pattern string matching algorithm, the risk prediction pattern matching algorithm is generally considered to be the best performance. There are many keyword patterns to match in content filtering and detection, so the risk prediction pattern matching algorithm needs to match each pattern separately [9]. The time and space complexity of the preprocessing stage of the risk prediction pattern matching algorithm is $O(m + n)$, the time and space complexity of the search stage is $O(mn)$, the worst-case comparison times of the non periodic pattern search is 3 N. The worst time complexity of risk prediction pattern matching algorithm is $O(mn)$, and the best time complexity is $O(n)$. The complexity of directly using risk prediction pattern matching algorithm to match multi-mode risk prediction pattern strings is $O(kn)$. Next, through the risk prediction pattern matching process of the construction project management system, the risk prediction pattern matching is realized.

2.2 Matching Risk Prediction Model

In the construction project management system, the matching of risk prediction pattern is to compare the information collected in the system with the rules in the existing pattern database, so as to judge whether the risk prediction pattern matches. The risk prediction pattern matching process of the construction project management system is shown in Fig. 1.

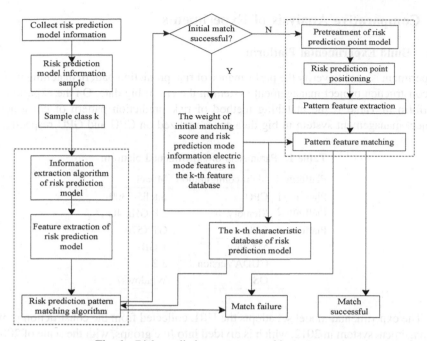

Fig. 1. Risk prediction pattern matching process

At present, the method of risk prediction pattern matching has become one of the most commonly used methods in the field of construction project management, which has been proposed for a long time. However, the survey shows that when using the risk prediction pattern matching algorithm in the era of big data, the pattern matching algorithm determines the detection efficiency and resource occupancy of the construction project management system to a large extent, and the current increase of network speed and network application makes the data in the network increase rapidly, which makes the pattern matching algorithm more and more important. It is precisely because For this reason, the research and exploration of the optimization method of pattern matching algorithm has never stopped, which is becoming more and more popular in today's network security [10, 11].

To sum up, in the era of big data, this paper analyzes the central idea of the risk prediction pattern matching algorithm, designs the specific implementation steps of the risk prediction pattern matching algorithm based on the description of the risk prediction pattern matching algorithm, completes the design of the risk prediction pattern matching algorithm, combines the risk prediction pattern matching process of the construction project management system, and realizes the risk prediction pattern Match of [12, 13].

3   Comparative Analysis of Experiments

3.1   Build Experimental Platform

Experiment is used to verify the performance of risk prediction pattern matching method of construction project management system in the era of big data. On the experimental platform of Table 1, the matching method of risk prediction pattern of construction project management system in big data era is realized on CPU and GPU respectively.

Table 1. Parameters of experimental platform

Platform	Main configuration	Model
Platform 1	CPU	Inteli5-2300
Platform 2	Memory	2.8 GHz 4G
Platform	GPU	GTX570
	Memory	1 GB
	CUDAVersion	3.2
	OS	Windows7

The experimental model set adopts the URL collected from the construction project management system in 2012, which is divided into five groups, with the scale of 20000, 40000, 80000, 160000 and 320000, as shown in Table 2. 12 MB URL of text data collected in the same way. The hit times in the table refer to the times that the URLs in the URL set can match in the total text data. Because there are duplicate URLs in the text data, the hit times include repeated cumulative URL matches. Taking URL set 1 as an example, there are 20000 URLs in the URL set, and 283256 URLs in URL set 1 are hit in all URL text data. The length of URLs in all 5 URL sets is 8–15 risk prediction modes.

Table 2. Test set

Test set	Number	Hit counts	Text set size
Test set 1	2000	283256	12 MB
Test set 2	4000	515649	12 MB
Test set 3	8000	994134	12 MB
Test set 4	160000	1921158	12 MB
Test set 5	320000	3966726	12 MB

3.2   Analysis of Experimental Results

As the WM based risk prediction pattern matching method of construction project management system has the characteristics of high matching efficiency and less memory, it is one of the classic multi pattern matching methods. In order to evaluate the risk prediction pattern matching method of construction project management system in the era of big data, this paper compares the model matching method of risk prediction of construction project management system based on WM algorithm. Due to the different requirements of platform 1 and platform 2 for the operation environment, this experiment will match the risk prediction pattern matching method of the construction project management system in the era of c-big data and the risk prediction pattern matching method of the construction project management system based on WM algorithm on the experimental platform 1, and match the risk prediction pattern matching method of the construction project management system in the era of g-big data and G- Based on the WM algorithm, the risk prediction pattern matching method of the construction project management system performs the matching task on the experimental platform 2. Four matching methods are used on their own platforms, and five sets of URL sets are used as pattern sets to match with the total data sets.

Figure 2 shows the preprocessing time comparison of four algorithms on five sets of URLs.

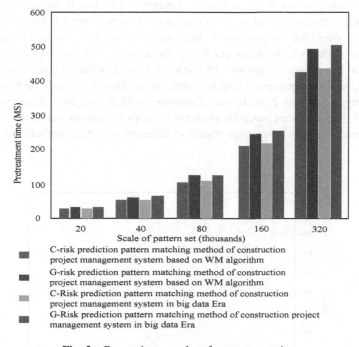

C-risk prediction pattern matching method of construction project management system based on WM algorithm

G-risk prediction pattern matching method of construction project management system based on WM algorithm

C-Risk prediction pattern matching method of construction project management system in big data Era

G-Risk prediction pattern matching method of construction project management system in big data Era

Fig. 2. Comparison results of pretreatment time

From the experimental results in Fig. 2, it can be seen that the preprocessing time of c-risk prediction pattern matching method of construction project management system based on WM algorithm on five sets of URLs is 26.528–423.037 ms, the preprocessing time of c-risk prediction pattern matching method of construction project management system in the era of big data is 27.114–435.947 ms, and the preprocessing time of c-risk prediction pattern matching method of construction project management system in the era of big data is 27.114–435.947 ms The preprocessing time of test pattern matching method is 2.2%–3.05% longer than that of c-wm-based risk prediction pattern matching method of construction project management system; the preprocessing time of g-wm-based risk prediction pattern matching method of construction project management system and g-big data-based risk prediction pattern matching method of construction project management system is longer than that of each other The original algorithm on platform 1. The preprocessing time of G-Risk prediction pattern matching method of construction project management system based on WM algorithm on five sets of URLs is 31.257–495.184 ms, that of G-Risk prediction pattern matching method of construction project management system in the era of big data is 32.116–504.996 ms, and that of G-Risk prediction pattern matching method of construction project management system in the era of big data is 32.116–504.996 ms It is 1.98%–2.75% more than that of G-Risk prediction pattern matching method of construction project management system based on WM algorithm, and the preprocessing time of G-Risk prediction pattern matching method of construction project management system in the era of big data is 15.84%–18.45% more than that of c-risk prediction pattern matching method of construction project management system in the era of big data.

The preprocessing time of the matching method on platform 2 is longer than that of the original algorithm on platform 1. Because there are fewer logic operation components on platform 2, the preprocessing of the algorithm in the early stage is carried out on platform 1, and the memory of platform 1 and platform 2 cannot be addressed uniformly, so the preprocessed data structure on platform 1 can only be transferred to the memory of platform 2 one by one. Compared with the original algorithm based on platform 1, the algorithm based on platform 2 needs to transfer data time.

The matching time comparison results of different matching methods are shown in Fig. 3.

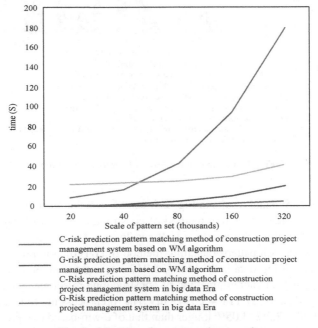

C-risk prediction pattern matching method of construction project management system based on WM algorithm

G-risk prediction pattern matching method of construction project management system based on WM algorithm

C-Risk prediction pattern matching method of construction project management system in big data Era

G-Risk prediction pattern matching method of construction project management system in big data Era

Fig. 3. Matching time comparison results

From the experimental results in Fig. 3, it can be seen that (a) on the large pattern set (more than 80000 items), the performance of the risk prediction pattern matching method of the construction project management system in the era of c-big data is better than that of the risk prediction pattern matching method of the construction project management system based on the WM algorithm, and the larger the pattern set, the better the performance. This can be attributed to two reasons: one is that the conflict rate of c-wm-based risk prediction pattern matching method of construction project management system increases with the increase of the scale of pattern set, resulting in the reduction of algorithm efficiency; the other is that the larger the pattern set, the more obvious the advantage of c-big data Era construction project management system risk prediction pattern matching method in the structure parallelism.

(b) Among the two matching methods based on platform 2, the performance of G-Risk prediction pattern matching method of construction project management system and G-Risk prediction pattern matching method of construction project management system based on WM algorithm are better than their respective original algorithms, and the efficiency of risk prediction pattern matching method of construction project management system in g-big data Era is far away Compared with the g-wm based risk prediction pattern matching method, the larger the pattern set, the more significant the advantage. In addition to the two reasons mentioned in (a), there are three reasons for this result: G-Risk prediction pattern matching method of construction project management system in the era of big data makes full use of the multi-layer storage structure of platform 2 to improve data processing speed, while due to the structure problem of matching method, G-Risk prediction pattern matching method of construction project management system based on WM algorithm It is difficult to use platform 2 high-speed on-chip memory shared memory to speed up processing speed on large mode set.

The comparison results of memory consumption of different matching methods are shown in Fig. 4.

From the experimental results in Fig. 4, it can be seen that with the increase of the number of URLs, the memory occupied by different matching methods increases correspondingly. Moreover, the memory occupied by the matching method of risk prediction mode of construction project management system in the era of c-big data is less than that of the matching method of risk prediction mode of construction project management system based on WM algorithm, and the memory occupied by the matching method of risk prediction mode of construction project management system in the era of g-big data is less than that of the matching method of risk prediction mode of construction project management system based on WM algorithm Match method. This is because in addition to suffix table and prefix table, c-risk prediction pattern matching method of construction project management system based on WM algorithm also needs to build jump table, while in the era of c-big data, risk prediction pattern matching method of construction project management system only needs to build multiple extended bloomfilters based on pattern set, and does not need extra memory space.

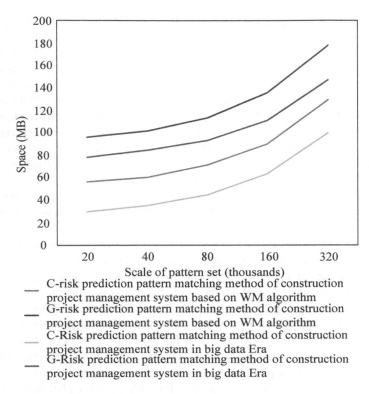

Fig. 4. Comparison results of occupied memory

The throughput comparison results of different matching methods are shown in Fig. 5.

From the experimental results in Fig. 5, we can see that with the increase of the size of the URL pattern set, the throughput of different matching methods decreases gradually, and the efficiency of all matching methods tends to bottleneck. This is because the pattern set is close to the upper limit of GPU memory. In contrast, in the era of g-big data, the throughput of risk prediction pattern matching method of construction project management system is always higher than that of other matching methods, and it is 2–5 times higher than that of risk prediction pattern matching method of construction project management system based on WM algorithm. In the era of c-big data, when the scale of pattern set is less than 40000, the throughput of risk prediction pattern matching method of construction project management system based on WM algorithm is lower than that based on c-big data. When the scale of pattern set increases to 80000 or more, the throughput of risk prediction pattern matching method of construction project management system in the era of c-big data is higher than that of risk

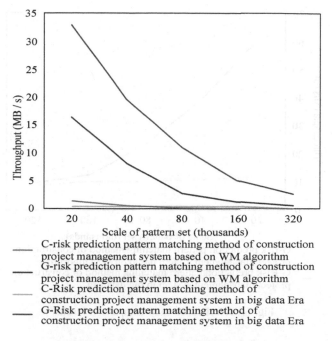

C-risk prediction pattern matching method of construction
project management system based on WM algorithm
G-risk prediction pattern matching method of construction
project management system based on WM algorithm
C-Risk prediction pattern matching method of
construction project management system in big data Era
G-Risk prediction pattern matching method of
construction project management system in big data Era

Fig. 5. Throughput comparison results

prediction pattern matching method of construction project management system based
on WM algorithm. This is because of the small scale of pattern set, the conflict rate of
c-risk prediction pattern matching method of construction project management system
based on WM algorithm is smaller, and the efficiency is higher than that of c-risk
prediction pattern matching method of construction project management system in the
era of big data. With the rapid increase of pattern set, c-risk prediction pattern matching
method of construction project management system based on WM algorithm is rushed
The burst rate increases and the matching efficiency decreases.

This paper also compares the acceleration ratio (serial time/parallel time) of the risk
prediction pattern matching method of the construction project management system in
the era of g-big data on different pattern sets with the risk prediction pattern matching
method of the construction project management system based on the WM algorithm, as
shown in Fig. 6.

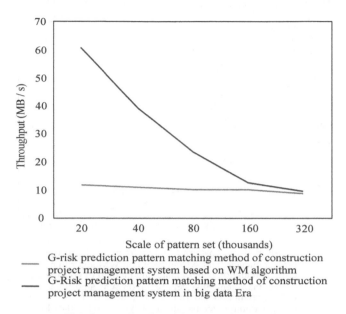

Scale of pattern set (thousands)
___ G-risk prediction pattern matching method of construction
project management system based on WM algorithm
___ G-Risk prediction pattern matching method of construction
project management system in big data Era

Fig. 6. Acceleration ratio comparison results

From the experimental results in Fig. 6, it can be seen that the acceleration ratio of the risk prediction pattern matching method of the construction project management system in the era of g-big data is always higher than that of the risk prediction pattern matching method of the construction project management system based on the WM algorithm. At the same time, it decreases with the increase of the pattern set, and gradually matches with the risk prediction pattern matching formula of the construction project management system based on the WM algorithm The acceleration ratio of the method tends to be the same. This is because the GPU's memory limit is close at this time. When the memory is large enough, the difference between the acceleration ratio of the two is very obvious. In the era of g-big data, the acceleration ratio of risk prediction pattern matching method of construction project management system can reach 60 times in the best case and nearly 10 times in the worst case. The acceleration of g-wm-based risk prediction pattern matching method for construction project management system is only about 12 times of the best case. This fully shows that in the era of c-big data, the degree of structural parallelism of risk prediction pattern matching method of construction project management system is far greater than that of c-wm based risk prediction pattern matching method of construction project management system, which is more suitable for parallel pattern matching [14, 15].

Based on the above experimental results, the performance of risk prediction pattern matching method in construction project management system is far better than other matching methods, no matter on platform 1 or platform 2.

4 Conclusion

In this paper, the risk prediction pattern matching method of construction project management system in the era of big data is proposed. In the background of big data era, the risk prediction pattern matching algorithm is designed. Combined with the risk prediction pattern matching process, the risk prediction pattern matching of construction project management system is realized. The results show that the risk prediction pattern matching method of construction project management system in the era of big data has higher accuracy.

Due to the urgency and importance of big data processing, big data technology has been highly valued in the global academic circle. In this paper, the pattern matching of risk prediction of construction project management system has been completed under the background of big data, but the implementation of project management after matching still needs further research, which will also be the focus of future research.

Acknowledgements. This research is funded by (Research on the application of cloud computing technology in the construction project management system in the era of big data, with No. JAT190908).

References

1. Hassan, G.M.: Discontinuous and pattern matching algorithm to measure deformation having discontinuities. Eng. Appl. Artif. Intell. **81**(2), 223–233 (2019)
2. Santiago, Cláudio P., Lavor, C., Monteiro, S.A., Kroner-Martins, A.: A new algorithm for the small-field astrometric point-pattern matching problem. J. Glob. Optim. **72**(1), 55–70 (2018). https://doi.org/10.1007/s10898-018-0653-y
3. Shin, H.S., Turchi, D., He, S., et al.: Behavior monitoring using learning techniques and regular-expressions-based pattern matching. IEEE Trans. Intell. Transp. Syst. **20**(4), 1289–1302 (2019)
4. Golan, S., Kopelowitz, T., Porat, E.: Streaming pattern matching with d wildcards. Algorithmica **81**(5), 1988–2015 (2019). https://doi.org/10.1007/s00453-018-0521-7
5. Korman, S., Reichman, D., Tsur, G., et al.: Fast-match: fast affine template matching. Int. J. Comput. Vision **121**(1), 111–125 (2017). https://doi.org/10.1007/s11263-016-0926-1
6. Wang, L., Zhiwen, Y.U., Guo, B., et al.: Mobile crowd sensing task optimal allocation: a mobility pattern matching perspective. Front. comput. Sci. China **12**(2), 231–244 (2018). https://doi.org/10.1007/s11704-017-7024-6
7. Gandotra, E., Singla, S., Bansal, D., et al.: Clustering morphed malware using opcode sequence pattern matching. Recent Pat. Eng. **12**(1), 30–36 (2018)
8. Ai, L., Ramaswamy, L., Luo, S.: Impact vertices-aware diffusion walk algorithm for efficient subgraph pattern matching in massive graphs. IEEE Access **7**, 44555–44561 (2019)
9. Lu, M., Liu, S., Kumarsangaiah, A., et al.: Nucleosome positioning with fractal entropy increment of diversity in telemedicine. IEEE Access **6**, 33451–33459 (2018)
10. Kociumaka, T., Pissis, S.P., Radoszewski, J.: Pattern matching and consensus problems on weighted sequences and profiles. Theory Comput. Syst. **63**(3), 506–542 (2019). https://doi.org/10.1007/s00224-018-9881-2
11. Asadi, P., Rezaeian Zeidi, J., Mojibi, T., et al.: Project risk evaluation by using a new fuzzy model based on Elena guideline. J. Civ. Eng. Manag. **24**(5), 284–300 (2018)

12. Fu, W., Liu, S., Srivastava, G.: Optimization of big data scheduling in social networks. Entropy **21**(9), 902 (2019)
13. Shuai, L., Weiling, B., Nianyin, Z., et al.: A fast fractal based compression for MRI images. IEEE Access **7**, 62412–62420 (2019)
14. Liu, S., Li, Z., Zhang, Y., et al.: Introduction of key problems in long-distance learning and training. Mobile Netw. Appl. **24**(1), 1–4 (2019)
15. Liu, S., Sun, G., Fu, W. (eds.): eLEOT 2020. LNICST, vol. 340. Springer, Cham (2020). https://doi.org/10.1007/978-3-030-63955-6

Intelligent Performance Evaluation Method of Assembly Construction Project Management Based on Cloud Computing Technology

Qiu-yi Li[✉]

Fuzhou University of International Studies and Trade, Fuzhou 350202, China
lqyl3799475957@163.com

Abstract. In order to better evaluate the performance of construction project management in Jinzhou, an intelligent evaluation method based on cloud computing technology is proposed. Combined with the practice of construction project management, the performance evaluation index system of construction project is established. Using the formula of relative membership degree to standardize the data, using linear programming to determine the weight of index factors, vector projection formula to find the approach degree between the actual evaluation value and the target evaluation value to evaluate the construction project management performance. Finally, it is proved by experiments that the performance intelligent evaluation method of assembly construction project management based on cloud computing technology has higher accuracy and timeliness in the practical application process, and fully meets the research requirements.

Keywords: Cloud computing · Construction engineering · Engineering management · Performance evaluation

1 Introduction

With the rapid development of China's infrastructure construction, in the process of urbanization, construction engineering has become the backbone of the steady development of national infrastructure. In the implementation of construction projects, engineering management has become the focus of attention of construction project practitioners, especially audit of project cost budget and settlement, which has become an important link in the control of project cost in construction project management, and plays a decisive role in project risk, cost execution, management of project quality and guarantee of progress [1]. In such a new era of construction project management environment and management needs, to explore more application ways of project cost budget and settlement audit has become the key to ensure the efficient implementation of construction projects and the consensus of practitioners in the industry [2]. Construction engineering has the characteristics of large investment, long construction period, involving many subjects, many uncertain factors and high risk.

© ICST Institute for Computer Sciences, Social Informatics and Telecommunications Engineering 2021
Published by Springer Nature Switzerland AG 2021. All Rights Reserved
S. Liu and L. Xia (Eds.): ADHIP 2020, LNICST 347, pp. 425–437, 2021.
https://doi.org/10.1007/978-3-030-67871-5_38

Construction project performance evaluation is an important part of construction project management. The intelligent performance evaluation method of prefabricated construction project management based on cloud computing technology can not only provide theoretical guidance for the practice of construction enterprises, but also provide a good basis for managers to grasp the project operation in the process of the project, so as to provide real-time monitoring and control.

2 Intelligent Performance Evaluation Method of Assembly Construction Project Management

2.1 One Point One Weight Algorithm of Attribute Index in Assembly Building Engineering

In the process of project price budget and settlement audit, there are many methods applied, in order to fully deal with the long construction period and large construction period The impact of the amount of capital investment and the changeable market environment on the application of the cost budget and settlement audit, generally, the cost budget and settlement audit will be carried out in a variety of audit methods, such as screening audit, analogy audit, key audit, sub audit and so on [3]. Based on this need, the algorithm of attribute index weight is optimized.

There are two factors influencing the comprehensive evaluation of construction project performance:

The first level of evaluation index set is a = (asset growth rate A_{21}, project profit rate A_{22}, cost reduction rate, B_3 project funds in place rate is B_5) * agility + accounting. B_4 and further combined with operational profitability A_{46}, customer satisfaction A_{31}, cooperation unit closeness A_{20};

The second level evaluation index set is table = (duration control capability A_{ij}, risk control capability A_{12}, contract performance rate B_{13} and response time B_4),

If the project schedule completion rate is A_{31}, the project contracting capacity is A_{32}, the cash flow turnover time A_{33}, the inventory turnover rate is A_{34}, the current asset turnover rate is B_5, the labor productivity is A_{36}, the safety production control ability is A_{37}, the owner satisfaction rate A_{40}, the quality target compliance rate is A_{42}, the completion punctuality rate is A_{44}, the rework rate is B_4, the owner complaint rate A_{45}, the design change control ability is B_6, and the environmental protection control ability is B_7. Based on this, the evaluation index grades of construction project management are divided as follows (Table 1):

Table 1. Classification of construction project management evaluation indexes

Division of primary and secondary factors of evaluation	Evaluation index of building materials	Evaluation criteria and equivalence division results
Main evaluation grade classification standard	Main sources	Subsidiary indicators affecting the evaluation results of the model
Classification standard of secondary evaluation grade	Toughness	The key index influencing the evaluation results of the model is only second to the loss degree of building materials
Main evaluation grade classification standard Classification standard of secondary evaluation grade	Management intensity	The key indicators influencing the evaluation results of the model are much more important than other factors
	Pressure bearing capacity	Subsidiary indicators affecting the evaluation results of the model
Main evaluation grade classification standard	Pricing range	Under the premise of the change of evaluation standard, the pricing range of building materials will also change

Due to the different units of each evaluation index, the data will be incommensurable. If the calculation is conducted directly, the results will be unreasonable. Therefore, before the aggregation, the evaluation value and the target value provided by experts should be standardized respectively. The specific algorithm is as follows:

Benefit index weight algorithm:

$$B_{ij} = \frac{A_{ij} - 1}{A_{ijmax} - A_{ijmin}} \tag{1}$$

The index algorithm of construction project contracting is as follows:

$$S_{ij} = \frac{B_{ij}}{2(A_{ijmax} - A_{ijmin})} + 1 \tag{2}$$

Where I is the I first level indicator and j is the j second level indicator. Considering both subjectivity and objectivity, the two methods are integrated, that is, the weight is given by experts and owners in the form of interval range according to the industry characteristics and experience of construction engineering [4]. Obviously, the determination of a reasonable attribute weight vector $w_{ij} = (w_{i1}, w_{i2}, w_{i3}, \ldots, w_{in})$ should make the total deviation between its performance evaluation result and the target value provided by the expert smaller, as well as the better. In this way, the expert's opinion is fully considered and the data itself information is fully utilized [5]. Therefore, in order to determine the weight of secondary evaluation indexes, the following linear programming model is established:

$$\min G = \sum \sum A_{ij}(w_{ij} - B_{ij}) \tag{3}$$

Use formula (3) to calculate the weighted value of performance evaluation of actual data and target data respectively:

$$P_{ij} = \prod (B_{ij} + w_{ij})k/Q_i \qquad (4)$$

There, k are two vectors. The larger the projection value of vector Q_i coincidence on the vector, the closer the vector coincidence and A are, the closer the evaluation result of engineering practice is to the target value given by experts, the better [6]. Further, the numerical calculation table of each secondary index calculated by the directional principle is as follows (Table 2):

Table 2. Secondary index specification for construction project management

B_{10}	B_{11}	B_{12}	B_{13}	B_{14}
Actual evaluation value	82	85	88	84
Target evaluation value	12	11	14	10
Rights information provided by experts	[0, 12, 6, 24]	[3, 12, 8, 16]	[2, 16, 4, 20]	[5, 16, 2, 13]
Actual data specification value	0.12	0.23	0.16	0.20
Target data specification value	12.16	13.41	16.46	15.42
Secondary weight	10.46	9.558	9.76	9.32
Weighted comprehensive actual value	0.13	0.46	0.35	0.28
Weighted combined target value	0.230	0.220	0.168	0.167
Weighted marginal target value	0.236	0.245	0.231	0.264

2.2   Evaluation Model of Construction Project Management

In order to strengthen the supervision of large-scale public buildings, indemnificatory housing, commercial housing and key projects of municipal Party committee and municipal government, and ensure the project quality. Strengthen the solid quality supervision of foundation and main structure engineering. Adhere to the supervision focus on the foundation, main structure and key parts affecting the use safety and use function, and implement the mandatory standards of engineering construction to ensure the safety of engineering structure [7]. Strengthen the management of completion acceptance of housing construction projects to prevent unqualified projects from flowing into the society. According to the relevant national and provincial regulations, supervise the organization form of completion acceptance, the implementation of acceptance procedures, the implementation of acceptance scheme, review the completion acceptance data, and supervise the implementation of national mandatory standards, specifications and quality evaluation results for the completion acceptance of the project, so as to ensure the quality of the completion acceptance.

Furthermore, the excessive dependence on the contract by the flexible management will not only weaken the mutual trust between the cooperative participants, but also lead to opportunistic behavior. The performance management evaluation model is a flexible project management model based on the contract signed and the agreement made by the cooperative participants through consultation as the basic code of conduct [8, 9]. Combined with the principle of cloud computing, based on the engineering management protocol, with management as the means and goal realization as the core, the management mechanism system includes cooperation mechanism, coordination mechanism, communication mechanism, trust mechanism and incentive mechanism. The specific management mechanism framework model is shown in the Fig. 1.

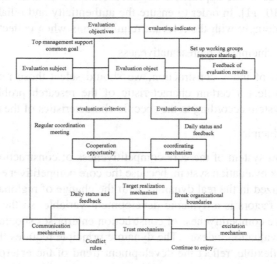

Fig. 1. Framework model of construction project management mechanism

In order to evaluate the core competitiveness of construction enterprises more objectively and effectively, and to provide reference for the cultivation of competitiveness of later construction enterprises, we should abide by the following principles when formulating the evaluation index system of core competitiveness of construction enterprises:

(1) Objective principle

The main purpose of building the index system is to analyze and evaluate the core competitiveness of construction enterprises, which needs to consider the industry characteristics and development background of construction enterprises as well as the relevant theories in the process of index selection.

(2) The principle of combining qualitative and quantitative analysis

From the concept point of view, the core competitiveness of construction enterprises is a more abstract concept. In order to reflect it more comprehensively, the qualitative and quantitative expression methods are used in the construction of index system. Quantitative analysis reflects the attributes of things through specific quantity, and qualitative analysis explores the essence of things through observation, experiment and analysis. The combination of qualitative analysis and quantitative analysis can ensure that the evaluation results of core competitiveness indicators of construction enterprises are more real and objective.

(3) Follow the principle of data authenticity

The construction of indicator system cannot be separated from the support of data. Scientific and authentic data can effectively reflect the rationality of each indicator and help the next analysis process. On the contrary, if the collected data is not authentic enough, it will lead to the appearance of unreasonable indicators and affect the final analysis results [10, 11]. In order to ensure the authenticity and reliability of the data, we must strictly comply with the relevant requirements when collecting the data.

(4) Follow the principle of representativeness

In the process of index construction, we should select those representative indicators that can reflect a certain characteristic of the research problem, and build a scientific index system according to the specific characteristics of the research [12, 13].

(5) Dynamic principle

The evaluation system of the core competitiveness of construction enterprises is a relatively complex evaluation system, because the core competitiveness of construction enterprises is realized in the real dynamic, with the change of regional, environmental, policy and other factors is a dynamic development variable, so the dynamic characteristics of the core competitiveness of construction enterprises should be considered in the selection of evaluation indexes. The dynamic principle requires that the evaluation index should be flexible, reflect the development trend of the enterprise, adapt to the changes of the market environment, and reflect the characteristics of the core competitiveness in a long period of time. The dynamic evaluation index can help the enterprise to respond to the changes of the market environment and make adjustments quickly [14, 15].

2.3   Implementation of Project Management Performance Evaluation

Combined with the previous research results, the knowledge management ability is divided into four dimensions: knowledge acquisition ability, knowledge integration ability, knowledge creation ability and knowledge application ability. Through reading the relevant literature at home and abroad, appropriate items are selected to measure it. In this study, several documents were collected for a brief introduction: Shi Jiangtao defined the concept of knowledge integration and put forward the factors that affect the ability of knowledge integration. The restrictive factors of enterprise knowledge creation ability are also a multi-level dynamic system, and the factors related to knowledge creation ability are numerous and complex. In order to better realize the transfer evaluation of project management performance, the evaluation system of project management performance is constructed as follows (Fig. 2):

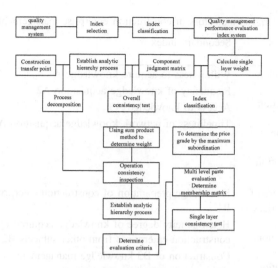

Fig. 2. Project management performance evaluation system

It plays an important role in the process of construction project management. The work content is closely related to the project contractor, the owner, the construction unit and the supervisor. Each relevant party must perform its own responsibilities, strengthen the project change management, and avoid problems in the change management caused by its own factors, which will affect the whole process The quality of individual projects results in the loss of economic interests of all parties. The initial index system is divided into three levels: the first level is the core competitiveness of the enterprise; the second level is the ability to acquire knowledge, the ability to integrate knowledge, the ability to create knowledge and the ability to apply knowledge; the third level is obtained by reading relevant literature and combining the collaborative innovation network of construction enterprises. The measurement indicators of the core competitiveness of construction enterprises are the quality rate of construction projects, the proportion of professional and technical personnel of enterprises and the brand awareness of enterprises. The impact indicators of the management system of construction enterprises are analyzed. The specific indicator system is shown in the Table 3:

Table 3. Impact indicators of construction enterprise management system

First level indicators	Secondary index	Exponential property
Knowledge acquisition ability a Knowledge integration capability B Knowledge creation ability C Knowledge application ability D	Willingness to acquire knowledge A1	qualitative
	Frequency of knowledge acquisition A2	qualitative
	Access to knowledge A3	ration
	Timeliness of network knowledge acquisition A4	qualitative
Core competitiveness of construction enterprises x First level indicators Knowledge acquisition ability a Knowledge integration capability B	Knowledge preparation of construction enterprise B1	ration
	Heterogeneity degree of knowledge acquired by construction enterprises from other subjects B2	ration
	Construction of B3 knowledge management system in construction enterprises	directional
	Trust degree between construction enterprises and other network subjects B4	qualitative
Knowledge creation ability C Knowledge application ability D Core competitiveness of construction enterprises x First level indicators	Incentive intensity of construction enterprises to knowledge creation C1	ration
	Proportion of construction enterprises participating in knowledge creation C2	qualitative
	The importance of construction enterprise to staff training	ration
	Cooperation degree between construction enterprises and other network entities C4	ration
Knowledge acquisition ability a Knowledge integration capability B Knowledge creation ability C Knowledge application ability D	The degree of using collaborative innovation network to develop new products in construction enterprises D1	qualitative
	Number of new patents of construction enterprises in recent years D2	qualitative
	Number of new construction methods developed by construction enterprises and network members in recent years D3	ration
	New technology adoption rate of construction enterprise D4	directional
Core competitiveness of construction enterprises x	Proportion of technical personnel in No.1 industry x1	qualitative
	Excellent rate of enterprise construction project x2	directional
	Enterprise brand awareness x3	qualitative

(1) Willingness to acquire knowledge A1

Construction enterprises expand the channel of knowledge acquisition by building collaborative innovation network. For enterprises, the increase of knowledge source does not mean the increase of knowledge. Only when enterprises realize the importance of knowledge and have the desire to acquire knowledge urgently, the ability of knowledge acquisition of enterprises will be enhanced.

(2) Frequency of knowledge acquisition A2

The frequency of knowledge acquisition in the collaborative innovation network not only reflects the communication between the network subjects, but also reflects the rate of knowledge update of the network subjects. Generally, the higher the frequency of knowledge acquisition, the faster the knowledge flow in the collaborative innovation network, and the easier it is for enterprises to acquire the knowledge they want.

(3) Convenience of acquiring knowledge A3

The main purpose of enterprises to acquire knowledge is to hope that the acquired knowledge can bring economic effects to enterprises, and the purpose of enterprises is to pursue profits, which also determines that enterprises have various difficulties in acquiring key knowledge. The convenience of acquiring knowledge from network members determines the transfer efficiency of key knowledge between enterprises.

(4) Timeliness of knowledge acquisition A4

Through the construction of collaborative innovation network, the channels for enterprises to acquire knowledge are expanded. While acquiring knowledge, enterprises must ensure the timeliness of knowledge. Only in this way can knowledge have value. Knowledge with low timeliness may mislead enterprises to make wrong decisions and affect the development of enterprises.

(5) Reserve level of construction enterprise B1

To a certain extent, the knowledge reserve of an enterprise not only reflects the quality level of employees, but also reflects the degree of knowledge accumulation of an enterprise. Generally, the knowledge that an enterprise reserves is the knowledge that can bring actual benefits to the enterprise after being tested. At the same time, these knowledge can improve the efficiency of knowledge integration.

(6) Heterogeneity degree of construction enterprises B2

In the collaborative innovation network, different network subjects often have different organizational culture, which also determines that the knowledge owned by enterprises often has their own characteristics. Generally, the closer the enterprise nature and corporate culture are to knowledge integration, the easier it is.

(7) Completion degree of knowledge management system of construction enterprise B3

Knowledge management system is an information system that collects and processes all knowledge in an organization. Enterprises must classify and store the acquired knowledge information effectively for knowledge integration. A complete knowledge management system can help enterprises to call the information they need quickly and greatly promote knowledge integration.

(8) Trust degree of construction enterprises B4

The higher the degree of trust among the subjects in the collaborative innovation network, the faster the knowledge flows in the network, and the stronger the efficiency of knowledge integration. At the same time, the establishment of trust relationship strengthens the cooperation among network members, which is more conducive to the development of knowledge integration activities.

(9) Incentive strength of construction enterprises C1

By setting up various incentive mechanisms and incentive measures, enterprises can stimulate employees' enthusiasm for knowledge creation and promote the development of knowledge creation activities. At the same time, the greater the support of enterprises for knowledge creation, the more new knowledge they create.

(10) Proportion of construction enterprise participants C2

People are the main body of knowledge creation, and knowledge can produce value because of people's use. The higher the proportion of people who have knowledge creation, the easier it is for enterprises to create new knowledge.

(11) Importance of construction enterprises on management C3

Knowledge is constantly updated. The enterprise often trains its employees so that they can realize that only through continuous learning can they adapt to the development of the company and enhance their sense of crisis. At the same time, through training, they can increase the number of talents in the enterprise and lay the foundation for knowledge creation.

(12) Cooperation degree of construction enterprises C4

Knowledge creation needs the full cooperation between the network subjects, and the full communication can promote the production of new knowledge by gathering the advantageous resources of each subject together. In this process, the higher the degree of cooperation between the subjects, the stronger the efficiency of knowledge integration of enterprises.

3   Analysis of Experimental Results

In order to verify the actual application effect of the intelligent performance evaluation method of assembly construction project management based on cloud computing technology, this paper takes the general contractor of construction project as the core enterprise as a case to learn from the concept of supply chain, and expounds the establishment of the performance evaluation method and evaluation effect of construction project. Experiment on Windows platform. Establish database, input and organize through Excel 2003. SPSS13.0 is used for statistical analysis, mainly descriptive statistical analysis; in the use of analytic hierarchy process to determine the weight of the index system and the use of fuzzy comprehensive evaluation method to comprehensively evaluate the performance score, the algorithm of matrix operation is mainly used. The number of sample cases shall be determined according to the sampling proportion of 5%. A multi-stage, proportional sampling method is used to extract the target data of the survey case experts under the overall goal of the project planning. Based on the above experimental environment comparison detection, the specific detection results are as follows:

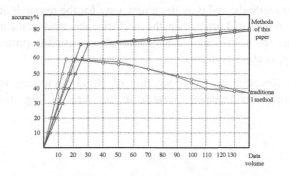

Fig. 3. Comparison of accuracy of project management performance evaluation

Further, in the same environment, the time-consuming situation of the two methods is compared and tested, and the test results are recorded, as follows:

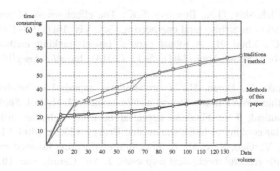

Fig. 4. Time consuming test of project management performance evaluation

It can be seen from the inspection results in Fig. 3 and Fig. 4 that, compared with the traditional evaluation methods, the intelligent evaluation method based on cloud computing technology proposed in this paper has higher evaluation accuracy in the practical application process, and the evaluation time of construction project management performance is significantly reduced in the same environment. It is proved that the performance intelligent evaluation method based on cloud computing technology is more effective and accurate, which fully meets the research requirements.

4 Conclusion

This paper constructs a performance evaluation model of construction project based on vector projection, quantifies the effect of construction project, and compares the effect of specific project management with the ideal target parameters, so as to help construction enterprises effectively classify key indicators and clarify the key points of performance management. The partial weight and information problem model proposed by the author can effectively avoid affecting subjective Human factors. The empirical study shows that the method is clear and can effectively avoid the shortcomings of other methods, and provides a new way for the research in this field. In the future research, we should strengthen the performance management of construction projects, innovate the sustainable development of construction projects, and improve their practical application performance.

Fund Projects. Research on the application of cloud computing technology in the construction project management system in the era of big data (Project No.: JAT190908).

References

1. Aydın, A., Tiryaki, S.: Impact of performance appraisal on employee motivation and productivity in Turkish forest products industry: a structural equation modeling analysis. Drvna Industrija **69**(2), 101–111 (2018)
2. Lee, J.S., Keil, M.: The effects of relative and criticism-based performance appraisals on task-level escalation in an IT project: a laboratory experiment. Eur. J. Inf. Syst. **27**(5), 551–569 (2018)
3. Globa, A.A., Ulchitskiy, O.A., Bulatova, E.K.: The effectiveness of parametric modelling and design ideation in architectural engineering. Sci. Vis. **10**(1), 99–109 (2018)
4. Liu, S., Liu, G., Zhou, H.: A robust parallel object tracking method for illumination variations. Mobile Netw. Appl. **24**(1), 5–17 (2018). https://doi.org/10.1007/s11036-018-1134-8
5. Ali, T., Mazdak, A., Daren, H.: A probabilistic appraisal of rainfall-runoff modeling approaches within SWAT in mixed land use watersheds. J. Hydrol. **56**(4), 476–489 (2018)
6. Indora, S., Kandpal, T.C.: Financial appraisal of using Scheffler dish for steam based institutional solar cooking in India. Renewable Energy **135**(5), 1400–1411 (2019)
7. Hu, P., Cheng, Y., Guo, X., et al.: Architectural engineering inspired method of preparing Cf/ZrC–SiC with graceful mechanical responses. J. Am. Ceram. Soc. **102**(1), 70–78 (2019)
8. Liu, S., Fu, W., He, L., Zhou, J., Ma, M.: Distribution of primary additional errors in fractal encoding method. Multimedia Tools Appl. **76**(4), 5787–5802 (2014). https://doi.org/10.1007/s11042-014-2408-1
9. Mittal, Y.K., Paul, V.K., Sawhney, A.: Methodology for estimating the cost of delay in architectural engineering projects: case of metro rails in India. J. Inst. Eng. (India) **100**(2), 311–318 (2019)
10. Abdi, A., Taghipour, S., Khamooshi, H.: A model to control environmental performance of project execution process based on greenhouse gas emissions using earned value management. Int. J. Project Manag. **36**(3), 397–413 (2018)

11. Liu, S., Lu, M., Li, H., et al.: Prediction of gene expression patterns with generalized linear regression model. Front. Genet. **10**, 120 (2019)
12. Mehmood, R., Lee, M.H., Hussain, S., et al.: On efficient construction and evaluation of runs rules-based control chart for known and unknown parameters under different distributions. Qual. Reliab. Eng. Int. **35**(2), 582–599 (2019)
13. Liu, S., Sun, G., Fu, W. (eds.): eLEOT 2020. LNICST, vol. 340. Springer, Cham (2020). https://doi.org/10.1007/978-3-030-63955-6
14. Liu, S., Yang, G. (eds.): ADHIP 2018. LNICST, vol. 279. Springer, Cham (2019). https://doi.org/10.1007/978-3-030-19086-6
15. Ma, G., Wu, M.: A big data and FMEA-based construction quality risk evaluation model considering project schedule for Shanghai apartment projects. Int. J. Qual. Reliab. Manag. **37**(1), 18–33 (2019)

Research on Host Intrusion Detection Method Based on Big Data Technology

Lei Ma$^{(\boxtimes)}$ and Hong-xue Yang

Beijing Polytechnic, Beijing 100016, China
malei235@tom.com

Abstract. When the host runs a large number of applications at the same time under normal activities, the abnormal probability value of the host after the fusion of evidence is large, resulting in false alarms, resulting in a reduction in the final detection accuracy of the detection method. A host intrusion detection method based on big data technology. Using big data processing intrusion detection index weight, sliding window is introduced. According to the number of times of host resource availability anomaly in the time window, the value of anomaly probability is controlled, the index anomaly closed value is determined, and the availability anomaly threshold is set to realize host intrusion detection. The experiment builds a data collection platform and compares the two traditional detection methods with the detection methods studied in the paper. The results show that the detection accuracy of the proposed detection method is about 98%, and the detection of host intrusion behavior is more accurate and the detection time is shortened.

Keywords: Big data technology · Intrusion behavior · Outlier probability · Detection accuracy

1 Introduction

With the rapid development of Internet technology, a lot of data information is produced in all aspects of production and life, which makes the data sources more and more extensive. At the same time, network security management is also facing severe challenges. There are more and more hacker attack channels, more Trojan horses and virus technologies. The speed of network security analysis data increases exponentially and the lag of data analysis speed increases A network security vulnerability [1]. Big data technology has the ability to quickly obtain valuable information from a large number of data with complex structure and various types. It can reveal the content and change trend that can't be seen by traditional means. It is a hot topic in the current academic, industrial and even national governments. Big data technology brings new opportunities and challenges to the development of the information security industry. In order to better play the role of big data and reduce the damage of data caused by cyber attacks, the research of intrusion detection technology based on big data is urgent, so that big data The safe and stable interaction of information data in this era.

In various industries and related fields, relying on the background of the network era, big data provides a good platform and effective channels for resource sharing and

S. Liu and L. Xia (Eds.): ADHIP 2020, LNICST 347, pp. 438–449, 2021.
https://doi.org/10.1007/978-3-030-67871-5_39

data exchange. However, with the increase of data volume and centralized storage of data, the security protection of massive data becomes more difficult, and a large number of centralized storage and processing of data inevitably increases the number of users According to the risk of leakage. The targets of hackers are more related to data clusters such as financial institutions, large companies, campus networks, etc. Once successfully attacked, hackers will obtain huge wealth from them. This year's ransomware attacked finance, government, and enterprises [3].

With the advent of the era of big data, hackers' attack methods are also changing, showing the development trend of covert attack methods. Analysis of its source is mainly due to the sudden increase in the amount of data and the increasingly close relationship between the data, which not only makes it difficult to detect hacker attacks, but also brings a wider range of harm through the connection between data and information. Traditional intrusion detection technology collects some network status information by placing multiple detectors in the network, and then the central control center analyzes and processes the information. However, in the face of large-scale, heterogeneous network environment and distributed cooperative attack, it is not enough. The main reasons are: first, the workload of the management control center is too large, and there are problems with the operation of the system; second, there is a certain delay in network transmission, and the transmitted information data cannot be transmitted to the management control center in time; third, there are platform differences in heterogeneous networks, making the analysis system face many difficulties. In view of these reasons, the distributed intrusion detection system came into being, and once became a hot spot in the field of intrusion detection [3]. Therefore, this paper proposes a host intrusion detection method based on big data technology.

2 Research on Host Intrusion Behavior Detection Method Based on Big Data Technology

2.1 Index Weight of Intrusion Detection in Big Data Processing

Big data technology intrusion behavior detection index weights adopt the objective weight assignment method to form the actual data of each host resource characteristic index in the decision plan. It has the advantages of objectivity and strong mathematical theoretical basis. It is determined by the analytic hierarchy process After the subjective weights of various indicators of the host's resources, the entropy weight method is used to determine its objective weights. When the information entropy takes the maximum value, the probability of the corresponding set of states appearing has an absolute advantage $H(X)$. The description formula of $H(X)$ is:

$$H(X) = -\sum_{i=1}^{n} p_i \log p_i (0 \leq p_i \leq 1, \sum_{i=1}^{n} p_i = 1) \tag{1}$$

In the above formula: p_i represents the extreme value of entropy and i represents the quantity. The information entropy solves the problem of measuring the amount of information and is a measure of the uncertainty of the system state. That is, the amount of information required to understand the uncertainty can be used to eliminate uncertainty. How much to express, entropy is a commonly used indicator to measure the regularity of disordered data. Entropy weight is the relative intensity of each index in the sense of competition when all kinds of detection indexes are determined after given factor set and measurement set [4]. If the degree of variation of the index value is smaller, the corresponding information entropy value is larger, the amount of information provided by the index is smaller, and the weight of the index is also smaller. Therefore, information entropy is an objective method to allocate the weight, which is determined by calculating the entropy value. In essence, it is to select the best factor to reflect the availability of host resources. The application of entropy weight method can eliminate the artificial interference in the calculation of each index weight in AHP, and make the evaluation result more real.

For the abnormality of the i index R, denoted by r, and then give a quantitative definition of the abnormality, divided into 4 levels, as shown in the following Table 1:

Table 1. Quantification level of abnormality degree

Abnormal degree	Abnormality	Quantized value
r_1	The indicator usage increment exceeds 0–15%	0.1
r_2	The use increment of this indicator is more than 16–50%	0.25
r_3	The use increment of this indicator is more than 51–70%	0.7
r_4	The use increment of this indicator is more than 71–100%	1

According to the grade index shown in the table above, and according to the calculation formula of entropy, after normalizing the relative importance of index R_i, the calculation is expressed as:

$$e_i = -\frac{1}{\ln n}\sum_{i=1}^{n} r_i \log r_i \qquad (2)$$

In the above formula, when the values of $r_i(i = 1, 2, \ldots, n)$ are equal, the entropy value e_i is at most 1, so when $0 \leq e_i \leq 1$, $\ln n$ is the maximum entropy value when all are equal. When the entropy value is maximum, the contribution of this indicator to the detection result is the smallest, so the weight of the detection factor R_i can be determined using the $1 - e_i$ metric process. Therefore, it is normalized to obtain the objective weight of the detection factor R_i:

$$w_i = \frac{1 - e_i}{n - \sum_{i=1}^{n} e_i}, 0 \leq w_i \leq 1, \sum_{i=1}^{n} w_i = 1 \qquad (3)$$

In the above formula, w_i represents objective weight. The weight obtained by AHP reflects the subjective preference of decision-makers, and the weight calculated by entropy weight method reflects the objective relationship between each host resource index. In order to give consideration to the expert's experience judgment of attributes, at the same time, it strives to reduce the subjective randomness of subjective weight. Use the linear weighting method to determine the comprehensive weight of the attribute, namely:

$$w_i = \mu w_{si} + (1 - \mu)w_{oi}, (i = 1, 2, \ldots n) \tag{4}$$

In the formula, w_i is the comprehensive weight, w_{si} is the subjective weight of the index determined by the analytic hierarchy process, and w_{oi} is the objective weight of the index determined by the entropy weight method. μ is the subjective preference coefficient, $1 - \mu$ is the objective preference coefficient, and $0 \leq \mu \leq 1$. When μ is less than 0.5, the proportion of subjective weight in the overall weight is relatively large, and vice versa. In general, the specific value of μ is given by the decision maker according to preferences, and it can be proved that μ can obtain the optimal value and optimize the change results, as shown in the following Fig. 1:

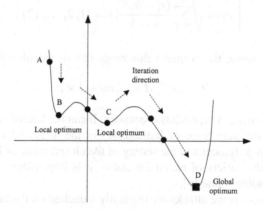

Fig. 1. Optimize change

According to the change process shown in the figure above, suppose S represents the host information system, w_{si} and w_{oi} are the subjective weight and objective weight of indicator R_i respectively, and w_i represents the combined weight of the two. An optimization model is established for the above indicators, and the calculation formula is as follows:

$$\min\left\{\sum_{i=1}^{n} [\mu(\frac{1}{2}(w_i - w_{st})^2) + (1 - \mu)(\frac{1}{2}(w_i - w_{oi})^2)]\right\} \tag{5}$$

Using the above formula to optimize the above weight value calculation process, according to the intrusion behavior detection index weights, determine the abnormal closed value of each index to realize the detection of host intrusion behavior.

2.2 Determination of Abnormal Closed Value of Each Index

When determining the abnormal closed value of each index, first of all, you need to obtain the host resource availability profile under normal circumstances, and use this to determine the abnormal value of the host resource index [5]. We collect data samples on several hosts in the laboratory. Host resource availability indicators are collected once a minute, 1440 times in a continuous 24 h. In order to ensure the objectivity of experimental data, the normal operation of the host computer is ensured in the process of data acquisition. For each index data collected, we first calculate the average m and standard variance c, and then calculate the corresponding normal value range H average and standard variance calculation formula of the index:

$$\begin{cases} m_i = \sum_{i=1}^{n} \frac{r_i}{n} \\ c_i = \sqrt{\sum_{i=1}^{n} \frac{(r_i - m_i)^2}{n-1}}, i = (1, 2, \ldots, 12) \end{cases} \tag{6}$$

In the above formula, the normal value range can be calculated as follows:

$$H_i = m_i - d \times c_i, m_i + d \times c_i \tag{7}$$

In the above formula, A represents d constant. It can be determined according to the actual situation of different resource indicators. For the threshold value of target host resource availability judgment and the setting of the closed value of host security status exception, due to the variety of exception status, it is impossible to accurately determine the corresponding value [6].

In actual detection, some attacks are manually launched on the target host, and then the corresponding threshold is calculated according to the sampled data. Of course, the corresponding threshold can also be set based on the experience of the administrator to determine the pairwise comparison matrix, where the target layer judgment matrix is A, and the matrix can be expressed as:

$$A = \begin{bmatrix} 1 & \frac{1}{2} & 4 & 3 & 3 \\ 2 & 1 & 7 & 5 & 5 \\ \frac{1}{4} & \frac{1}{7} & 1 & \frac{1}{2} & \frac{1}{3} \\ \frac{1}{3} & \frac{1}{5} & 2 & 1 & 1 \\ \frac{1}{3} & \frac{1}{5} & 3 & 1 & 1 \end{bmatrix} \tag{8}$$

In the indicator layer, there is only one indicator of CPU utilization and the indicator matrix of network resources is B_3, so it directly inherits the result of the judgment matrix of the target layer. The calculation matrix can be expressed as:

$$B_3 = \begin{bmatrix} 1 & 3 \\ \frac{1}{3} & 1 \end{bmatrix} \tag{9}$$

Based on the above formula, the storage resource index is B_2, and the process/thread index matrix is B_5. The values of the two indexes are equal. The calculation formula is as follows:

$$B_2 = B_5 = \begin{bmatrix} 1 & 3 & 5 \\ \frac{1}{3} & 1 & \frac{1}{5} \\ \frac{1}{5} & \frac{1}{3} & 1 \end{bmatrix} \tag{10}$$

The resource matrix can be recorded as B_4:

$$B_4 = \begin{bmatrix} 1 & 1 & 1 \\ 1 & 1 & 1 \\ 1 & 1 & 1 \end{bmatrix} \tag{11}$$

Use the comparison matrix given above based on expert experience to calculate the weight vectors of the hierarchical single ranking, and after the consistency test, use big data technology to build a recognition framework, that is, filter out security events per unit time and make corresponding statistics and records, Construct the evidence collection:

$$\bigcup E_i (i = 1, 2, \ldots, n) \tag{12}$$

According to the above formula, we use the evidence BPA method to allocate the credibility of the data in the host, calculate the trust function and likelihood function of each data, and use the Dempster synthesis method to calculate the joint credibility evaluation function, trust function and likelihood function of the known evidence. If there is a new intrusion in the detection cycle, continue the data fusion process until the end of the detection cycle, build a distribution function according to the comprehensive trust level, calculate the trust level of the intrusion host data, and use the small value of the trust level value as the index abnormal closed value. Use these abnormal closed values to realize the detection of host intrusion behavior [7].

2.3   Implement Host Intrusion detection

There are many parameters that can be used for the statistical characteristics of host resource availability. Drawing on previous work, and through a large number of experimental analysis, five primary indicators are selected to form a host resource availability measurement system. Each primary indicator also includes a corresponding secondary Indicators [8, 9]. Because the change of each indicator can better reflect the dynamic characteristics of the application program's demand for the availability of host resources, and the incremental information can also reduce the amount of calculation and memory utilization, all the secondary indicators in the host resource indicators selected by this method, except the CPU utilization, represent the increment in unit time. "Increment" refers to the absolute value of the change of the indicator relative to the last unit time. The five first-level indicators are the total statistics of storage resources, computing resources, bandwidth resources, processes/threads, and IO resources. Corresponding to its index level and meaning in the host, as shown in the following Table 2:

Table 2. Host availability index and meaning

First level indicators	Secondary index	Variable representation	Specific meaning
Computing resources	CPU	CPUUse	CPU Usage rate
Storage resources	physical memory	phyMem	Physical memory usage increment
	Virtual Memory	virtMem	Virtual memory usage increment
	Memory usage	memUse	Approved memory usage increment
Internet resources	Send traffic	bandSend	Send packet increment
	Receive traffic	bandReci	Receive data packet increment
Process/thread count	Process	Process	Process number increment
	Thread	Thread	Thread increment
	Handle	Count	Handle number increment
IO resources	IO reading	IO read	IO read increment
	IO write	IO write	IO write increment
	IO others	IO other	IO other increments

According to the above weight determination process, each index shown in the above table is assigned to the first and second level indexes, and the mathematical model of resource availability evaluation of the target host is established, with the calculated abnormal closed value as the evaluation index, Set the threshold of the total intrusion behavior data of the switchboard to L. When the abnormal closing value is greater than or equal to L, there is an abnormal intrusion and a sliding window is introduced, According to the number L of occurrences of host resource availability abnormalities in the time window, it is determined whether the current host security status is abnormal, that is, the abnormal status of the host at the current time is determined according to the number of host resource availability abnormalities within the latest window. among them:

$$p = \sum_{t-\Delta t}^{\Delta t} c \tag{13}$$

Among them, c is the number of host resource intrusions per unit time, Δ is the length of the time window, t is the intrusion time, $A(s)$ is used to indicate the current state of the host, 1 is abnormal, and 0 is normal, then the security status frequency is closed φ:

$$A(s) = \begin{cases} 1, p \geq \varphi \\ 0, p < \varphi \end{cases} \tag{14}$$

In the above formula, when $p < \varphi$, it means that the host is safe and there is no intrusion behavior; when $p \geq \varphi$, there is intrusion behavior. Based on the threshold value of host resource availability exception, judge the host resource availability exception within the window time, and judge the host security status exception at the current time according to the closed value of the host security status exception frequency [10, 11]. The thresholds L and φ are determined by the network administrator according to the experience and the requirements of the host security state. The research on the host intrusion detection method based on big data technology is finally completed.

3 Simulation Experiment

3.1 Experiment Preparation

Three computers with the same parameters are prepared for the experiment. The parameters of the computers are as follows (Table 3):

Table 3. Experimental computer parameters

No. Name parameter	No. Name parameter	No. Name parameter
1	Memory	8 GB DDR4 RAM
2	Storage	Solid state drive SSD
3	Processor	Amd Ruilong 5 3580u mobile processor integration
4	Graphics card	Amd radeon TM Vega graphics card
5	interface	Usb-c, usb-a, surface connect port

According to the parameters shown in the table above, build a data collection platform, as shown in the following Fig. 2:

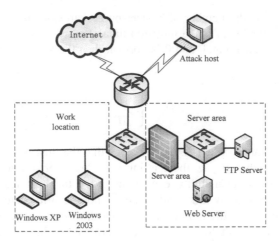

Fig. 2. Data acquisition platform

According to the platform structure shown in the figure above, simulate host intrusion behavior, deploy intrusion detection points, collect all host data and intrusion data, as shown in the table below (Table 4):

Table 4. All data and intrusion data in the host

Number of layers	32-64-256-128-64-32	32-1152-2
1	8.0365	0.8286
2	4.8646	0.7035
3	3.0567	0.6154
4	6.8237	0.5525
5	3.0342	0.5080
6	4.5420	0.4864
7	7.3056	0.4797
8	7.1946	1.3818
9	6.1387	0.9705
10	5.1188	0.5335

In the table above, 32-64-256-128-64-32 represents the data in all levels of the host, and 32-1152-2 represents the external intrusion data of the host. Network intrusion detection method based on improved majorcluster clustering (traditional behavior detection method 1) and host intrusion detection method based on data mining (traditional behavior detection method 2) are used respectively, and the proposed big data technology-based intrusion detection method proposed in this paper Host intrusion detection methods are tested to compare the performance of the three methods [12, 13].

3.2   Analysis of Results

Based on the above processing, the host intrusion detection time is obtained by iterating the data shown in the table above, as shown in the following figure:

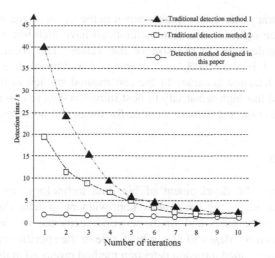

Fig. 3. Comparison of test time results

According to Fig. 3, with the increasing number of iterations, the detection time of the three detection methods gradually decreases. The host intrusion detection time of the traditional behavior detection method 1 is reduced from 40 s to 3S, and the host intrusion detection time of the traditional behavioral behavior detection method 2 is reduced from 15 s to 3S. However, the host intrusion detection time of the proposed method is relatively stable, which is reduced from 3S to 2S The time was the lowest among the three methods.

In this experimental environment, the detection accuracy of the three detection methods is calculated, and the experimental results are as follows Fig. 4:

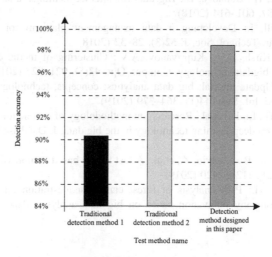

Fig. 4. Experimental results of accuracy of three detection methods

From the accuracy calculation results shown in the figure above, we can see that the three host intrusion behavior detection methods all have high test accuracy, but there are still certain differences in the size of the values. The accuracy of traditional detection method 1 is about 90%, the accuracy of traditional detection method 2 is about 92%. The detection accuracy of the test method studied in this paper is about 98%. This method has high sensitivity to host intrusion data to ensure the accuracy of intrusion data detection and is more suitable for the actual host intrusion detection.

4   Conclusion

In accordance with the development of big data technology, with the continuous expansion of data scale, the continuous improvement of data throughput and the requirements of multi hosts and multi network segments, the intrusion detection technology based on big data will play a huge role in the specific application platform in this field. The distributed intrusion detection method proposed in this paper improves the shortcomings of traditional intrusion detection system. It not only makes DIDS technology develop rapidly, but also provides safe and reliable protection for large-scale network.

References

1. Lakshmanaprabu, S.K., Shankar, K., Rani, S.S., et al.: An effect of big data technology with ant colony optimization based routing in vehicular ad hoc networks: towards smart cities. J. Cleaner Prod. **217**, 584–593 (2019)
2. Chen, M.-Y., Lughofer, E.D., Polikar, R.: Big data and situation-aware technology for smarter healthcare. J. Med. Biol. Eng. **38**(6), 845–846 (2018)
3. Ge, M., Bangui, H., Buhnova, B.: Big data for internet of things: a survey. Future Gener. Comput. Syst. **87**, 601–614 (2018)
4. Turner, C., Gill, I.: Developing a data management platform for the ocean science community. Mar. Technol. Soc. J. **52**(3), 28–32 (2018)
5. Rycarev, I.A., Kirsh, D.V., Kupriyanov, A.V.: Clustering of media content from social networks using bigdata technology. Comput. Opt. **42**(5), 921–927 (2018)
6. Watson, H.J.: Update tutorial: big data analytics: concepts, technology, and applications. Commun. Assoc. Inf. Syst. **44**(1), 364–379 (2019)
7. Zeng, W., Xu, H., Li, H., et al.: Research on methodology of correlation analysis of sci-tech literature based on deep learning technology in the big data. J. Database Manage. **29**(3), 67–88 (2018)
8. Shuai, L., Weiling, B., Nianyin, Z., et al.: A fast fractal based compression for MRI images. IEEE Access **7**, 62412–62420 (2019)
9. Xue, Y., Feng, H.: Path analysis of forest carbon sequestration on poverty alleviation papermaking company innovation based on big data analysis. Paper Asia **35**(1), 28–32 (2019)

10. Fu, W., Liu, S., Srivastava, G.: Optimization of big data scheduling in social networks. Entropy **21**(9), 902 (2019)
11. Liu, S., Liu, D., Srivastava, G., et al.: Overview and methods of correlation filter algorithms in object tracking. Complex Intell. Syst. (2020). https://doi.org/10.1007/s40747-020-00161-4
12. Huda, M., Maseleno, A., Atmotiyoso, P., et al.: Big data emerging technology: insights into innovative environment for online learning resources. Int. J. Emerg. Technol. Learn. **13**(1), 23–36 (2018)
13. Lu, M., Liu, S.: Nucleosome positioning based on generalized relative entropy. Soft Comput. **23**(19), 9175–9188 (2018). https://doi.org/10.1007/s00500-018-3602-2

Intelligent Control Method for Load of Multi-energy Complementary Power Generation System

Shi-hai Yang[1], Xiao-dong Cao[1], Wei-guo Zhang[2(✉)], and Feng Ji[1]

[1] Jiangsu Electric Power Company Research Institute, Nanjing 210000, China
ysh09800@126.com
[2] NARI Technology Co. Ltd., Nanjing 211106, China

Abstract. In order to reduce environmental pollution, the load of power generation system has become an important basis for the balance of power supply and demand. Therefore, an intelligent load control method for multi energy complementary power generation system is proposed. Firstly, the intelligent control scheme of multi energy complementary power generation system is formulated, the circuit principle of the control method is determined, and the frequency control parameters of power grid, frequency regulation of generator set and control mode of generator set are controlled. The experimental results show that the load intelligent control method of the multi energy complementary generation system can effectively control the load of the power generation system, and the multi energy complementary generation can smooth the randomness of single energy, the intermittent fluctuation of energy and the control effect of energy storage, and reduce the impact on the power grid, which is very suitable for distributed grid connected operation.

Keywords: Multi-energy · Complementarity · Power generation · Control

1 Introduction

Wind energy, solar energy, marine energy and other renewable energy have huge reserves, wide distribution, and no pollution in development and utilization. How to make use of these resources to transform the electric energy which is needed by the modern information society has been paid more and more attention by all countries [1]. Some islands and remote areas are far from the main power grid due to their geographical characteristics, or the power grid is difficult to set up, coupled with the lack of energy conversion technology, the living standards of residents are backward, and it is difficult to carry out industrial production, while the natural resources of the islands are often rich, so the development and utilization of renewable resources of the islands are many at one stroke. However, these natural resources are very sensitive to the climate, with small energy density and strong randomness, which can not be realized as a stable energy output or at a considerable cost [2]. Renewable natural resources have the characteristics of small energy density, wide distribution and strong randomness. According to the research of hydrodynamics, the randomness and uncertainty of wind

S. Liu and L. Xia (Eds.): ADHIP 2020, LNICST 347, pp. 450–459, 2021.
https://doi.org/10.1007/978-3-030-67871-5_40

energy and water energy will lead to the following changes of the frequency and voltage of the output power of wind turbine, so it is necessary to control and adjust the power in the actual operation [3]. At the same time, because of the complexity of generators and some power electronic devices, and the control technology of some system models is only suitable for some specific power systems. In addition, the wind turbine generator units are usually in remote areas, offshore or island, generally unattended, and can only be monitored remotely, which puts forward higher requirements for the matching and reliability of the control system of the generator units.

According to the current situation and development trend of wind power generation, according to the planning of the national energy administration, large-scale wind power bases are planned to be built in areas rich in resources, so as to accelerate the decentralized development of wind energy in other regions. By 2020, wind turbines will reach 200 GW. China's wind energy is distributed in the northeast, Xinjiang, Inner Mongolia and other remote areas. In this process, the stability of power grid frequency becomes the key to affect the quality of power supply [4]. Active power is closely related to frequency. The balance between supply and demand of active power under rated frequency is the basic premise of ensuring frequency quality and one of the basic methods of frequency control. However, this is a constrained nonlinear multi extremum optimization problem, which is difficult to deal with by traditional methods. Therefore, an intelligent load control method for multi energy complementary power generation system is proposed. The optimal control of generator active power output can not only realize the balance of supply and demand of active power, but also ensure the frequency stability of power generation system. It can also make the power generation system run economically. Frequency regulation is divided into one, two and three regulation. Primary regulation is the inherent characteristic of generator, and differential regulation is usually used. In order to ensure the frequency safety of the system, it is necessary to adjust the frequency twice by using the power frequency static characteristics of the generator synchronizer. In order to improve the economy of power generation system, the generator is usually optimized and adjusted, which can not only ensure the safety of system frequency, but also reduce the generation cost of power generation system.

2   Intelligent Control Method for Load of Multi-energy Complementary Power Generation System

2.1   Design the Structure of Intelligent Load Control Method for Multi-energy Complementary Power Generation System

The multi-energy storage joint generation system shares a single inverter topology as shown in Fig. 1:

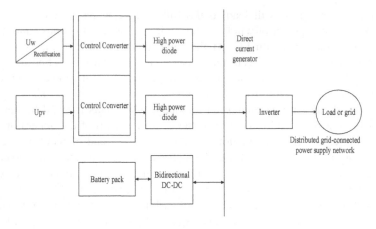

Fig. 1. Intelligent load control for multi-energy complementary power generation system

Two V2-controlled Boost-Buck converters (DC-DC) are designed on the DC side of grid-connected or inverters. The output of wind power generation converted by AC-DC rectifier and photovoltaic power generation are controlled respectively by parallel connection of high power diodes and DC bus [5]. By using common inverters, the switching between distributed grid-connected power supply and isolated off-grid power supply is realized. The multi-energy storage combined power generation system based on this topology has the advantages of low cost, suitable for edge area, good stability of fluctuation, suitable for grid-connected operation, stable DC bus voltage and simple control strategy [3].

The model is composed of three parts: strong power flow loop, weak current sampling display control loop and control program module. The discharge circuit designed in the high power flow circuit is used to adjust the multi-energy output deviation not to be too large [6].

2.2   Introduction Control Circuit Principle

When the Boost and Buck circuits are used separately, they have only a single step-up or step-down function. A Kv and A1 module are added to the Boost-Buck circuit, and A3 module is used to achieve both the effect of voltage rise and fall and the constant voltage control of the output double loop feedback (Kv-A1, outer loop Kv-A3-A2-A1) of the Boost-Buck circuit. The inner loop control can improve the transient response speed of the Boost-Buck circuit, while the outer loop control can improve the control accuracy of the Boost-Buck circuit and keep the output voltage basically constant. In practical application, the input is added and the output filter reduces the pulsation of the current [7]. The V2 control method uses the voltage of the output filter capacitor C of the Boost-Buck converter to replace the inductor current in the peak current control. As the peak voltage detection input of comparator A1 and output by single chip computer, the PWM signal compensated by skew compensation circuit is connected to JK flip-flop to modulate IGBT work. Rate tube conduction angle duty cycle signal, thus changing the inductance current and output voltage.

2.3   Calculation of Power Network Frequency Control Parameters

When the grid frequency and power change, they will have a dynamic relationship. When the power difference P, the frequency deviation f of the power grid will have the following transfer function:

$$\frac{\Delta f(s)}{\Delta P(s)} = \frac{1}{M_i s + D_i} \tag{1}$$

Formula M_i is the inertia time coefficient of the generator set, and $M_i = 50$–10 s is the general one. Formula D_i is the frequency regulation effect coefficient of the load (also known as the model load damping constant), then $D_i = \frac{\Delta P_{pt}}{\Delta f} (MW/Hz)$.

When D_i is replaced by a standard unitary value, it represents the percentage of load active power change caused by 1% frequency change. Take $D_i = 1$ as an example, that is, when frequency +0.01 (+0.5 Hz), the load value will be +0.01 (load 1%) and $D_i = 1$–3.

In primary frequency modulation of generating units, the curve of static relationship between output power and frequency of generating units becomes the static characteristics of power and frequency of generating units [8]. The slope of power-frequency static characteristic curve can be expressed as $\frac{\Delta P}{\Delta f} = K_g$.

Governor is an important part of generator set. Its main function is to maintain the load distribution and rotation speed between units. The mathematical function of steam turbine governor is $1/T_{ni}S + 1$, Of which Is the governor time factor, which is usually 0.05–0.25 s. Steam turbines convert steam stored energy into mechanical energy, which is then converted by generators into electrical energy. The transfer function of reheat steam turbine is $\frac{SCT_{Ri} + 1}{(ST_{Ri} + 1)(ST_{ti} + 1)}$. It needs to consider the time delay of steam flow in reheat system. C is the proportion of power generated before steam enters reheater, which is generally 25%–30%, TRi is the time coefficient of reheater, which is generally 5–10 s.

2.4   Regulating the Frequency of a Generator Set

Stand-alone control area: there is only one non-reheat thermal generator unit in the control area with an installed capacity of 600 MW (rotating standby capacity of 100 MW), with a load of 500 MW. at a frequency of 50 Hz We take the firepower unit's Di equal to 1Mi equal to 10. After several seconds of reduced amplitude oscillation, it finally stabilizes near the calculated value of theoretical analysis.

The wind power generation control mode and the photovoltaic power generation control mode belong to the single power supply constant voltage control mode, can be divided into two kinds of situations:

The time period of non-light radiation. The wind speed is enough to control the control mode of single Boost-Buck converter using V2 of wind power generation;

No wind speed period. Enough optical radiation to control a single Boost-Buck converter using a PV-based V2 converter;

The multi-energy complementary power generation control mode belongs to the dual power supply deviation control and constant voltage control mode, specifically: there is both light radiation and wind speed time, when wind speed, light radiation is enough, but wind, When the output voltage difference between the two photogeneration exceeds the allowable range of the Boost-Buck converter, the PWM signal is output by the bias control module of the single-chip microcomputer to make the wind power discharge circuit work, and the difference of the output voltage between the two is adjusted to the allowable value. The V2 control dual-Boost-Buck converter which starts the wind-light combined power generation works simultaneously [9, 10].

2.5   Control Generation Control Mode

The multi-energy storage combined generation control mode includes multi-energy storage and distributed grid-connected fluctuation and fuzzy control, multi-energy storage island off-grid load control and fuzzy control. The multi-energy storage and distributed grid-connected control mode is to control the switch-on direction of the bidirectional DC-DC converter under the distributed grid-connected power supply mode, according to whether the output power fluctuation of the multi-energy power exceeds the national standard [11, 12]. By controlling the start-up and stop time of energy storage or discharge, the energy storage can reduce the power wave momentum of multi-energy and reduce the impact of the power network.

The active power of multi-energy complementary generation is sampled according to a certain sampling period, and the on-off direction of switch is controlled in the circuit of bi-directional DC-DC converter. When the sampling difference between two adjacent points is positive and the absolute value is greater than the national standard, Starting energy storage to store remaining energy; When the sampling difference between the two points is negative and the absolute value is greater than the national standard, the starting energy storage releases the storage energy.

2.6   Control Load Network Load

The strategy multi-energy storage island type off-grid control mode is that the multi-energy storage combined power generation under the isolated off-grid direct load operation mode, according to the multi-energy generation power and the load demand power match or not. The on-off direction of switch is controlled in the circuit of bi-directional DC-DC converter, and the start-up and stop time of energy storage or discharge is controlled. The purpose of energy storage is to increase the follow-up of multi-energy and load and to reduce the abandoned multi-energy quantity.

For the off-grid operation model of multi-energy storage island, the multi-energy storage power in the model should be matched with the load demand power at every moment. Therefore, the multi-energy combined output is divided into three regions according to the installed capacity: peak, middle waist and low valley. The load is also divided into three regions according to the maximum load, namely, high load, medium load and low load. According to the following principles, a multi-energy storage and nine-house district control strategy based on tracking load fluctuation is formulated.

(1) When the multi-energy output and load level are in the low-valley and low-load region, the middle-waist load region and the peak-high load region, respectively, it is considered that the multi-energy combined output is in the state of natural tracking load fluctuation (positive peak-shaving), and the output of the energy storage model is set to 0 at this time.

(2) When the multi-energy output level is in the low valley region and the load level is in the high load region, or when the multi-energy output level is in the high peak region and the load level is in the low load region, it is considered that the multi-energy power output is in the state of unable to track the load fluctuation (inverse peak-shaving). Set the energy storage model to:

When $P_{ES(i)} \leq \left| P_{wg(i)} - P_{L(i)} \right| - P_F$, $P_{wg(i)} - P_{L(i)} < 0$, low valley and high charge energy storage discharge; When $P_{ES(i)} \leq \left| P_{wg(i)} - P_{L(i)} \right| - P_F$, $P_{wg(i)} - P_{L(i)} > 0$, peak low charge energy storage charge. Among them, $P_{L(i)}$ is the total load of the model at I time, PF is the setting threshold, and its setting amount is determined by 10% of the storage capacity, considering that the storage should not be filled frequently.

3 Results

According to the above-mentioned principle and conception, Xinjiang University has developed the hardware module of V2 control double Boost-Buck transform controller and the software modules of constant voltage control, fluctuation control, load control, fuzzy control and deviation control for multi-energy storage combined power generation. Physical hardware and multi-energy storage networking test platform.

3.1 Experimental Preparation

(1) In the 0:00–7:00, 20:00–24:00 non-light radiation period and multi-energy generation differential voltage greater than the allowed range of 7:00–8:00, 19:00–20:00 wind-generated V2 control of a single Boost-Buck converter control mode. The test results are qualified and meet the requirements.

(2) When the differential voltage of multi-energy generation is less than the allowable range of 7:00–19:00 time, the wind-light combined power generation V2 control dual-Boosck converter control mode, the test results are qualified voltage and meet the requirements.

Based on the control strategy of reducing multi-energy fluctuation and tracking load, positive peak shaving, flat peak shaving and reverse peak shaving frequency are used to calculate the effect of energy storage after energy storage. Table 1 gives the variation and effect of multi-energy electric power fluctuation, following load and economic index before and after adding energy storage.

It can be seen that the effect is better in stabilizing the fluctuation of multi-energy electricity, improving the characteristic of following load, reducing the amount of multi-energy electricity abandonment, increasing the utilization of resources and improving the economic benefit of multi-energy electricity.

Table 1. Economic indicators of multi energy electric energy storage

Economic indicators	Time	Fan power/kW	Photovoltaic power/kW	Photoelectric complementary power/kW
Multi energy power fluctuations	0	386.55	0	387
Subsequent load	5	290.89	0	291
Before increasing energy storage	10	398.54	703.20	1102
After increasing energy storage	15	298.10	742.40	1041

3.2   Load Comparison of Power Generation Control Mode

The multi-energy complementary charging mode and the traditional charging mode are used to charge the battery pack composed of 12 V/9 A·h battery series and parallel respectively. The dynamic curve of charging current in the two modes is shown in Fig. 2.

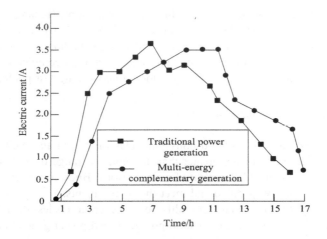

Fig. 2. Two charging current variation curves

(1) The charging time of traditional charging mode is nearly 17 h, the charging time of multi-energy complementary generation is nearly 15.5 h, and the charging time is shortened.

(2) Considering the safety problem, the maximum charging current of the battery is 4 A, the maximum current is 3 A in the traditional charging mode, and the maximum charging current is close to 3.5 A in the multi-energy complementary charging mode, indicating the fuzzy control mode. Ability to automatically identify maximum charge current.

3.3 Test of Inverter Output

In the simulation test of multi-energy storage test platform, the output voltage waveform of the inverter is compared with that of the actual field multi-energy grid-connected bus voltage recording, as shown in Fig. 3.

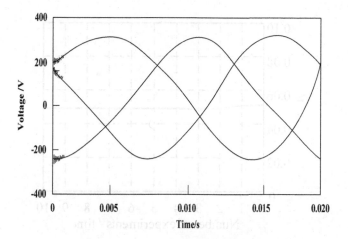

Fig. 3. Inverter output voltage ripple

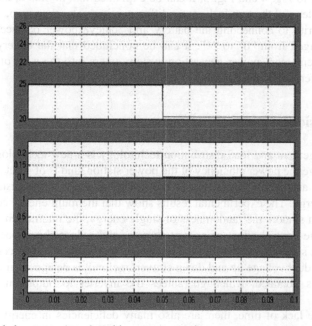

Fig. 4. Power balance results of multi energy complementary generation system when the battery is not saturated

When the battery of multi energy complementary power generation system is not full, the regulation method of load tracking will consider the state of the battery. Assuming that the current power is reduced from 25 kW to 22 kW, the SOC of the battery is 0.6. The simulation curve is shown in Fig. 4. The load tracking decision changes in 0.05 s as shown in Fig. 5.

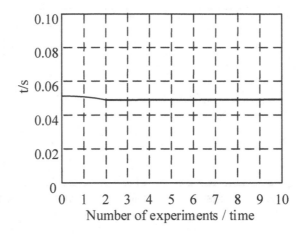

Fig. 5. Load tracking decision changes at 0.05 s

According to Fig. **4** and Fig. 5, it can be explained that in 0.05 s, the input power becomes smaller and the terminal power difference is greater than zero. The controller will give priority to connecting the battery and achieve power balance by charging. Decision changed from "0" to "1". It can be concluded that the battery of the multi energy complementary power generation system enters into the state of charge when the power difference is near zero.

4 Conclusions

Energy management and load distribution technology is a new technology, which can greatly improve the power efficiency of the power station, and is of great significance to the economic and reliable operation of the power station and the sustainable development of energy. The experimental results show that the multi energy complementary generation can smooth the randomness of single energy, the intermittent fluctuation of energy and the control effect of energy storage, and reduce the impact on the grid, which is very suitable for distributed grid connected operation. The constant voltage control of the dual voltage up and down converter circuit is designed, and the ideal DC bus voltage is obtained, which guarantees the inverter to output high-quality AC current.

Due to the lack of time, there are also many deficiencies in energy management. Some problems and solutions need to be further studied. Therefore, some suggestions and ideas are put forward for the research work in this paper: due to the lack of a large number of field test data, the knowledge base of fuzzy control system can not be trained and optimized by using fuzzy neural network. In the future work, the controller can be modified manually according to the membership function and control rules, so that the performance of the controller can reach a good level.

References

1. Jiang, Z., Hao, R., Ai, Q.: Research on the interaction mechanism of industrial parks based on multi-energy complementarity of cooling, heat and power. Electr. Power Autom. Equip. **37**(6), 260–267 (2017)
2. Lu, J.J., He, Z.Y., Zhao, C.Y.: MMC three-terminal active power regulation system for multi-energy cooperative operation. Electr. Power Autom. Equip. **37**(6), 245–252 (2017)
3. Tang, H., Yang, G.H., Wang, P.Z.: Optimal capacity configuration of multi-energy complementary generation system based on CO_2 emissions. Power Constr. **38**(3), 108–114 (2017)
4. Zeng, M.: Research on system planning method of regional energy supply and service network based on benders decomposition optimization. J. North China Electr. Power Univ. Nat. Sci. Ed. **44**(1), 89–96 (2017)
5. Wei, L.J., Huang, M., Sun, J.J., et al.: Nonlinear behavior analysis of photovoltaic-storage hybrid power generation system with constant power load. J. Electr. Technol. **32**(7), 128–137 (2017)
6. Liu, S., Yang, G. (eds.): ADHIP 2018. LNICST, vol. 279. Springer, Cham (2019). https://doi.org/10.1007/978-3-030-19086-6
7. Pan, F., Cheng, F.H., Luo, C.: Research on charge and discharge control strategy of microgrid hybrid energy storage system. Northeast Electr. Power Technol. **417**(3), 21–26 (2018)
8. Ye, L., Qu, X.X., Mo, Y.X., et al.: Intra-day time scale operation characteristics analysis of multi-energy water and multi-energy complementary power generation system. Power Syst. Autom. **42**(4), 158–164 (2018)
9. Zheng, P., Shuai, L., Arun, S., Khan, M.: Visual attention feature (VAF): a novel strategy for visual tracking based on cloud platform in intelligent surveillance systems. J. Parallel Distrib. Comput. **120**, 182–194 (2018)
10. Guo, F.T., He, S., Wang, W.Q.: Research on optimal dispatch of power load in microgrid of power generation system. Comput. Simul. **35**(09), 99–103+184 (2018)
11. Liu, S., Li, Z., Zhang, Y., et al.: Introduction of key problems in long-distance learning and training. Mob. Netw. Appl. **24**(1), 1–4 (2019)
12. Liu, S., Lu, M., Li, H., et al.: Prediction of gene expression patterns with generalized linear regression model. Front. Genet. **10**, 120 (2019)

Simulation of Multi-area Integrated Energy for Cooling, Heating and Power Based on Large Data Analysis

Feng Ji[1(✉)], Shi-hai Yang[1], Xiao-dong Cao[1], and Yong-biao Yang[2]

[1] Jiangsu Electric Power Company Research Institute, Nanjing 210000, China
jf987006@126.com
[2] Southeast University, Nanjing 210096, China

Abstract. Because the current power supply does not take into account the regional climate cold and hot issues, leading to some areas of power energy supply is greater than demand, waste a lot of unnecessary electricity, power utilization rate is low. To solve the above problems, a multi-area integrated energy dispatching method based on large data analysis is proposed. Firstly, the data warehouse method is used to integrate the supply and demand data of multi-region related comprehensive energy sources, then the integrated data is processed, and then the power demand level is divided based on fuzzy clustering. Finally, the multi-region power is reasonably supplied according to the level, and the comprehensive energy dispatch of cooling, heating and power is completed. The experimental results show that the multi region integrated energy scheduling method based on big data analysis can reduce the average power supply power by 353.534 kW, and improve the utilization rate of electric energy.

Keywords: Largedata analysis · Integrated energy of cooling · Heating and power · Integrated energy dispatch · Simulation

1 Introduction

Under the double pressure of energy crisis and environmental pollution, the construction of regional comprehensive energy system has attracted wide attention all over the world. Compared with the traditional cold, hot and power distribution system, the regional comprehensive energy system has higher energy utilization efficiency and emission reduction value [1]. In China, the construction of regional comprehensive energy system is still in its infancy, and the scientific planning and optimal allocation of regional comprehensive energy supply system is of great significance to its optimal operation and overall performance improvement, which is an important problem to be solved in the further development of regional comprehensive energy system in China. The regional comprehensive energy system is composed of two parts: energy station and heat and cold transmission pipeline network. Integrated energy station is the core of regional integrated energy system. Its design process can be divided into three parts: basic architecture design, equipment combination optimization and decision-making based on operation simulation. The basic structure design is to determine the type of the

S. Liu and L. Xia (Eds.): ADHIP 2020, LNICST 347, pp. 460–471, 2021.
https://doi.org/10.1007/978-3-030-67871-5_41

alternative cold and heat power conversion device and its connection relationship according to the primary energy status, site, energy supply type and demand of the user demand and other elements available in the area. The process of equipment combination optimization is similar to that of power supply planning. Generally, operation simulation is used to optimize the comprehensive energy equipment type, capacity and number combination to achieve the optimal goal. When considering the multi-objective of economy and environmental protection, it is also necessary to obtain the recommended scheme through the comprehensive decision-making process.

With the continuous complexity of the structure and function of the power grid, the requirements for the economy, environmental protection and reliability of the power system are getting higher and higher. How to improve the energy utilization rate, improve the energy structure of the power system, alleviate the contradiction between energy demand and energy shortage, energy utilization and environmental protection, and make the power grid move towards a clean, safe, efficient and reliable development path has become a difficult problem and key to power operation [2]. China has a vast territory and occupies many temperature zones. Therefore, there are some differences in temperature in each region. Therefore, in summer indoor cold supply and winter indoor heat supply, according to the regional temperature, is one of the main ways to solve the above problems [3]. However, how to realize the rational dispatch of multi-regional integrated energy resources of cooling, heating and power, the large data analysis in the early stage is very important, and it is the basis of realizing the rational distribution of power energy. In the process of big data analysis, it mainly includes data acquisition related to regional energy supply, then processing the collected power data, dividing the power supply and consumption of multi-region into several levels according to the actual demand of local power, and finally realizing the optimal distribution of power according to the level, reasonable dispatch, in order to reduce unnecessary consumption of power energy and improve the energy structure. In order to validate the effectiveness of the multi-region integrated energy dispatching method based on large data analysis in this study, a simulation experiment was carried out. The results show that the total energy supply in North China has been greatly reduced after using this method to dispatch the integrated energy of the region. This shows that the utilization rate of power energy has been improved and the energy shortage has been alleviated.

2 Multi-area Cooling, Thermoelectricity and Integrated Energy Dispatch

With the rapid development of social science and technology, human living and development space is more and more vulnerable to population growth, energy shortage, environmental pollution and ecological destruction [4]. In order to improve energy efficiency, reduce environmental pollution and improve energy structure, countries around the world have made the exploration of alternative energy sources and the sustainable development of energy as a top priority. With the rapid economic growth and social development in China, the contradiction between energy demand and energy shortage, energy utilization and environmental protection has become increasingly prominent. With its cleanliness and renewability, electric energy has become the focus of research. It has more and more advantages in improving energy utilization efficiency, developing new energy and alleviating energy demand. It is of great significance to solve the contradiction between economic development and environmental constraints in China [5]. At present, as one of the core technologies in the field of power utilization, there are still many problems to be solved. On the one hand, to meet the requirements of power system security, reliability and economy, it is the key and difficult point in the field of power supply to study the optimal dispatching methods and strategies [6]. On the other hand, with the increase of the permeability of renewable power generation, it is of great practical significance to reduce the impact of randomness and fluctuation of renewable power generation on the system by using power energy optimal dispatching technology.

2.1 Data Integration of Supply and Demand for Integrated Cooling, Thermal and Power Energy Based on Multi-regional Relevance

In the power regulation business, there are many business systems, such as power information collection, geographic information, distribution automation, dispatching automation, production management, weather forecast, marketing and so on. These systems describe the information of an object from different perspectives. At present, the information is independent, that is, multiple files of an object. In order to extract more valuable information from regulatory data, it is necessary to integrate descriptive information from multiple perspectives of an object and conduct joint analysis to realize one object, one file, one device, one file, one user, one event and one file. For this reason, it is necessary to adopt a unified data model, data structure and association relation oriented to power grid object, and integrate and manage these data to support large data analysis and calculation. In order to solve this problem, this chapter studies data integration [7].

Data integration is the integration of data from several different data sources into a unified data set, which may be integrated logically or physically. The core task of data integration is to integrate related distributed heterogeneous data so that users can access them transparently. Common data integration methods include federated database system, middleware integration method and data warehouse method. In this paper, the data warehouse method is used to integrate the supply and demand data of multi-region related integrated cooling, heating and power energy, so this method is only described below [8].

The data warehouse method is to establish a data warehouse to store data. As shown in Fig. 1, what it actually does is copy data from each data source to the same place, that is, data warehouse.

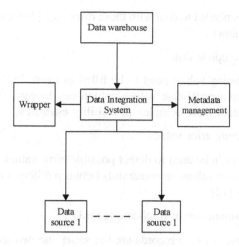

Fig. 1. Data warehouse

In order to extract the existing data sources and organize them into the form of comprehensive data that can be used for decision analysis, the basic architecture of a data warehouse should have the following basic components:

(1) Data source. Operational database systems and external data that provide the lowest level of data for data warehouses [9].
(2) Monitor. Responsible for sensing changes in data sources and extracting data according to the needs of data warehouse.
(3) Integrator. The data extracted from the operational database are transformed, computed, synthesized and integrated into the data warehouse.
(4) Data warehouse. Store data that has been converted according to the enterprise view for analysis and processing. According to different analysis requirements, data are stored in different degrees of integration. Metadata should also be stored in the data warehouse, which records the data structure and any changes in the data warehouse to support the development and use of the data warehouse.
(5) Customer applications. Provides a tool for users to access and query data in data warehouse, and expresses the results of analysis in an intuitive way [10].

2.2 Data Processing of Supply and Demand for Integrated Energy of Cooling, Heat and Power

Data integration is followed by data processing steps, and data cleaning operations are usually required in actual business processes. Usually, dirty data exists in business systems, including incomplete data, wrong data and duplicate data. Therefore, data cleaning technology is needed to deal with these dirty data [11]. Common data cleaning methods are given below:

(1) Solutions to Incomplete Data

In most cases, missing values need to be filled in manually; some missing values can also be derived to approximate values, which can be replaced by average, minimum, maximum or other more complex probability estimates.

(2) Method of detecting error values

Statistical analysis can be used to detect possible error values, or simple rule bases can be used to detect data values, or constraints between different attributes can be used to detect error values [12].

(3) Detection and elimination of duplicate records

If all attribute values of both records are the same, the two records are duplicated. The purpose of eliminating duplicate records can be achieved by retaining one of the records.

(4) Detection and elimination of inconsistent data

By defining integrity constraints, inconsistent data can be detected, and connections can be found by analyzing data. Cleaning can be done by specifying simple transformation rules or using domain-specific knowledge [13].

2.3   Power Demand Classification Based on Fuzzy Clustering

Fuzzy clustering analysis is a method of clustering objective things according to their characteristics, degree of affinity and similarity [14]. Its characteristics are that the conclusion of fuzzy clustering does not mean that the objects belong to a certain category absolutely, but that the objects belong to a certain category relatively with clear values. To a certain extent, it belongs to another category. In the fuzzy clustering analysis, the first step is to calculate the fuzzy similarity matrix, and different fuzzy similarity matrices will produce different classification results; even if the same fuzzy similarity matrix is used, different thresholds will produce different classification results [15, 16]. The basic steps of power demand classification based on fuzzy clustering are shown in Fig. 2 below.

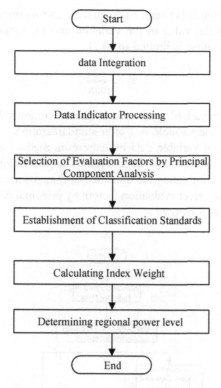

Fig. 2. Basic steps of power demand classification based on fuzzy clustering

The first step is to process the statistical indicators of each representative point, in order to make the data of each indicator comparable, the dimensionless processing is carried out. Dimension is an important concept in physics. In theoretical calculation and numerical calculation, it is often necessary to do dimensionless processing (in fact, the main characteristic quantities of the system are used as the units of the corresponding physical quantities). By doing so, the theoretical calculation is simple, the numerical calculation is convenient, and the physical equation can be transformed into a specific mathematical equation, which is convenient for mathematical processing. The extremum method is chosen here to be dimensionless.

There are three ways to choose:

$$x_i' = \frac{x_i}{\max - \min} \tag{1}$$

That is to say, every variable is divided by the total distance of the value of the variable, and the value range of each variable after standardization is limited to $[-1,1]$.

$$x_i' = \frac{x_i - \min}{\max - \min} \tag{2}$$

That is to say, the difference between each variable and its minimum value is divided by the total distance of the value of the variable, and the range of the value of each variable after standardization is limited to [0,1].

$$x'_i = \frac{x_i}{\max} \tag{3}$$

That is, the value of each variable is divided by the maximum value of the variable, and the maximum value of the variable is 1 after standardization.

Dimensionalization of variable data by extremum method is to transform original data into data bounded by a specific range through the maximum and minimum values of variable values, thus eliminating the influence of dimension and order of magnitude.

The second step is to select evaluation factors by principal component analysis. The process is shown in Fig. 3 below.

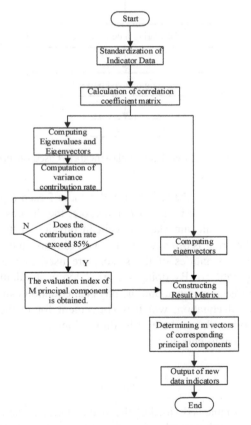

Fig. 3. Selection of evaluation factors by principal component analysis

The third step is to establish the classification standard. Table 1 below is the power demand classification standard.

Table 1. The power demand classification standard

Power demand level	Electricity demand (KW)
Level 1	[500,1000)
Level 2	[1000,2000)
Level 3	[2000,3000)
Level 4	[3000,4000)
Level 5	[4000,5000)

Next, we need to convert the standard of fuzzy disaster degree classification to Table 1. The purpose is to establish the membership function of the fuzzy set of disaster loss grade, and to unify the standard of dividing the loss grade of natural disasters by using different indicators.

Here A, B, C, D and E are grade 1, grade 2, grade 3, grade 4 and grade 5, respectively. For convenience, we calculate the logarithmic function with 10 as the base, convert the corresponding values of each index into natural numbers, and get the criterion of fuzzy disaster grade division.

The fourth step is to calculate the weight of evaluation index by using analytic hierarchy process. Firstly, a hierarchical structure is established based on the above-mentioned analysis indicators. Then, a judgment matrix is constructed by comparing the nine-level scoring system. Then, the maximum eigenvectors and eigenvectors in the judgment matrix are calculated. According to the calculated results, the consistency test of the constructed judgment matrix is carried out. Finally, after the consistency test is passed, the weight set of the indicators can be obtained.

The fifth step is to calculate the comprehensive evaluation result vector. By choosing the appropriate synthesis operator, the weight set A and membership matrix R are combined to get the fuzzy comprehensive evaluation result vector B of the evaluation object, that is, the membership degree of the evaluated object to each level of fuzzy subset as a whole.

The sixth step is to determine the regional power level. The principle of weighted average comprehensive average is used to determine the regional power level. The principle of this method is that the level of the object is a continuous relative position, using the numerical value 1,2,. N denotes each rank in turn, and these values are called ranks of each rank. Then, the values B J (j = 1,2) in the comprehensive evaluation vector B are used. (n) The ranks of each rank are weighted to get the relative position of the evaluated object.

The principle of weighted average can preserve the original data and process data more intact, avoid the loss of data information, and quantify the grade of the evaluated object, making the evaluation results more intuitive.

2.4   Realization of Reasonable Dispatch of Comprehensive Energy for Cooling, Thermoelectricity and Power

(1) Determining decision variables

Decision variables mainly refer to the generation capacity and power demand level, and there is a certain relationship between them. For areas with low power demand level, the generation capacity is very small, and for areas with high power demand level, the generation capacity is large.

(2) Analysis of objective function

The main factors involved in the objective function are generation capacity and power demand level. With the further development of energy-saving concept, higher requirements are put forward for power supply scheduling. The level of force demand in the objective function determines the amount of electricity generated. The development of objective function in the direction of power dispatching can promote the optimization of power dispatching decision-making model and make Low-carbon Science and technology innovate and develop continuously.

(3) Model constraints

At present, many power enterprises do not pay attention to the classification of power demand levels, resulting in huge waste of power in the actual supply process. In view of this situation, the government has issued an order for the rational distribution of electricity, which has become a constraint for the operation of many large power enterprises. However, in order to achieve the balance between supply and demand levels, it is not enough to rely on the efforts of the government alone. Power enterprises need to improve their own power dispatching mode and decision-making model.

3   Simulation Experiment

In order to prove the effectiveness of the multi-area integrated energy dispatching method for cooling, heating and power based on large data analysis, a simulation experiment is needed. Firstly, an experimental simulation platform is established by using the simulation tool CloudSim. The experimental data are selected from the power supply data of several areas in North China from 2015 to 2019. Now use the method of this study to conduct reasonable dispatch. After the dispatch is completed, the power supply is counted. The results are shown in Table 2 below.

Table 2. Statistical results of power supply

Year	Actual result (KW)	After reasonable dispatch (KW)	Differ (KW)
2015	2536.46	2047.35	489.11
2016	3011.68	2834.87	176.81
2017	3517.21	3247.98	269.23
2018	4563.78	4179.54	384.24
2019	5217.23	4768.95	448.28

From Table 2, it can be seen that the power supply decreases by 489.11 KW, 176.81 KW, 269.23 KW, 384.24 KW and 448.28 KW respectively, with an average decrease of 353.534 KW, which shows the effectiveness of this method.

In order to further verify the reliability of the method in this paper, the time required for the comprehensive energy classification of cooling, heating and power in reference [2, 3] is compared, the results are shown in Fig. 4.

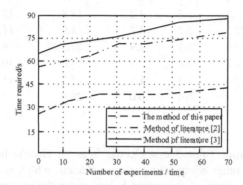

Fig. 4. Comparison of time required for different methods of comprehensive energy classification

It can be seen from Fig. 4 that the time required by the methods in reference [2, 3] in the classification of comprehensive cold, heat and power energy is relatively long, while the time required by the method in this paper is relatively short. The reason is that the dimensionless treatment is carried out in the process of comprehensive energy classification of cooling, heating and power, and the calculation is simple, and the numerical calculation is convenient. The physical equation can be transformed into specific mathematical equation, which is convenient for mathematical processing. The extremum method is used to dimensionless, the original data and processing data are retained completely, the loss of data information is avoided, the grade of the evaluated object is quantified, and the evaluation result is more intuitive.

The main parameters of the optimization algorithm are: the initial population size is 50, the population dimension is 10, the search space is 0–10000 kW, and the maximum number of iterations is 1000. When the inertia weight is large, the global search ability is strong, but the local search ability is poor. Previous studies have shown that the value of inertia weight W should be less than 1 to ensure the local search ability of the algorithm. Therefore, according to experience, the value range of inertia weight W is 0.9 maximum and 0.2 minimum. Learning factors affect the change speed of particles. The existing algorithm research shows that learning factors usually select constants in [0,4], and the number of particles retained in the front end is 8. Equipment capacity configuration corresponding to 8 non inferior schemes is shown in Table 1 (equipment with capacity configuration result of 0 is omitted).

Table 3. Configuration capacity optimization results

Programme	Configure capacity/MW						
	Gas engine	Waste heat boiler	Gas fired boiler	Voltage refrigerator	Absorption refrigerator	Heat storage irrigation	Cold storage tank
1	9.8	13.2	8.9	8.4	11.1	15.6	20.9
2	12.5	17.2	7.7	6.9	14.5	15.3	27.5
3	15.6	21.3	5.5	4.8	18.8	12.6	36.5
4	16.4	22.5	5.2	4.7	18.9	11.5	28.9
5	20.2	25.6	3.3	2.9	21.3	9.2	47.2
6	21.2	28.2	2.7	2.4	24.6	10.5	50.6
7	22.6	29.3	2.3	1.8	25.3	11.2	52.4
8	22.9	30.2	0.0	0.0	28.9	4.6	60.5

It can be seen from Table 3 that electric heating, electric hot water boiler, gas turbine, battery and other equipment are not selected, which indicates that the cost performance of these types of equipment is low under the example price and parameters, and the economic competitiveness is not high. There are heat storage and cold storage devices in each scheme, and the configuration capacity is high, which shows that the configuration of heat storage and cold storage equipment can effectively improve the economy of the comprehensive energy supply system, and has a good value of carbon emission reduction, so it should be given priority. Absorption refrigeration equipment can reduce carbon emission, but its economy is not as good as electric refrigeration equipment. In addition, the comparison of the comprehensive energy supply scheme of this example shows that the larger the capacity of the electric refrigeration equipment is, the higher the overall economic benefit is.

4   Conclusions

In summary, this study explores some problems of power scheduling. At present, regional differences are not taken into account in power system dispatching, which leads to low power utilization rate and serious energy waste. In view of the above situation, a multi-region integrated energy scheduling method based on big data analysis is studied. The simulation results show that this method is effective, solves the problem of unnecessary power consumption and achieves the purpose of energy saving.

The influence of new energy generation forms, including wind power and photovoltaic, on the planning of integrated energy system is not taken into account, and further study is needed. With the deepening of the national energy system reform and the vigorous development of the energy Internet, the integrated energy system is bound to have a larger and larger display stage, and the rational and efficient planning of the integrated energy system will also be the focus of future research.

References

1. Lv, Y., Yang, X., Zhang, G., et al.: Experimental research on the effective heating strategies for a phase change material based power battery module. Int. J. Heat Mass Transf. **128**(15), 392–400 (2019)
2. Ehsan, M.M., Guan, Z., Klimenko, A.Y.: A comprehensive review on heat transfer and pressure drop characteristics and correlations with supercritical CO2 under heating and cooling applications. Renew. Sustain. Energy Rev. **92**(12), 658–675 (2018)
3. Wang, S., Wu, Z., Yuan, S., et al.: Method of multi-objective optimal dispatching for regional multi-microgrid system. Proc. CSU-EPSA **29**(5), 14–20 (2017)
4. Deng, J., Ma, R., Hu, Z., et al.: Optimal scheduling of micro grid with CCHP systems based on improved particle swarm optimization algorithm. J. Electr. Power Sci. Technol. **33**(2), 45–48 (2018)
5. O'Malley, C., Erxleben, A., Kellehan, S., et al.: Unprecedented morphology control of gas phase cocrystal growth using multi zone heating and tailor made additives. Chem. Commun. **56**(42), 25–31 (2020)
6. Fanpeng, B., Shiming, T., Fang, F., et al.: Optimal day-ahead scheduling method for hybrid energy park based on energy hub model. Proc. CSU-EPSA **29**(10), 123–129 (2017)
7. Mykhaylo, L., Serge, N.N., Sivakumar, P., et al.: A concept of combined cooling, heating and power system utilising solar power and based on reversible solid oxide fuel cell and metal hydrides. Int. J. Hydrogen Energy **43**(40), 18650–18663 (2018)
8. Du, K., Calautit, J., Wang, Z., et al.: A review of the applications of phase change materials in cooling, heating and power generation in different temperature ranges. Appl. Energy **220**(15), 242–273 (2018)
9. Lu, M., Liu, S.: Nucleosome positioning based on generalized relative entropy. Soft. Comput. **23**(19), 9175–9188 (2018). https://doi.org/10.1007/s00500-018-3602-2
10. Liu, S., Liu, D., Srivastava, G., et al.: Overview and methods of correlation filter algorithms in object tracking. Complex Intell. Syst. (2020). https://doi.org/10.1007/s40747-020-00161-4
11. Baum, M., Dibbelt, J., Pajor, T., et al.: Energy-optimal routes for battery electric vehicles. Algorithmica **82**(5), 1490–1546 (2020). https://doi.org/10.1007/s00453-019-00655-9
12. Fu, W., Liu, S., Srivastava, G.: Optimization of big data scheduling in social networks. Entropy **21**(9), 902 (2019)
13. Pham, T.T., Maréchal, A., Muret, P., et al.: Comprehensive electrical analysis of metal/Al2O3/O-terminated diamond capacitance. J. Appl. Phys. **123**(16), 161–172 (2018)
14. Chaudhary, R., Aujla, G.S., Kumar, N., et al.: Optimized big data management across multi-cloud data centers: software-defined-network-based analysis. IEEE Commun. Mag. **56**(2), 118–126 (2018)
15. Wang, F., Zhang, J., Xu, X., et al.: A comprehensive dynamic efficiency-enhanced energy management strategy for plug-in hybrid electric vehicles. Appl. Energy **247**(1), 657–669 (2019)
16. Liu, S., Bai, W., Liu, G., et al.: Parallel fractal compression method for big video data. Complexity **2018**, 1–6 (2018)

Research on Key Performance Evaluation Method Based on Fuzzy Analytic Hierarchy Process

Shi-han Zhang[1(✉)] and Gang Qiu[2,3]

[1] Shenyang Institute of Technology, Fushun 113122, China
zhangshihan1324@163.com
[2] Department of Computer Engineering, Changji College,
Changji 831100, China
[3] College of Computer Science and Technology, Shandong University,
Jinan 250100, China

Abstract. Under the traditional performance evaluation method, the financial analysis process lays too much emphasis on the integrity and the weight of performance factors is not clear, which leads to the poor aggregation degree of performance evaluation data. A key performance evaluation method based on fuzzy analytic hierarchy process (FAHP) is proposed. The data flow of performance indicators is established according to the current situation of the enterprise on the basis of the three indicators of work efficiency of financial staff, utilization of financial funds and overall financial operation efficiency. By applying data envelopment analysis and static tree analysis, a comprehensive analysis model is established. The index data flow is sampled according to the boundary performance value and the weight of each factor is calculated by using the idea of fuzzy hierarchy. After quantification, the final fuzzy evaluation is obtained and the key performance evaluation is realized. The experimental results show that compared with the traditional key performance evaluation method, the data aggregation degree of the designed key performance evaluation method is improved by 29%, and the overall scientific nature is stronger.

Keywords: Fuzzy hierarchy · Evaluation results · Data aggregation · Sampling extraction

1 Introduction

At the beginning of reform and opening-up in the last century, China was backward in society, weak in economic foundation, low in scientific and technological level, workers'cultural literacy, backward in ideology, low in international status, less exchanges with foreign countries, lack of experience in market economy, and imperfect in various economic systems, laws and regulations. In recent years, with the stability and prosperity of China's social economy and finance, some economists in China have begun to emphasize financial economic management. The core issue is how to prepare effective financial formation management policies and how to rationally allocate existing financial resources in the financial management of enterprises in China. The

S. Liu and L. Xia (Eds.): ADHIP 2020, LNICST 347, pp. 472–481, 2021.
https://doi.org/10.1007/978-3-030-67871-5_42

solution of this problem has an important impact on the rational management of financial economy of enterprises in China. In order to ensure the smooth progress of the financial work of enterprises and carry out financial management with the highest efficiency, it is necessary to establish a proprietary financial key performance evaluation index system, which can not only ensure financial security, but also improve work efficiency. Facts have proved that only by using new management means to ensure the standardization of financial information management process, can the risk be minimized, some duplicate financial work content be avoided, and the efficiency of financial information management is gradually improved. The method of fuzzy analytic hierarchy process (FAHP) is to express the fuzziness of the evaluation index on the impact degree of the final evaluation target parameters in the form of a fuzzy set, form an evaluation matrix that can be directly evaluated, and obtain the evaluation result of a fuzzy set through a fuzzy transformation, and evaluate the final evaluation target parameters according to the size of membership. By introducing the Fuzzy Analytic Hierarchy Process (FAHP) evaluation scheme, we can establish a unique evaluation matrix through the objective weight of the indicators, and realize the evaluation of key financial performance [1].

2 Design of Key Performance Evaluation Method

2.1 Establishment of Performance Indicators Data Flow

The extraction and establishment of financial key performance data flow can be simply regarded as a combination of organizational viewpoints and thinking modes of financial sharing. The organization of data viewpoint is the primary problem in designing an effective method of financial key performance evaluation. It needs to use mathematical logic to influence data system integration. Through the follow-up Fuzzy Analytic Hierarchy Process (FAHP), an analysis matrix is established for comprehensive calculation.

In the establishment of data flow of financial performance indicators, different levels of financial information need to correspond to different value systems and financial advantages, and finally introduce different rating weights. The whole data system of performance indicators is divided into policy level and strategy level according to the level of financial information. Its core indicators include three main aspects: the efficiency of financial personnel, the utilization rate of financial funds, and the overall operational efficiency of finance [2].

The work efficiency of financial personnel needs to be determined by means of per capita financial operation quota. Its algorithm is the ratio of the total operation quota to the total number of financial personnel. This value reflects the intrinsic relationship between human resources input and output. The larger the ratio, the stronger the purchasing economy. In addition, it is necessary to determine the value of the per capita procurement cost, that is, the ratio of the average monthly salary of the procurement personnel to the public financial expenses of the procurement department. The smaller the ratio, the higher the procurement efficiency of the procurement department.

The determination of the index of the utilization rate of financial funds needs the support of three data: budget preparation rate, procurement budget completion rate and procurement control rate. Budget preparation rate is the proportion of purchasing funds to the overall budget, which reflects the efficiency of purchasing departments' management of financial funds. The procurement budget completion rate is the ratio of the procurement amount to the procurement budget. This ratio defines the effect of the procurement budget implementation. The higher the proportion, the more efficient the financial budget implementation is, and the better the procurement goal can be achieved. Purchasing control rate can reflect the degree of capital savings of purchasing department. However, high control rate is not the ultimate goal in procurement, because the accuracy of procurement budget preparation and budget precision can also determine the control rate to a certain extent [3].

The overall operational efficiency of finance is the proportion of enterprise financial expenditure in the total amount. On the one hand, this proportion shows the importance of financial application of enterprises, on the other hand, it also reflects the impact of economic operation of enterprises. Under certain conditions of the overall budget, the larger the ratio, the smaller the corresponding proportion of other forms of economic management. Detailed economic indicators are as follows: Table 1:

Table 1. Factor analysis table

Level	Scope	Status quo	Methods
Policy level	Organizational model	Risk analysis	Performance improvement
	Financial facilitation	Project performance status	Economic project data
	Financial missiveness	Improve potential	Project execution data
	Financial internal information division	Comprehensive risk	Economic data
	Overall financial information standards	In-project roles	Economic planning data
Strategic level	Mission of the position	Financial situation	Data system
	The main duties and responsibilities	Performance improvement	Training design
	Economic enforcement conditions	Economic information analysis	Incentive plan
	Economic punishment data	Regional economic data	Factors data
	Economic analysis data	Positive data	Choose the part of data
	Economic development data	Negative data	New member data

The above-mentioned evaluation index system mainly aims at the actual evaluation scheme of current enterprise financial performance index data flow. Through evaluation and analysis, numerous data streams can be formed, which can provide data base for subsequent data sampling and extraction [4].

2.2   Indicator Data Flow Sampling

The establishment of data flow in cloud Accounting Financial Sharing Center needs to aim at the completion rate of the above evaluation indexes and influencing factors. The actual calculation of data flow needs to consider many reasons. After the data flow summary work is completed, we can extract the data indicators samples of the current key data of enterprise finance through the designed extraction template. In order to ensure the scientificity and accuracy of data extraction, the design introduces two kinds of data logic analysis methods: data envelopment analysis and static tree analysis, and combines the current financial indicator data for theoretical quantification [5].

Data Envelopment Analysis (DEA) is a comprehensive logical structure analysis method derived from the intersection of modern financial operations management and mathematical economics. It is mainly through the use of mathematical logic planning to convert multiple evaluation indicators or data flow information into output departments or units (also collectively referred to as decision units), and then judge whether each decision unit is DEA according to the effectiveness between each decision unit. Effective, in simple terms, is to analyze whether each decision-making unit is in efficient production, and its effective data is extractable data.

In the field of statistics and financial economics, efficient production is a prerequisite for generating production functions. DEA method can be used to determine the distribution structure of integrated production frontiers, so DEA can also be regarded as an unconventional parametric statistical evaluation method. Because there are many decision-making units involved, the designed analysis model introduces the static tree structure into the DEA data envelopment analysis method, and establishes a static tree analysis system, as shown in Fig. 1:

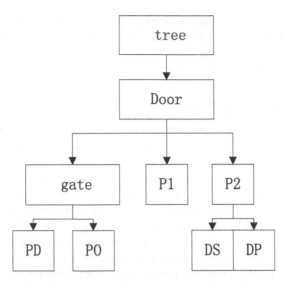

Fig. 1. Schematic diagram of static tree analysis

The static tree analysis method can use the linear specification to obtain the production frontier boundary of each economic data analysis gate of the current enterprise, and directly calculate the completion efficiency and input and output of the decision unit under each gate. In the actual calculation, the decision unit should be comprehensively selected first, the total number is N; m and S respectively refer to the performance rate input and output of each decision unit, and assume the ith input variable of the Kth decision unit, denoted by the symbol Xik (i = 1, 2, ... m), corresponding to the Jth output in this series of decision units expressed by Yjk (i = 1, 2, ... s), relevant The calculation model is as follows:

$$s,t \begin{cases} \sum_{k=1}^{n} X_k H_k + s^- = \theta X_t \\ \sum_{k=1}^{n} X_k H_k - S^+ = Y_t \\ s^- \geq 0, \ s^+ \geq 0, \ H_k \geq 0, \ k = 1, 2, \ldots n \end{cases} \tag{1}$$

In Eq. 1, s^- and s^+ respectively represent the economic verification coefficients of each decision gate; H and θ represent the decision-making influence parameter variables of each decision gate; where θ is the performance comprehensive efficiency value of each financial information influence gate [6].

The efficiency obtained by formula (1) can be represented by λ. In order to distinguish the value range, each part of the value range can be divided into: A1, A2, A3, A4. In performance rating, the theoretical efficiency is assumed to be a fixed amount, then each range variable and corresponding output are:

$$A_1 = A_2 = A_3 = A_4 = 3.5 \times 10^{-8} \qquad (2)$$

$$U_1 = U_2 = U_3 = U_4 = \frac{1}{4}H \qquad (3)$$

Where H is the Solomon efficiency coefficient and is based on the actual situation.

Before the calculation, the static tree analysis can be transformed into a static analysis chain, and the data extraction convenience rate is improved as shown in Fig. 2.

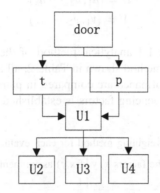

Fig. 2. Static tree analysis chain

Based on Fig. 2 and Eqs. (2) and (3), the decision-making unit input and output analysis matrix is established to obtain the boundary performance value C [7].

$$C = \begin{bmatrix} -(U_1+A_1) & A_1 & A_2 & 0 & 0 \\ U_1 & -(U_1+A_2) & 0 & U_2 & U_3 \\ U_1 & 0 & -(U_1+A_3) & 0 & 0 \\ 0 & U_2 & 0 & -(U_1+A_4) & U_4 \\ 0 & 0 & U_4 & 0 & -(U_2+A_1) \end{bmatrix} \qquad (4)$$

The boundary performance value can represent the completion efficiency of each desired work within the indicator system. Using DEA analysis, the input and output values of all decision units are converted to boundary performance values, projected into the geometric space, with the lowest input or highest output as the boundary [8, 9]. When the decision-making unit (DMU)'s boundary performance refers to being at the spatial boundary, the DMU is considered to be efficient and efficient, with a relative efficiency value of 1, which means that the DMU can no longer reduce input or increase production under the same conditions. The data is extractable data. If the DMU is within the bounds, the DMU is an inefficient unit and gives a value between 0 and 1. The performance metric between 1 indicates that if the output is unchanged, the input or investment remains unchanged. It shows that if you continue to add inputs, you can increase the output, and the data is non-extractable data [10].

2.3   Achieve Performance Evaluation

According to the sampled data extracted above, based on the fuzzy analytic hierarchy process, the final performance target can be evaluated. The principle steps are as follows:

A set of factors U composed of factors to be evaluated that affect the set of evaluation results of the target and a set of comments v composed of comments that judge each factor are set. Can be expressed as:

$$U = (u_1, u_2, \cdots u_n)$$
$$V = (v_1, v_2, \cdots v_n)$$

(5)

Firstly, according to the 1:1 analytical method of the analytic hierarchy method, according to the evaluation method shown in Table 2, all the factors in the factor set U that affect the financial performance are compared in pairs, and the fuzzy quantitative assignment between the influencing factors is established [9].

Table 2.   Fuzzy weighting method for each evaluation factor weight

The importance of the factors x, y	f(x,y) assignment	f(x,y) assignment
Equally important	1	1
X is slightly important	3	1/3
X is important	5	1/5
X is very important	7	1/7
X is extremely important	9	1/9

By comparing the two factors, we can get an n-direction positive matrix C. The application formula (4) of fuzzy assignment of each factor in matrix C is transformed to obtain the weight coefficient of single factor. On this basis, the weight vector composed of multi-factor weight coefficients is obtained, and the weight vector is expressed as n. Then, according to the evaluation in the single factor evaluation set v, the discretized values are applied to the normal distribution according to the grading criteria, and the fuzzy comprehensive evaluation matrix R is constructed according to the interval membership degree of the function. After determining the weight vector and the comprehensive evaluation matrix, the result of the comprehensive evaluation Y can be obtained by the fuzzy operation, where Yj is the subordinate of the corresponding fuzzy comment corresponding to the judgment object, and generally needs to analyze and evaluate the result according to the principle of maximum membership degree. It is considered to be the last fuzzy comment [11, 12].

3 Experimental Results and Analysis

The simulation data set is taken as an experimental sample to illustrate the basic process of determining the key performance value by applying fuzzy comprehensive evaluation method and verifying the rationality of the method. Applying the 1–9 ratio method, according to the relative importance of different factors to evaluate the target parameters, the fuzzy assignment is performed, and the normalization transformation is applied to determine the single factor index weight coefficient. For example, the evaluation of the efficiency of financial application, the main factors affecting the financial structure, the status quo of enterprises. According to the factors, the two factors are compared to determine the importance of the factors.

Commonly used, good, medium, poor, and poor five-level evaluation criteria to fuzzy evaluation of a good or bad thing. The calculation method and evaluation standard of single factor index in the evaluation system are mainly determined based on the results of ministerial standards and digital model theory. For example, the plane permeability coefficient of variation coefficient is good, good, medium, poor, and the corresponding range is less than 0.5 0.5–0.6 0.6–0.7 0.7–0. 8 and greater than 0.8. You can directly seat the number according to the size of the parameter. For the criteria of new fiscal factors, irregular factors and other indicators, the application of mathematical model theory is determined. If the fund utilization rate standard is determined, the typical theoretical model established by the application simulates the influence degree of different capital utilization efficiency, and the five-level evaluation standard is established according to the simulation result. The end evaluation index in the evaluation system has a clear boundary value, and the single factor evaluation matrix can be obtained according to the specific value of the parameter and the single factor evaluation standard, and the normal distribution function is used to obtain the interval membership degree. The upper level does not have an explicit evaluation index of the boundary value, and the single factor evaluation matrix can be obtained by using the one-factor evaluation matrix of the next-level evaluation index and the weighting set determined by the analytic hierarchy process. On the basis of the single factor evaluation matrix, the fuzzy evaluation matrix of each level can be obtained. The matrix values are extracted to determine the final fuzzy evaluation results, which are compared with the traditional performance analysis methods. The results are shown in Fig. 3.

As can be seen from the experimental results in Fig. 3, the data aggregation degree of both the traditional method and the proposed method fluctuated to a certain extent during the whole experimental process. However, in terms of the average degree of perfect fit, the proposed method is significantly higher than the traditional method, with the highest data aggregation index value up to 0.026, indicating that the proposed method has better application effect and obvious advantages.

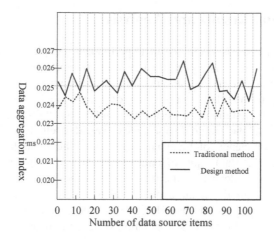

Fig. 3. Data aggregation degree comparison chart

4   Conclusion

Combining with the actual financial characteristics and the important problems faced by the authentic company, this paper applies the research template of the Fuzzy Analytic Hierarchy Process (FAHP) to make up for the influence of financial factors neglected in the past financial performance evaluation on the results of performance evaluation, and constructs a scientific, reasonable and comprehensive financial performance evaluation system of the company. This paper also has some limitations. Firstly, the data of financial indicators selected in this paper come from the current stage of our country. Although financial audit and performance evaluation need to audit the annual report of the company, the assurance provided by auditor is limited to reasonable assurance, so it is impossible to guarantee the authenticity and integrity of the annual report disclosure data 100%. Secondly, the selection and revision of the index system in this paper combines the analysis and judgment of the characteristics of the current financial enterprises. It has certain subjectivity. Different evaluation subjects may have different opinions on the selection and revision of the index. To some extent, the conclusion of this study is not unique. Finally, because the change of external environment will affect the performance evaluation system of enterprises, and the conclusion of this paper is based on the historical data disclosed in the annual reports of listed companies from 2014 to 2016, so it reflects the past financial performance performance, and fails to accurately predict the future development trend of financial performance of Listed Companies in construction industry. It is hoped that in the follow-up study, we can pay attention to the changes of the external environment and adjust the evaluation system appropriately to adapt to the new situation of industry development.

References

1. Zhang, P., Ling, W., Zheng, Y., et al.: Research on evaluation method of distribution automation operation based on fuzzy analytic hierarchy process. Electr. Meas. Instrum. **53** (22), 72–77 (2016)
2. Hao, C.S., Sheng, J.K., Fan, X.M., et al.: Comprehensive evaluation model of mine safety based on fuzzy analytic hierarchy process. Coal Technol. **35**(2), 234–236 (2016). (in Chinese)
3. Cui, G., Wu, F., Li, M., et al.: Research on safety risk assessment of highway and bridge construction based on fuzzy analytic hierarchy process. Chin. Market **2**(41), 46–49 (2016)
4. Guo, L., Yang, S., Zhou, L., et al.: Comprehensive evaluation of highway traffic safety based on fuzzy analytic hierarchy process. Traffic Sci. Technol. **19**(2), 8–12 (2017)
5. Liu, S., Fu, W., He, L., et al.: Distribution of primary additional errors in fractal encoding method. Multimed. Tools Appl. **76**(4), 5787–5802 (2017)
6. Qin, B., Li, Q., Tan, J.: Safety evaluation of masonry and wooden structure ancient buildings based on fuzzy analytic hierarchy process. J. Civ. Eng. Manage. **34**(5), 52–59 (2017)
7. Wen, M., Ma, R.: Analysis on the construction of key performance indicator system for public policy evaluation – a case study of mass innovation space performance evaluation system piloted in X city. Adm. BBS **53**(52), 70–71 (2017)
8. Liu, S., Lu, M., Li, H., et al.: Prediction of gene expression patterns with generalized linear regression model. Front. Genet. **10**, 120 (2019)
9. Dong, B., Peng, J.: Simulation of abnormal data mining algorithm in complex network data stream. Comput. Simul. **33**(1), 434–437 (2016)
10. Zheng, X.: Research on power wireless private network networking and security protection technology based on TD-LTE. Electron. Des. Eng. **25**(7), 83–86 (2017)
11. Liu, S., Liu, D., Srivastava, G., et al.: Overview and methods of correlation filter algorithms in object tracking. Complex Intell. Syst. (2020). https://doi.org/10.1007/s40747-020-00161-4
12. Lu, M., Liu, S.: Nucleosome positioning based on generalized relative entropy. Soft. Comput. **23**(19), 9175–9188 (2018). https://doi.org/10.1007/s00500-018-3602-2

Research on Dynamic Assignment of Distributed Tasks Based on Improved Contract Network Protocol

Zhi-li Tang[iD] and Jing-long Wan[(⊠)][iD]

School of Aeronautics, Northwestern Polytechnical University,
Xi'an 710072, Shanxi, China
`wanjl@mail.nwpu.edu.cn`

Abstract. This paper introduces the basic idea and operation mechanism of contract network. In view of its shortcomings, it introduces an agent mental model and proposes a distributed task allocation algorithm based on improved contract network protocols. The algorithm improves the bidding process and the bidding process, reduces the system communication volume and fully considers the status information of the drone itself. Finally, a simulation experiment is designed to compare and analyze the advantages of the improved contract network over the traditional contract network.

Keywords: Contract Network Protocol (CNP) · Distributed task allocation · Combat mission modeling

1 Introduction

In the information battlefield, the various arms and operating entities distributed in different locations implement distributed interconnection and interoperability based on the operational network to realize dynamic allocation and coordination of distributed tasks. The battlefield situation is complex and changeable and the pre-war plan is very easy to be disrupted. Therefore, it is required that the accusation center be able to coordinate the battlefield resources and entities on the battlefield in real time. In a dynamic environment, reform the action plan for the new battlefield situation and select the appropriate battle The entity performs the corresponding combat mission. Before receiving an order from a superior, it is required that combat entities be able to autonomously implement combat task allocation and resource coordination in accordance with corresponding rules, so as to avoid jeopardizing fighters and causing mission failures. Therefore, it is of great significance to study the dynamic assignment of distributed tasks. This paper is based on the Contract Net Protocol (CNP) task allocation model. It analyzes the deficiencies and shortcomings in the traditional contract network,

© ICST Institute for Computer Sciences, Social Informatics and Telecommunications Engineering 2021
Published by Springer Nature Switzerland AG 2021. All Rights Reserved
S. Liu and L. Xia (Eds.): ADHIP 2020, LNICST 347, pp. 482–497, 2021.
https://doi.org/10.1007/978-3-030-67871-5_43

introduces the Agent mental model, improves the bidding and bidding stages, and performs simulation verification and analysis.

2   The Basic Theory

2.1   Basic Theory of Contract Network

The Contract Net Protocol (CNP) [1,2] is a method proposed by Smith and Davis in 1980 to solve distributed problems. In the contract network protocol model, it consists of multiple agents that can transmit messages to each other. According to the different responsibilities for each agent, the agents are divided into three types: bidding agent, bid agent, and winning agent.

As a blind bidding method, the contract net protocol can solve the single task assignment. However, the UCAV fleet needs to face multi-tasking concurrent situations. The negotiation process using the traditional contract net protocol has the following disadvantages [3,4]:

The bidding agent conducts bidding in the form of broadcast and all the agents that receive the message can participate in the bid, which will generate a large communication load, waste resources in the system and have low negotiation efficiency. The bid agent needs to communicate with the bidding agent multiple times during the evaluation stage, which leads to an increase in the communication volume of the system, and the negotiation process is long and cumbersome. The bid document evaluation mechanism is incomplete. In the contract net protocol, the bidding agent only judges the winning bidder by the value of the bid and does not evaluate the actual task amount of the bid winner.

The task assignment process of the entire contract net protocol is a distributed dynamic task assignment method [5,6], which depends on the independent decision-making ability and control strategy of each agent.

2.2   Agent Mental Model

The Agent mental model [7–9] records the capability information of the Agent, the relationship between the Agents, the mental state of the Agent and the changes in the system environment. The mental model consists of the following parameters: ability, trust, familiarity, positivity, risk, and busyness.

Agent mental model is defined as a two-group: $<Relation\_Agent, Parameter\_Metal>$. $Relation\_Agent = <Agent_i, Agent_j>$ represents the relationship between two agents and $Parameter\_Metal = <A, B, P, F, RT, Cl, BD>$ represents the mental parameters of the Agent mental model.

Ability A. In the contract net protocol, the bidding Agent tends to find Agent with stronger task execution capabilities to cooperate. The capability parameter update of the agent does not depend on the completion of the task, but depends on the capability of the agent. This paper considers the capability value of UCAV from the task load and weapon load. The parameter update function is:

$$A(Agent_i, T_k) = \frac{Load\_Task_i}{Load\_Task_{max}} + \frac{Load_i}{Load_{max}} \tag{1}$$

In the formula, $Load\_Task_i$ represents current mission load; $Load\_Task_{max}$ represents maximum mission load; $Load_i$ represents current weapon load; $Load_{max}$ represents maximum weapon load.

Believability B. Trust believability represents $Agent_i$'s evaluation of $Agent_j$'s completed tasks and the parameter update function is:

$$B(Agent_i, T_i) = \begin{cases} B_{init}, \\ \quad i = 1, 0 <= B_{init} <= 1 \\ min(B(Agent_i, T_{i-1}) + G_s * \xi, 1), \\ \quad i > 1, 0 < G_s <= 1, \xi > 0 \\ max(B(Agent_i, T_{i-1}) + G_f * \zeta, 1), \\ \quad i > 1, 0 < G_f <= 1, \zeta > 0 \end{cases} \tag{2}$$

In the formula, G_s represents $Agent_j$ completes the completion degree of task T_{i-1} and the higher the degree of completion, G_s the closer to 1; G_f represents $Agent_j$ completes the failure degree of task T_{i-1}, the higher the failure degree, the closer to 1; ξ represents the reward coefficient for completing the task; ζ represents the penalty coefficient of the task failure.

Familiarity F. Familiarity represents that the number of bidding tasks for $Agent_j$ completion $Agent_i$ accounts for the proportion of each other's bidding tasks. The parameter update function is:

$$F(Agent_i, Agent_j, T_k) = \frac{N_{ij}}{N_{ij} + N_{ji}} \tag{3}$$

In the formula, N_{ij} represents the number of bidding tasks of $Agent_i$ completed by $Agent_j$; N_{ji} represents the number of bidding tasks of $Agent_j$ completed by $Agent_i$.

Positivity P. The positivity is the proportion of the number of bidding tasks $Agent_j$ participating in $Agent_i$ to the total bidding task of N_{lm}. The higher the proportion, the higher the enthusiasm of $Agent_i$ participating in the bidding. The parameter update function is:

$$P(Agent_j, T_k) = \frac{N_{lm}}{N_l} \tag{4}$$

Risk RT. The risk tolerance indicates the degree to which $Agent_i$ can bear the risk. The parameter update function is:

$$RT(Agent_i, T_k) = \begin{cases} RT_{init}, \\ i = 1, 0 <= RT_{init} <= 1 \\ min(RT(Agent_i, T_{k-1}) + G_s * \psi, 1), \\ i > 1, 0 < G_s <= 1, \psi > 0 \\ max(RT(Agent_i, T_{k-1}) + G_f * \phi, 0), \\ i > 1, 0 < G_f <= 1, \phi > 0 \end{cases} \quad (5)$$

Busyness BD. The busyness indicates the current busyness of the $Agent_i$. The parameter update function is:

$$BD(Agent_i, T_k) = \frac{N_{b\_used}}{N_{b\_total}} \quad (6)$$

3 Distributed Combat Mission Modeling

3.1 Task Performance Indicator Function

In distributed operations, each UCAV is regarded as an independent agent and the entire combat system forms a multi-agent system. Each agent has a high degree of autonomy, fully shares intelligence and performs task assignment and coordination through mutual negotiation. Given a UCAV set $V = \{V_1, V_2, ...V_{N1}\}$ and a task set $T = \{T_1, T_2, ...T_{N2}\}$, each UCAV can complete one or more tasks T. Task performance is defined as the revenue of the task completion minus the corresponding cost. This paper mainly constructs the task performance index function from the perspective of benefit and cost and establishes the mathematical model of the index function. The task performance indicators are analyzed from the following aspects, including attack mission revenue, voyage cost, airtime cost, and fleet fitness.

Suppose the decision variable x_{ij} is:

$$x_{ij} = \begin{cases} 1, V_i performs T_j task \\ 0, V_i don't perform T_j task \end{cases} \quad (7)$$

Attack Mission Revenue. The benefits of an attack mission obtained by a drone depend on the capabilities of the drone performing the mission and the value of the mission. The capability of the drone is determined jointly by factors such as its comprehensive capabilities and the weapons it is mounted on. The value of the task is given by the accusation center according to certain rules before the task is performed. The expected reward function $Reward_{ij}$ of the attack task before executing the task. The expression is:

$$Reward_{ij} = \sum_{i=1}^{N_v} \sum_{j=1}^{N_t} \frac{x_{ij} * P_{Dij} * T\_Value(T_j)}{N_T * T\_Value_{max}} \quad (8)$$

In the formula, N_v represents total number of drones; N_t represents total number of task targets; P_{Dij} indicates the kill probability of the UAV V_i to mission target, calculated according to the weapon configuration on the aircraft and the enemy target information; $T\_Value(T_j)$ is the value of performing task objective T_j; $T\_Value(max)$ is the maximum value to perform the task.

Voyage Cost. Voyage costs consider UAV V_i fuel consumption from mission start to mission end. During the flight of the UCAV fleet, the pursuit of the least consumption and the shortest time to complete the flight process. Therefore, this paper simplifies the trajectory planning process and adopts a modified straight range method to perform fast calculations to adapt to the dynamic and complex flight environment.

$$PathCost_{ij} = \sum_{i=1}^{N_v} \sum_{j=1}^{N_t} \frac{x_{ij} * (D_{ij} + \overline{D_{ij}})}{N_T * D_{max}} \tag{9}$$

In the formula, D_{ij} indicates the straight flight of the drone to perform the mission; $\overline{D_{ij}}$ indicates the flight trajectory corrected after considering the threat source; D_{max} represents the maximum combat radius of UCAV.

Airtime Cost. While considering the UCAV flight range, we should also consider the flight time of the UCAV mission, balance the flight time of each UCAV in the formation, and avoid a dangerous event when an UCAV flight time is too long. Therefore, the airtime cost function is used to measure the flight time of the UCAV flight formation's execution mission.

$$\overline{T\_Cost_{ij}} = \frac{1}{N_V * max(T_i)} \sum_{i=1}^{N_v} T_i \tag{10}$$

Cluster Fitness. Cluster fitness refers to the ability of the UCAV to respond to uncertain risks and the ability to adapt to the environment in the face of complex and changing battlefield environments. The remaining combat power of the fleet is used to measure the fitness of the fleet. The stronger the remaining combat power is, the higher the fitness of the fleet is. The remaining ammunition and endurance flight capacity have greatly affected the remaining combat power of the fleet. The remaining amount of ammunition determines the attack capability of the UCAV fleet and the endurance determines the threat avoidance and continuous combat capability of the UCAV fleet.

Select the half-gradient distribution function as a single-item attribute function, expressed as:

$$\xi = \begin{cases} 0, f <= f_{min} \\ \dfrac{f - f_{m}in}{f_{max} - fmin,} fmin <= f <= f_{max} \\ 1, f >= f_{max} \end{cases} \tag{11}$$

The remaining combat power of each UCAV is weighted according to the weight of the individual factors. The weighted value is set by the importance of the corresponding attribute, and the remaining combat power K is defined as:

$$K = a\xi_1 + b\xi_2 \quad (a + b = 1) \tag{12}$$

In the formula, ξ_1 represents the value of the single-item attribute function of the remaining ammunition; represents the single-item attribute function value of the endurance; a

b is the corresponding weighted value of the remaining ammunition and endurance, and $a + b = 1$.

The remaining combat power variance is:

$$Var = \frac{1}{N_v} \sum_{i=1}^{N_v} (K_i - \overline{K})^2 \tag{13}$$

The cluster fitness function is expressed as:

$$Margin = \frac{Var}{max(K_i - \overline{K})^2} \quad (i = 1, 2 \ldots N) \tag{14}$$

The above four index functions have been normalized, so that each index has a unified dimension, which is convenient for constructing task effectiveness functions.

3.2   Task Assignment Multi-constraint Optimization Model

Multi-UCAV task performance function is expressed as:

$$\begin{aligned} E_{ij} = &a * Reward(T_i) + b * (1 - PathCost(T_i)) \\ &+ c * (1 - \overline{T_Cost(T_i)}) + d * Margin \end{aligned} \tag{15}$$

In the formula, a, b, c, and d are weighted values of each index value and the weighted value is input according to a preset setting $(a + b + c + d = 1)$.

According to the expression of the mission performance function, it can be obtained that the multi-UCAV ground attack requirements are: Constraint 1: When task assignment, all task targets are assigned to UCAV, satisfying

$$\sum_{i=1}^{N_v} X_{ij} >= 1 \tag{16}$$

Constraint 2: The UCAV fleet has the best overall performance after completing combat missions.

$$max\left(\sum_{i=1}^{N_v} \sum_{j=1}^{N_T} X_{ij} * E\right) \tag{17}$$

Constraint 3: The task load of each UCAV cannot exceed the maximum capacity constraint, and is set to the maximum load of UCAV.

$$\sum_{i=1}^{N_v} X_{ij} \leq Load_{max} \tag{18}$$

4 Dynamic Allocation of Distributed Tasks Based on Improved CNP

4.1 Bidding Strategy

In order to reduce the communication volume occupied by the large-scale release of bidding tasks, this paper adopts an acquaintance bidding strategy, based on the Agent mental model, and designs a bidding decision function as a two-tuple:

$$Call\_for\_bidder =< Relation\_Agent, D\_Limit > \tag{19}$$

It consists of the bidding decision acquaintance relationship function $Relation\_Agent$ and decision threshold D_Limit, which indicates the number of bids issued. This function sets the decision threshold according to the network load and task level of the system at that time, and selects the bidding information sending object.

$$Relation\_Agent(agent_i, T_k) = \lambda_1 * A + \lambda_2 * B + \lambda_3 * F + \lambda_4 * P \tag{20}$$

Before issuing the task message, the bidding agent first calls the acquaintance relationship data from the knowledge base and substitutes it into the acquaintance relationship function $Relation\_Agent$ to obtain the acquaintance relationship sorting. Finally, the task publishing object is selected according to the decision threshold D_Limit.

4.2 Bidding Process

After receiving the bid invitation, the Agent needs to evaluate its own situation and give the bid value. The bid value of Agent is composed of two parts: the gain and the cost of completing the task. In addition to the consumption of the task itself, there are certain costs due to its state consumption and external environmental impact and task risk. The cost function given in this paper is:

$$Cost(Agent_i, T_k, C) = \sigma_1 * \frac{1}{RT} + \sigma_2 * CL + \sigma_3 * C_l \tag{21}$$

In the formula, σ_1, σ_2 and σ_3 are the weight coefficients of risk tolerance, busyness and fixed cost C_l.

The benefit of a task is measured by the amount of change in the task's performance value. First calculate the maximum overall performance value obtained by inserting a task into its own task sequence after assuming a buy task.

$$Efficacy(S\{T_j\} \bigcup T_K) = max(Efficacy(S\{T_i, T_k\})) \tag{22}$$

Then calculate the change in overall task performance after buying the task

$$\begin{aligned} Efficacy^+(T_k) = {}& Efficacy(S\{T_j\} \bigcup T) \\ & - Efficacy(S\{T_j\}) \end{aligned} \tag{23}$$

If the task performance change amount is < 0, indicating that the performance after the purchase task is lower than before, the "reject" message is sent directly to the bidding agent. Otherwise, continue to calculate the bid value and make a tender. The bid value of the tender is:

$$\begin{aligned} Price(Agent_i, T_k, C_l, E_i) = {}& \alpha * Cost(Agent_i, T_k, C) + \\ & \beta * Efficacy^+(T_k) \end{aligned} \tag{24}$$

In the formula,α and β are the weighting factors for the cost and the benefit

4.3 Bid Stage

After reaching the preset bid deadline, all the bids received by the bidding agent are selected, and the bidding agent with the highest bidding value is selected as the winning bidder. The highest bid value for all bids is:

$$Price_{Ui}(T_k) = max(Price(Agent_i, T_k, C_l, E_i)) \tag{25}$$

In the formula, $Price_{Ui}(T_k)$ represents the bidding value of the drone Ui participating in the bidding task T_k.

When the bidding agent issues a task invitation to the winning agent, it also publishes its status information and waiting time together with the task information. After receiving the invitation, the winning bid agent needs to give feedback to the bidding agent. If it refuses, it will directly reply to the "reject task" message. If the bidding agent receives the "accept" message within the time limit, it will sign a contract with it and announce the signing success message to other agents. If the waiting time is exceeded, the bidder will re-select the winning Agent to sign the contract according to the rules.

4.4 Task Execution Phase

Until the bidding task is completed, the entire task allocation process based on the improved contract net protocol does not end. After signing the contract, both parties still need to be responsible for the completion of the task. The winning agent needs to feedback the completion of the task at a certain time. If the bidding agent has not received the feedback message of the winning agent within the time limit, it will be considered that the winning agent has failed to complete the task successfully and the task will be put into the auction sequence and auctioned again. This paper makes strict time series requirements for the task auction process, so that all task coordination processes can be coordinated and operated, which improves the stability of the system and ensures the smooth execution of the task auction process.

The state transition timing of the Agent in the task allocation process based on the improved contract network protocol is shown in Fig. 1:

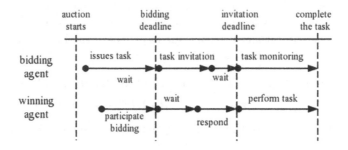

Fig. 1. State transition timing.

The task assignment and coordination mechanism designed in this paper has the following advantages:

Reducing the Waste of Time or Traffic. This paper determine the order of change of auction rights by using the principle of task priority, change the auction order according to the urgency and importance of the task, optimize the bidding process and reduce the waste of time or traffic;

Reducing Communication and Computing Load. In the bidding stage, the acquaintance bidding strategy is established by the capability, trust, enthusiasm and familiarity in the agent mental model, avoiding the blind broadcast method to waste system traffic, fully considering the subjective will of the candidate agent, and reducing the communication and computing load.

Combining Agent State and Task Efficiency Changes. In the bidding stage, the busyness and risk tolerance of the Agent mental model and the change of mission performance are combined to obtain the final bidding value, which fully considers the Agent's own state and the overall mission performance change of the mission.

Meet Real-Time Requirements. The entire task auction process has strict timing constraints to ensure the real-time requirements of the system. The process of distributed dynamic task allocation and coordination based on improved CNP [10] is shown in Fig. 2.

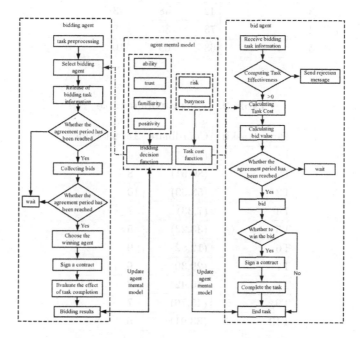

Fig. 2. Dynamic task allocation and coordination mechanism based on improved contract network protocol.

5   Algorithm Simulation and Results Analysis

In order to verify the distributed dynamic task assignment problem based on improved contract network protocol, a simulation experiment was designed to verify. The simulated battlefield environment is a space area of 100*100*20. There are 5 threat sources and 10 enemy targets in the battlefield. After the reconnaissance and detection of the battlefield, the combat command center dispatched 4 drones to attack the enemy targets.

5.1   Initial Data

Due to the limitations of the simulation environment, it is necessary to preset the battlefield environment, including the initial data such as the UCAV mental state, the target value and the damage probability of the UCAV to the target. The specific data is shown in Table 1, 2, 3, 4, 5.

Table 1. Threat source information table

Threat SourceID	X Axis	Y Axis	Threat radius
1	20	25	7
2	12	60	5
3	42	48	13
4	73	36	8
5	68	69	10

Table 2. Target attribute table

Target taskID	Coordinate	Target value
T1	(71,51)	4
T2	(80,61)	8
T3	(53,69)	10
T4	(7,33)	7
T5	(33,22)	5
T6	(12,82)	9
T7	(29,30)	6
T8	(86,42)	8
T9	(45,76)	7
T10	(33,91)	5

5.2 Simulation

In the following, two experimental scenario hypotheses will be carried out, including initial task assignment, emergence of new threat sources, and comparison with traditional contract networks to verify the effect of task allocation based on improved contract network protocol on battlefield emergencies.

Experiment 1: Firstly, the task is randomly assigned to the drone, and the task assignment is based on the improved contract network protocol. The distribution result is shown in Fig. 3 and the task performance change curve is shown in Fig. 4.

The task allocation process based on the improved contract network tends to be stable in the auction 21 rounds, while the traditional contract net tends to be stable in the auction 26 rounds. The results show that the distributed task assignment based on the improved contract network protocol is efficient and stable.

Experiment 2: A new threat source is suddenly detected on the UCAV4 mission execution route. If the original task sequence {T3,T6} is continued, it will be greatly threatened by security. Therefore, UCAV4 auctions the task

Table 3. UCAV attribute table

UCAV ID	Coordinate
U1	(0,70)
U2	(0,20)
U3	(10,0)
U4	(50,0)
U5	(80,0)

Table 4. Candidate UCAV mental status table when task T1 is executed

Agent parameter	U1	U2	U3	U4	U5
Ability A	0.1	0.3	0.7	0.4	0.5
Believe ability B	0.3	0.1	0.5	0.4	0.2
Familiarity F	0.6	0.3	0.4	0.5	0.2
Positivity P	0.7	0.6	0.4	0.2	0.6
Risk tolerance RT Ability A	3.0	1.5	5.0	4.5	10
Busyness BD	0.1	0.5	0.2	0.1	0.4

Table 5. Weight setting table

Weight name	Value
Tender decision	$\lambda_1 = 0.4, \lambda_2 = 0.2, \lambda_3 = 0.3, \lambda_4 = 0.1$
Decision function D_limit	4,3,2,2,3
Cost function weight	$\sigma_1 = 0.4, \sigma_2 = 0.4, \sigma_3 = 0.2$
Bidding value weight coefficient	$\alpha = 0.4, \beta = 0.6$

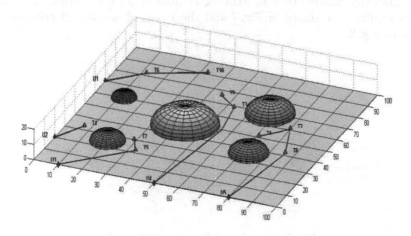

Fig. 3. Experiment 1 assignment results.

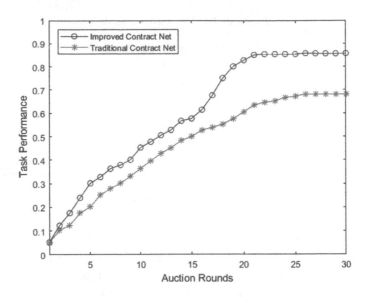

Fig. 4. Experiment 1 performance curve.

sequence, re-allocates and coordinates the task, and gets the assignment. The result is shown in Fig. 5 and the task performance curve is shown in Fig. 6.

UCAV4 hands over the task sequence to UCAV2 through auction, and through task negotiation, UCAV5 hands task T1 to UCAV4. The UCAV4 avoids the threat and obtains new task execution, and also avoids the waste of combat resources. After five auction rounds, the mission performance reaches a steady state.

Experiment 3: Multiple UCAV formations received new combat mission instructions {T11,T12} during the execution of the mission. The new missions were randomly handed over to existing combat units for auction. The distribution results were shown in Fig. 7 and the curve of mission effectiveness was shown in Fig. 8.

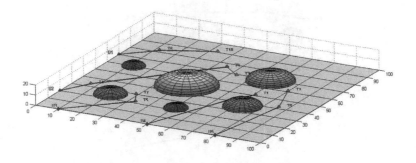

Fig. 5. Experiment 2 assignment results.

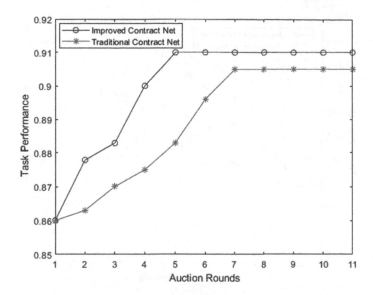

Fig. 6. Experiment 2 performance curve.

After the auction, the new task T11 was assigned to UCAV5, and the new task T12 was assigned to UCAV1, which affected the result of the original task assignment. Through resource coordination, task T2 was assigned to UCAV4 for execution. After 9 rounds, the task performance reached stable state.

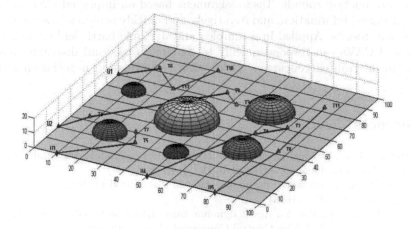

Fig. 7. Experiment 3 assignment results.

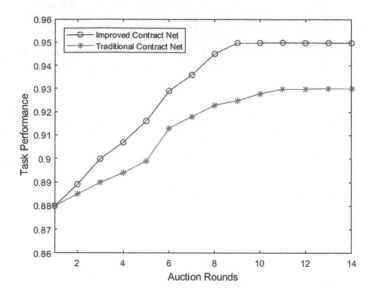

Fig. 8. Experiment 3 performance curve.

6   Conclusion

Facing the unexpected situation in the execution task, compared with the traditional contract network, the distributed task assignment based on the improved CNP can achieve a reasonable distribution result through a relatively small number of task auction rounds. Task assignment based on improved CNP is based on local shared information, and iteratively and quickly achieves reasonable task assignment results. Applied in a complex and dynamic battlefield environment, multiple UCAVs can respond quickly to dynamic tasks and distribute assignments to distributed concurrent tasks with a coordinated negotiation mechanism.

References

1. Smith, R.G.: The contract net protocol: high-level communication and control in a distributed problem solver. Control Decis. 15–64 (1964)
2. Chen, W.-K.: Frameworks for cooperation in distributed problem solving. Linear Netw. Syst. 123–135 (1993)
3. Yang, P., Liu, Y., Pei, Y.: Agent dynamic task allocation based on improved contract net protocol. J. Fire Control Command Control **36**(10), 77–80 (2011)
4. Li, X.-L., Dai, Y.-W.: A task allocation algorithm base on improved contract net protocol under the dynamic environment. J. Sci. Technol. Eng. **13**(27), 8014–8019 (2013)
5. Pang, Q.-W., Hu, Y.-J., Li, W.-G.: Research on multi-UAV cooperative reconnaissance mission planning methods: an overview. J. Telecommun. Eng. **59**(6), 741–748 (2019)

6. Fu, G.-Y., Li, Y., Fu, W.-Y.: Application of improved contract net on pursuit assignment allocation for multiple mobile robots. J. Sichuan Ordnance **40**(3), 98–102 (2019)
7. Zhao, Y., Wu, J.-F., Gao, Y.-P.: Information fusion method of hypersonic vehicle based on multi-agent navigation. Syst. Eng. Electron. **42**(2), 405–413 (2020)
8. Qin, J.-F., Zeng, F.-M.: Research on cooperation mechanism of multi-agent system based on improved contract net. J. Transp. Sci. Eng. **38**(5), 1065–1069 (2014)
9. Li, M., Liu, W., Zhang, Y.: Mulit-Agent dynamic task allocation based on improved contract net protocol. J. Shandong Univ. Technol. **46**(2), 51–56 (2016)
10. Guo, Z.-J., Mi, Y.-L., Xiao, Y.: Application of improved contract net protocol on weapon target assignment of air defense combat. Mod. Def. Technol. **45**(4), 104–111 (2017)

Adaptive Adjustment Method for Construction Progress of Fabricated Buildings Based on Internet of Things

Qian He and Qian-sha Li[(⊠)]

School of Civil Engineering and Hydraulic Engineering, Xichang University,
Xichang 615000, China
HExixi0628@126.com, liqiansha335@163.com

Abstract. Traditional assembly-type construction progress projects have shortcomings such as long construction period, poor environmental benefits and high building energy consumption. In view of these problems, this paper proposes an adaptive adjustment method of prefabricated building construction schedule based on Internet of things. Utilize the advantages of the Internet of Things, collect information on the construction progress, and then effectively scan and bind according to the construction progress information. Then build the progress management data information model of the prefabricated building project, calculate the minimum value of the maintenance cost load in the construction progress, and realize the adaptive adjustment target of the project schedule. The results show that the adaptive adjustment method of the construction progress of the fabricated building based on the Internet of Things is superior to the traditional method in engineering error rate and construction safety, which greatly reduces the construction period and is more effective in use.

Keywords: Internet of Things · Prefabricated building · Construction progress · Self-adaption · Method of adjustment

1 Introduction

As an important pillar industry of China's economy, the construction industry is crucial for the development of its digitalization and integration. In recent years, as an important breakthrough for the transformation and upgrading of the construction industry, the prefabricated building is the key to realizing the continuous innovation and development of the construction industry and leading development [1]. Internet of Things technology is a hotspot technology used in the construction industry. It mainly uses RFID technology and traditional wireless networks, and combines BIM technology to build an information system. In foreign countries, SONG JC and others have realized the automatic tracking and real-time information acquisition of building materials by combining RFID and GIS technology to locate building materials. In China, Chang Chunguang and others applied RFID technology and BIM technology to the assembly building, and studied the application process based on these two technical systems [2]. However, the short data transmission distance, the anti-interference of the wireless

© ICST Institute for Computer Sciences, Social Informatics and Telecommunications Engineering 2021
Published by Springer Nature Switzerland AG 2021. All Rights Reserved
S. Liu and L. Xia (Eds.): ADHIP 2020, LNICST 347, pp. 498–507, 2021.
https://doi.org/10.1007/978-3-030-67871-5_44

communication network and the high network transmission cost have become obstacles to the application of the Internet of Things technology. In the past two years, a new type of ultra-long thermal distance and low-power data transmission technology based on 1 GHz and below, which is called LoRa technology, was released by Semtech. As a kind of low-power wide-area Internet of Things, it has revolutionized the Internet of Things technology. By using spread spectrum technology and special LoRa modulation, it has the advantages of long transmission distance, strong penetrating power, low cost and low power consumption at low transmission rate. At present, LoRa technology is mainly used in smart parking lots, remote wireless meter reading, smart agriculture, intelligent street lamps, etc. The construction industry will become a new application field of this technology. The solution in this paper aims to fill the application gap of LoRa wireless communication technology in prefabricated buildings, and make breakthroughs in the digitalization, networking and intelligence of the construction industry [3].

2 Adaptive Adjustment Method for Prefabricated Construction Progress

In order to obtain the actual progress of the construction site, on-site personnel are required to collect information, and then complete the data collection and then compare with the planned progress. This method can't check the construction progress anytime and anywhere, can't adjust the network plan in time, and can't control the construction speed in time, which is not conducive to refined management. The integration of RFID and BIM technology enables real-time control of the progress management of fabricated buildings [4].

2.1 Construction Progress Information Collection

First of all, the BIM technology is used to model the prefabricated building (the current stage is overturned), and each major is synthesized, collided and deepened. Secondly, after completing the deepening design, each component is coded on the model, and the coding is simple and easy to understand. The relevant information of the component is hooked up, and the location information and planning time information need to be written in advance [5]. Then, the components coded by BIM technology are automatically plotted, and the drawings are handed to the PC component processing factory for processing. The encoding, naming and planning information in the model tag is processed and stored on the RFID chip, which is embedded in the prefabricated components [6]. Once again, after completing the component processing, the components are transported according to the planned time, and the actual factory time, the actual entry time, the actual delivery time, the actual admission time, and the actual lifting time are recorded. Finally, the hoisting equipment such as tower crane is used for assembling. If the accuracy is qualified, the installation time is automatically read and recorded. If the installation accuracy does not meet the requirements, a warning is issued. The whole process of assembly design, production, transportation and hoisting is shown in Fig. 1:

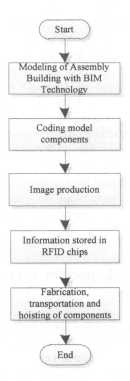

Fig. 1. The whole process of assembly component design, production, transportation and hoisting

The planning time of the different stage components can be linked to the component in advance, can be implemented in the modeling software, or can be implemented in BIM 5D or other progress simulation software. The actual time of the component can be imported into Revit through the secondary development software through the information of the reader, thereby realizing the information collection of the schedule. This "BIM + RFID" model satisfies the entire process of information management of component production and transportation, and can be reverse traced. It can not only ascertain the relevant time information of the component, but also track its responsible person and operator to facilitate the quality control of the component.

2.2 Effective Scanning and Binding Based on Construction Progress Information

Active tag (IoT transport module) saves the attribute information of the component in the module's memory before embedding. After the component is embedded in the active tag, the constructor can scan the RFID tag of the component through the RFID reader and upload the information to the cloud server. At the same time, the active tag is positioned by the GPS to locate the component, and the saved information in the tag is transmitted to the cloud server through the transmission module. The information of the passive tag and the active tag is then matched and bound in the cloud server. Through this matching method, the real-time information of the component can be saved to the cloud server through the transmission module of the active tag. And through the cloud server to update to the construction personnel's terminal equipment in real time, so as to record the installation location, arrival date, installation date and other information of the component in real time [7]. After the information is complete, the label is embedded on the prefabricated component. Since the reader installed at the construction site can read and recognize the component information held in the electronic tag without contact, the purpose of automatically identifying the component information is achieved. In this way, the components can be written to the actual time by hand scanner from production, storage to delivery [8]. On-site fixed readers can read a large number of label information, and have the characteristics of oil resistance, fast reading and writing speed, reliable information and timely. The collected information is connected to the computer through the reader, and the tag information read is the actual progress information of the component. This information is transmitted to the computer, and the simulation of the actual progress information is compared with the simulation of the planned progress to find out the cause of the impact [9, 10]. Using RFID technology to collect the actual production time of the component, the actual storage time, the actual delivery time, the actual admission time, the actual transportation time and the actual lifting time, the process is shown in Fig. 2. Make sure that the state of the component is visible at all times to prepare for subsequent call information.

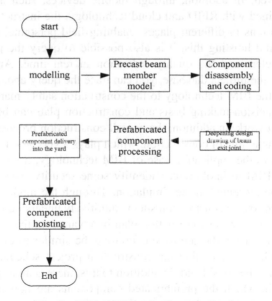

Fig. 2. BIM, RFID technology application

2.3   Constructing a Schedule Management Data Information Model for a Prefabricated Building Project

After selecting the best construction plan through the comparison of the plan, it is necessary to integrate the relevant information of the construction plan. And enter the information into the BIM model to establish a building model or planning network entity that matches the actual engineering conditions. For hoisting construction, the application of BIM technology is mainly reflected in the establishment of a construction model to simulate the implementation effect of the on-site construction plan. And according to the components selected for the construction of the project, the installation location and installation sequence of the components are clearly defined to facilitate construction guidance and scheduling. It should be noted that quality and performance inspections are required before the components enter the site, and the explanatory materials and quality certificates issued by the manufacturer are checked. After inspection, they can be put into use. At the same time, the command and dispatch management personnel can work through the on-site handheld equipment. The construction hoisting model is directly displayed on the handset. Field personnel can use handheld devices to scan the components and analyze the structure and related information of the components. Compare with the construction design plan, conduct in-depth research on the causes of the differences, improve the on-site construction work, and ensure the quality of component installation. Intuitive understanding, so that the construction team has an understanding of the building information, effectively avoiding the occurrence of installation errors, so that work efficiency and installation quality are improved. In addition, through mobile devices, such as tablets, mobile phones, etc. combined with RFID and cloud technology, the instructors can guide the construction conditions in different places, enabling field personnel to locate components smoothly. And hoisting this, it is also possible to query the parameters of the component parameters and the quality indication in real time. And uploading the completion data to the project database, you can trace the query about the construction quality and apply the BIM technology to the construction safety management. Firstly, the organization decision-making basis and construction plan can be provided to the safety management work. In addition, the overall construction process can be remotely monitored at the same time of construction, which plays a very good role in preventing safety hazards. From the application case of BIM technology, it can be analyzed that the simulation of BIM technology can identify some security risks. Each process of construction can be presented in the simulation. Through the model, not only can the visual characteristics of the construction site be intuitively understood, but the technical personnel can also inspect the construction plan in advance. In the simulation process, the emergency plan for the dangerous situation can be similar to the traditional engineering project. When constructing the construction project schedule, each key construction node should be considered. In addition to this, there is an important condition, that is, the order in which the prefabricated components are hoisted. Under normal circumstances, these tasks are done manually, but the manual error is large, and the plan to complete the entire lifting process is difficult to ensure accuracy. The introduction of BIM technology just makes up for this shortcoming and can make the lifting process efficient and quick. Then import the file into Navisworks and associate with the building BIM model, so that you can accurately organize the time and amount of engineering required for each project. Therefore, the specific construction plan is determined, and after all the information is collected, the 5D simulation work is carried out. At this point, the construction progress can be more clearly reflected, and the schedule can be more comprehensively organized to apply BIM technology to the construction schedule management to ensure that the process can be carried out in an orderly manner [11, 12]. At the same time, the overall view of managers should be strong. The important nodes and schedules should be compared to make the time and space allocation reasonable and make the best use of them. The building information model is the core, and the delay of the construction period can be avoided as much as possible. In terms of schedule management, the BIM model can be used as a basis for decision-making. While promoting construction, it is inevitable that some unforeseen unexpected situations will occur, resulting in disruption of the construction schedule. At this point, the construction schedule can be adjusted immediately according to the BIM model to avoid the construction schedule not being able to progress smoothly due to the untimely change, which leads to more errors.

2.4   Maintenance Cost Load Calculation in Construction Schedule

BIM-based 5D dynamic construction cost control is based on the 3D model, adding time and cost to form a 5D building information model. Through the virtual construction to see whether the material stacking, project progress, and capital input are reasonable, timely discover the problems existing in the actual construction process. Optimize the construction period, resource allocation, adjust resources and capital investment in real time, optimize the construction period and cost targets, and form an optimal building model to guide the next construction. Suppose a production line has II specifications, each specification produces $A = \{A_1, A_2, A_3, \cdots A_n\}$, s is the total number of products produced, and j represents a possible production sequence.

$$J = \{J_1, J_2, J_3, \cdots J_n\}, s = \sum_{i=1}^{n} A_i \text{ 伸} \tag{1}$$

There are a total of m stations in the production line, and the processing time of each product in each station is matrix p. p_{ik} represents the processing time of the $i = (1, 2, 3. \ldots \ldots n)$-th specification at the k-th station. For production plan J, the completion time of product J_j at station k can be expressed as $C(J_j, k)$, and the maximum completion time for the entire production plan is $C(J_s, m)$. Therefore, the most critical issue of the dynamic cost control problem is to solve the minimum $C(J_s, m)$, i.e., solve $\min C(J_s, m)$.

The same work station in the manufacturing industry can only process one product at a time, and the dynamic cost control in the assembly construction has its particularity. It mainly shows that multiple products can be cured at the same time in the dynamic cost control process, so the maintenance process is parallel. Secondly, considering that the product curing time is long, it is generally completed in non-working hours, and the start time of the next process after curing cannot be in non-working time, so it is postponed to the next working day. In addition, a series of processes of dynamic cost control cannot be interrupted and must be carried out continuously. The overtime of the process is OT. If the pouring operation is still not completed within the overtime hours, it is postponed until the next day, and other processes such as mold installation can be interrupted, so the overtime is $OT = 0$. The working time per day is TW, generally $TW = 8$ h, and non-working time is $TN = 24 - TW$. The completion time of each cost control task is as follows:

$$T = Max \begin{cases} C, (j_{j-1}, k), C(J_j, k - 1) + P_{jk} \\ Int(T/24) \end{cases} \tag{2}$$

$$C(J_j, k) = \begin{bmatrix} T \cdots if\,(T \leq 24D + TW + OT) \\ \vdots \\ 24(D+1) + P_{jk}, if\,(T \geq 24D + TW + OT) \end{bmatrix} \tag{3}$$

For concrete curing, since it can be processed in parallel,

$$T = C(J_j, k - 1) + P_{jk} \tag{4}$$

The objective function is $f(x) = min\,C(J_s, m)$, S is the total production task or order, and m is the number of 5D dynamic cost control or the number of stations in the assembly construction progress.

2.5   Realize the Progress Adjustment Target of Engineering Progress

Progress management is one of the three goals of project management. Using information technology to collect the actual progress of the fabricated components, so as to effectively arrange the construction tasks, so that the construction progress deviation is within the range. The use of RFID technology to achieve automated collection of actual progress saves a lot of manpower. Once the products of the building components are produced, they are irreversible. Once the error exceeds the controllable range, the components will be unusable, so the assembled buildings are highly demanded in the engineering design stage. Compared with the traditional construction method, the adaptive adjustment method of the assembly building schedule is more energy-saving and environmentally friendly, and is less affected by environmental factors. It can speed up the construction progress, make the project put into use as soon as possible, and exert production value. Realize no green formwork, no external scaffolding, no on-site masonry, no plastering green construction. Adopt "less specifications, multiple combinations" design principles to reduce construction costs and shorten construction period; use green technology to save later operating costs.

3   Simulation Experiment Demonstration and Analysis

In order to ensure the effectiveness of the adaptive adjustment method of the IoT-based fabricated building construction schedule designed in this paper, the simulation experiment demonstration analysis is carried out. Prefabricated component production tasks are generated by the production planning task entity and then processed by subsequent workstations one by one. When the prefabricated production components are processed on the workstation 9, the components are lifted, transported and stored in the component yard. The operational time of an activity is determined by its corresponding distribution function entity. In traditional prefabricated production, production is driven by production plans. In the simulation, queue entities are used to represent buffers between workstations. When a workstation is working, tasks are queued and waiting for a certain amount of time. Each queue entity can store multiple tasks. The linked files can be seen from the management project management link. The layout of the construction site is completed according to the imported CAD drawings. The optimized 3D model of the construction site layout is shown in Fig. 3.

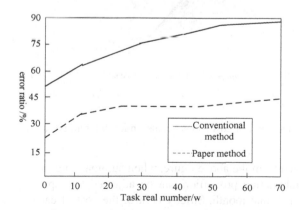

Fig. 3. Construction layout model

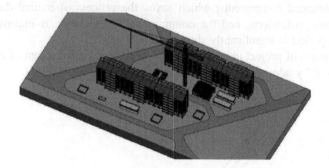

Fig. 4. Experimental comparison chart

In the process of data collection, due to unclear time definition, irregular statistical process and special conditions on site, it is easy to cause some data to be too large and too small. Therefore, first of all, the most basic data processing, the removal of abnormal data, and then the remaining data processing.

In order to ensure the effectiveness of the experiment, the traditional method and the adaptive adjustment method based on the IoT-based assembly building construction progress were compared, and the construction error rate of the two methods was statistically analyzed. The experimental results are shown in Fig. 4.

According to the analysis of Fig. 4, the engineering error rate of the adaptive adjustment method of the prefabricated construction progress based on the Internet of Things is far lower than that of the traditional method. Therefore, it can be concluded that the adaptive adjustment method of the prefabricated construction progress of the Internet of Things designed in this paper largely avoids the situation of on-site rework, waiting for work, secondary transportation, etc. He has obviously improved the quality of the project and also ensured the safety of the construction. Through the combination of the prefabricated component management system and the Internet of Things technology, the information management technology is used to optimize the management of the prefabricated components, which saves the process of mutual data exchange among various professions, and the communication efficiency is improved, and the construction period is significantly shortened.

In the process of project management, the integrated management of cost/schedule can be realized by adjusting the proportion of prefabrication (Fig. 5).

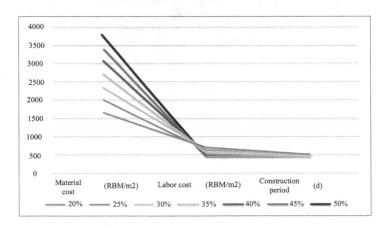

Fig. 5. Change trend of cost and schedule under different prefabrication proportion

As can be seen from the above figure, when summarizing the cost progress of each month, the prefabrication proportion can be adjusted appropriately according to the specific situation of that month. Strictly control the cost of each stage of the project, fully consider the possibility and trend of price changes in each stage of the project, and make timely adjustment measures according to the specific implementation situation, such as how to adjust the project schedule in case of natural disasters.

4 Conclusion

This paper analyzes and experiments on the adaptive adjustment method of the construction progress of the prefabricated building based on the Internet of Things. Based on the advantages of the Internet of Things, the information acquisition is carried out, and the design of the design and the information carried are obtained. After that, according to the design of the deepened design and the information carried, the production planning is completed, the production plan optimization of the precast concrete components is completed, and the construction schedule planning process is accurately controlled. The experimental results show that the adaptive adjustment method of the construction progress of the prefabricated building based on the Internet of Things is very efficient. When it progresses in the construction of the prefabricated building, it significantly improves the quality of the project, shortens the construction period, and ensures the safety of construction. It is hoped that the research in this paper can provide theoretical basis and reference for the adaptive adjustment method of assembly construction progress in China.

References

1. Jianguo, X., Jia, Z.: Prediction simulation of parallel transmission path of Internet of Things data. Comput. Simul. **35**(1), 172–175+231 (2018)
2. Xiaodan, L.: Study on construction process planning and control of assembly building. Doctorate thesis of Dalian University of Technology (2018)
3. Yang, X.: Research on application of BIM Technology in construction stage of assembly building. Master thesis of Wuhan University of Engineering (2017)
4. Liu, S., Li, Z., Zhang, Y., Cheng, X.: Introduction of key problems in long-distance learning and training. Mobile Netw. Appl. **24**(1), 1–4 (2018). https://doi.org/10.1007/s11036-018-1136-6
5. Bortolini, R., Formoso, C.T., Viana, D.D.: Site logistics planning and control for engineer-to-order prefabricated building systems using BIM 4D modeling. Autom. Constr. **98**, 248–264 (2019)
6. Jelena, N.: Building "with the systems" vs. building "in the system" of IMS open technology of prefabricated construction: challenges for new "infill" industry for massive housing retrofitting. Energies **11**(5), 1128 (2018)
7. Yuan, Z., Sun, C., Wang, Y.: Design for Manufacture and Assembly-oriented parametric design of prefabricated buildings. Autom. Constr. **88**, 13–22 (2018)
8. Dou, Y., Xue, X., Wang, Y., et al.: New media data-driven measurement for the development level of prefabricated construction in China. J. Clean. Prod. **241**, 118353 (2019)
9. Jrade, A., Jalaei, F.: Integrating building information modelling with sustainability to design building projects at the conceptual stage. Build. Simul. **6**(4), 429–444 (2013). https://doi.org/10.1007/s12273-013-0120-0
10. Zhao, L., Liu, Z., Mbachu, J.: Development of intelligent prefabs using IoT technology to improve the performance of prefabricated construction projects. Sensors **19**(19), 4131 (2019)
11. Fu, W., Liu, S., Srivastava, G.: Optimization of big data scheduling in social networks. Entropy **21**(9), 902 (2019)
12. Shuai, L., Weiling, B., Nianyin, Z., et al.: A fast fractal based compression for MRI images. IEEE Access **7**, 62412–62420 (2019)

Research on Structural Optimization of Reinforced Concrete Frame Based on Parallel Cloud Computing

Qian-sha Li[✉] and Qian He

School of Civil Engineering and Hydraulic Engineering,
Xichang University, Xichang 615000, China
LIqiansha335@163.com

Abstract. In order to solve the problems of excessive wear and poor bearing capacity of reinforced concrete frame, the optimization of reinforced concrete frame structure is studied with cloud computing method. The bearing capacity and related parameters of reinforced concrete structure are calculated by cloud computing method, and the support degree and potential structural wear displacement range of reinforced concrete frame structure are obtained. According to the calculation results, load parameters of building concrete structure are regulated and adjusted. The reinforced concrete frame with X-type braces is strengthened by combining with the standard parameters, and the reinforced concrete frame connection is established. The orthogonal intersection model of the bearing capacity of the structure can effectively complete the optimization study of the reinforced concrete frame structure. Finally, the experiment proves that the research of RC frame structure optimization based on parallel cloud computing has better bearing capacity and anti-wear performance than the traditional RC frame structure, and fully meets the research objectives.

Keywords: Parallel cloud computing · Reinforced concrete · Frame structure

1 Introduction

With the increasing prosperity and development of the construction industry, reinforced concrete frame structures are widely used in the construction industry because of their high bearing capacity and good ductility. But at present, there are relatively few studies on Optimization of prefabricated assembly structure of steel reinforced concrete frame joints splicing form, which is difficult to fully meet the building requirements. Therefore, based on cloud computing method, the reinforced concrete frame structure is optimized, and the stability of the frame structure is improved by adding prefabricated joints to the reinforced concrete frame [1]. By optimizing the joint combination in reinforced concrete frame, the ductility and bearing capacity degradation of the structure frame are avoided. Finally, the experiment proves that the method is more effective and scientific than the traditional method.

S. Liu and L. Xia (Eds.): ADHIP 2020, LNICST 347, pp. 508–517, 2021.
https://doi.org/10.1007/978-3-030-67871-5_45

2 Method

2.1 Calculation of Bearing Capacity of Reinforced Concrete Structures

In order to effectively achieve the goal of optimizing the structure of reinforced concrete miners, it is necessary to calculate the bearing capacity of concrete at first. Whether the reinforced concrete structure and the reinforcement part can be successfully fused becomes an important part of the overall structural optimization, and the effectiveness of this part depends on the bearing capacity of the joint at the joint site [2]. In the construction process, the bearing capacity of concrete joint is usually low, so it is easy to reduce the bearing capacity of the structure in the process of structural optimization. In order to solve the above problems and better complete the optimization design of reinforced concrete frame structures, it is necessary to fully consider the stress factors and characteristics of buildings, so as to establish a more standard load standard for reinforced concrete structures [3]. Secondary combination and optimization design of reinforced concrete structural parameters of complex network buildings are carried out, and the performance parameters of reinforced concrete materials in building structures are accurately estimated and recorded. The results are shown in Table 1.

Table 1. Bearing capacity of reinforced concrete materials

Type	Level	Yield strength/MPa	Ultimate strength/MPa	Modulus of elasticity/GPa
12 mm Steel skeleton	Q235b	270	430	620
20 mm Steel skeleton	Q235b	275	43.5	625
Longitudinal bars of beams and columns with diameter of 18 mm	HRB335	280	440	630
Longitudinal bars of beams and columns with diameter of 22 mm	HRB335	285	445	635
Beam and column stirrups with diameter of 18 mm	HRB325	290	450	640
Beam-column stirrups with diameter of 22 mm	HRB325	295	455	645

As shown in the table above, in order to avoid the torsion of reinforced concrete in the construction process and subsequent use, HRB buckling-resistant brace reinforcement frame is set up to ensure the stability of the structure in the process of optimization of reinforced concrete structure [4]. The parameters of HRB buckling-resistant brace structure are calculated. Due to the problems of over-limit axial compression ratio in reinforced concrete frame structures of complex buildings, it is necessary to increase the section of frame columns in order to improve their load capacity in the process of rebuilding complex network buildings [5]. The influence factor of

HRB buckling-proof frame structure stability is introduced into parallel cloud computing to calculate the structural support. The algorithm is as follows:

$$S_0 = \frac{1 + 4e_{\mathrm{n}}^2}{(1 - \frac{\beta^2}{R_n^2})^2} \tag{1}$$

In the formula: β is the spectrum parameter reflecting the load characteristics of steel bars, R is the strength parameter of building structures, the strength range of reinforced concrete structures is n and in the range of [0,1], e represents the damping frequency of the structural characteristics of the surface soil layer of building structures, A is the stationary parameter of building structures, then:

$$A = \sqrt{\frac{2}{\beta R_0^2}} \exp(-\frac{\beta^2 + e^2}{R_0^2}) \tag{2}$$

In the application of reinforced concrete frame in construction, the displacement of its structural angle is prone to occur, and it is difficult to ensure the stability and accuracy of the structural system and load capacity [6]. Therefore, combined with parallel cloud computing method, the minimum deflection angle of reinforced concrete lateral component is calculated and strengthened. By restraining the structure frame, the stress degree of the structure frame is weakened, and the problems of displacement and angle deflection are avoided [7]. The least square method is used to fit the minimum deflection angle, and the algorithm of building reinforced concrete anti-variable function is obtained as follows:

$$S(\varpi) = \frac{A + 4\lambda_i^2 \alpha^2 / \alpha_i^2}{(A - \alpha^2 / \alpha_i^2) + \lambda_n^2 \alpha^2 / \alpha_i^2} S_0 \tag{3}$$

In the formula, α_n and α_m are the maximum and minimum load-bearing damage coefficients in the random process, and λ is the responsiveness. Based on the above formulas and the static elastic-plastic principle, the anti-destructive capacity under buildings is deduced and calculated [8]. In general, if there are m nodes in the structure, the load-carrying capacity of the structural nodes is divided into n categories by parallel cloud computing method, then the classification matrix of the structural load-carrying capacity can be obtained by combining the numerical model theory as follows.

$$W = \begin{bmatrix} w_{11}, w_{12}, \ldots, w_n \\ w_{21}, w_{22}, \ldots, w_{2n} \\ w_{21}, w_{22}, \ldots, w_{2n} \\ \cdots \quad \cdots \quad \cdots \\ w_{m1}, w_{m2}, \ldots, w_{mn} \end{bmatrix} \tag{4}$$

Among them,

$$0 \leq w_{mn} \leq 1 \quad (n = 1, 2, 3. \ldots, m = 1, 2, 3 \ldots)$$

According to the above requirements, the initial classification matrix is designed. In the matrix, a is the next column of molecules, so the initial classification matrix is generated. If M = 1, the concrete bearing capacity formula of parallel cloud computing is as follows:

$$\xi = \sum_{n=1}^{a} S(\varpi) * \left(\frac{\|a_n - \beta_n\|}{\|a_i - \beta_i\|} \right)^2 * \log A \frac{a_i}{\sum_{i=1}^{n} W} \tag{5}$$

Through the above algorithm, the bearing capacity of reinforced concrete structures in buildings can be accurately obtained, so that the parameters of frame structures can be standardized according to the bearing capacity, so as to achieve the goal of structural optimization.

2.2 Reinforcement of Reinforced Concrete Frames

In order to solve the problem of serious damage of reinforced concrete frame in the construction process, it is necessary to reinforce the reinforced concrete frame structure. Firstly, the standard load capacity of reinforced concrete frame in our country's current buildings is analyzed by using the previous algorithm, and the relevant planning processing is carried out. The normative data of the damage resistance parameters of the building concrete structure are obtained as follows (Table 2):

Table 2. Specification for damage resistance parameters of building concrete structures

Serial number	Load category	Resistance loss parameter	Serial number	Resistance loss parameter	standard value
1	A steel bar	30 KG/m^2	6	Weight of reinforcing bar per meter	0.00617*d*d
3	Concrete	2500 KG/m^3	8	Reinforced concrete weight	2200 KG/m^3
4	Exterior wall	2.0	9	Conversion thickness	550 px
5	Interior wall	2.5	10	SPECT	5.0

In order to control the yield section of reinforced concrete accurately in the support section, it is necessary to make sure that any part other than the support section is in the elastic range. Because the connection of the reinforced frame with HRB buckling-resistant brace can be processed by the method of embedded parts, combined with the data in the table above and the previous algorithm, the standard dimensions of the connection plate and the additional reinforced frame in the reinforced concrete frame can be obtained by calculation. In order to ensure the stability of the frame structure, the composite wall of the complex network building model has its unique characteristics, and according to the building added. Solid Estimation Model is used to analyze the stress process and characteristics of buildings in the environment [9, 10]. Distribution beams are loaded on the building by jacks, and hidden beams are loaded on the top after secondary distribution of complex network buildings. Support is set on both sides of the building wall to ensure the stability of the building wall plane [11, 12]. The reinforced concrete frame with X-braced reinforcement frame is shown in the following Fig. 1.

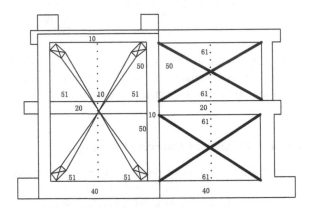

Fig. 1. Reinforced concrete frame reinforced with X-type braces

As shown in the figure, the reinforced concrete frame with X-brace reinforcement frame can effectively balance the internal force of reinforced concrete, balance the stiffness and the internal force that should be borne by the reinforcement design of the external structure. However, the vertical load of the structure is not considered for the external concrete frame. The external concrete frame, main frame and buckling-resistant brace bear the load and interlayer shear force together. In the process of design and treatment of reinforced concrete structures, the weight and bearing capacity of the building roof should be fully considered, and the weight and bearing capacity of the roof should always be kept in a moderate state to ensure that the ceiling is firm. In order to optimize the reinforced concrete frame structure, it is necessary to avoid the increase of ground load, ensure the stability and safety of the structure and reduce the bearing capacity of the members.

2.3   Structural Optimization of Reinforced Concrete Frame

According to the above ideas, the structure of the numerical model database system is standardized. After summarizing the existing structural criteria, the modal superposition method is proposed. Its iteration formula is as follows:

$$[A_\lambda^R]\Delta\overrightarrow{\mu}_\lambda = \xi(K^n - K_\lambda^{cn}) \tag{6}$$

Among them, K_λ^{cn} is the recovery load, $[A_\lambda^R]$ is the Jacobian matrix, and $\overrightarrow{\mu}_\lambda$ is the balanced iteration step. Before solving the problem, $[A_\lambda^R]$ is estimated by non-equilibrium load linearity and the convergence is checked. If the calculation results do not satisfy the convergence criterion, the deformation degree of reinforced concrete is studied. According to the results of deformation degree research of reinforced concrete frame structure, the displacement, velocity and acceleration of each layer of reinforced structure are geometrically non-linear. According to the elastic-plastic analysis, the ultimate load is calculated according to the previous calculation method. According to the calculation results, the above model is simplified to a unified orthogonal inter-section model of the bearing capacity of ideal reinforced concrete frame structure. For the convenience of display and understanding, the section of the model is simplified and drawn. The specific section structure is shown in the following Fig. 2.

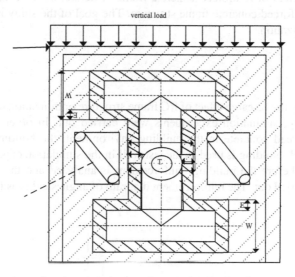

Fig. 2. Cross section of orthogonal intersection model for bearing capacity of reinforced concrete frame structure

According to the principle of load value and symmetry of building structure, only a quarter of the model is studied. Under lower loads, the model structure has not failed and can still bear loads, and the surrounding area is still in an elastic state. With the increase of transverse tension pressure, the plastic zone expands continuously. When the pressure reaches the limit value, the plastic zone almost extends to the whole range of reinforcement, and the building structure loses its bearing capacity. The information of steel bending degree under different pressure is as follows (Table 3):

Table 3. Ultimate moment of reinforced concrete frame structure

Number	K	D/r	Grade of concrete	Elevation	Bending moment
1–5	0.09	0.38549	C35	56.800	0.41021
6–10	1.00	0.51241	C40	48.600	0.54715
11–15	1.06	0.61254	C45	40.000	0.64852
16–20	1.12	0.77425	C50	28.900	0.74219
21–25	1.30	0.84573	C55	13.900	0.84528
26030	1.51	0.89135	C60	9.800	0.88452

In traditional buildings, with the increase of transverse tension and bearing pressure, the bending degree of steel bars in the same longitudinal section increases gradually, and the failure phenomenon is very easy to occur. The compressive strength and bearing capacity of reinforced concrete vertical members can be effectively improved by optimizing the frame structure with the ultimate bending moment specification parameters of reinforced concrete frame structure, so as to achieve the optimization of reinforced concrete frame structure. The goal of the study is to achieve the goal of modernization.

3   Result

In order to verify the research effect of RC frame structure optimization based on parallel cloud computing, simulation experiments were carried out. In order to simulate the stress of reinforced concrete in the building environment, the bottom wall of a real complex network building is used as the test wall in the simulation experiment, and the stress situation before the building reinforcement is analyzed and the secondary reinforcement is carried out. The reinforcement diagram of angle steel is as follows (Fig. 3).

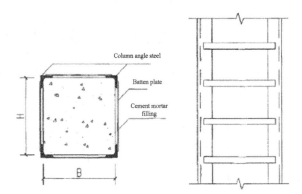

Fig. 3. Diagram of external angle steel reinforcement

The reinforced concrete columns strengthened by steel encasement can effectively improve the compression bearing capacity of the columns.

The reinforced structure is analyzed. Considering that the stiffness change of the member has some influence on the building structure, the method of converting section size is used to reflect it and then the structural analysis and detection are carried out. As the stiffness of frame beams changes little in building components, the effect of stiffness changes after strengthening of frame beams can be directly neglected. In the case of other conditions unchanged, the bearing capacity and bending degree of the traditional reinforced concrete frame structure and the optimized frame structure in this paper are compared and tested several times, and recorded. In order to facilitate the research, the traditional frame structure is set as A, and the optimized frame structure is recorded as B. At the same time, the multi-test results of two groups of structures are evaluated comprehensively, and the average parameters are drawn, and the following results are obtained (Fig. 4).

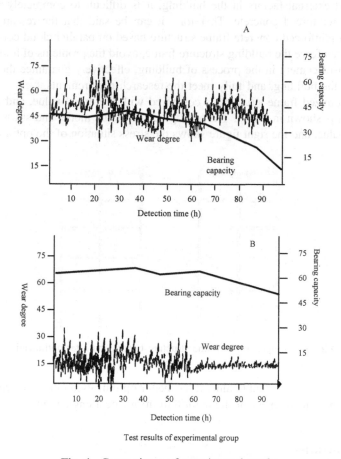

Fig. 4. Comparisons of experimental results

According to the analysis of the above test results, it is not difficult to find that the evaluation results clearly show that before the optimization of reinforced concrete frame structure, its bearing capacity is between 10% and 40%, the overall bearing capacity is relatively poor, and the wear degree is between 30% and 75%. This indicates that the worse wear degree of reinforced concrete frame is liable to cause linear deviation, which causes the torsion of building junction direction. Relatively speaking, the bearing capacity of the optimized frame structure is obviously increased, reaching 45%–60%. Although the anti-destructive reading of the structure is improved after optimizing the ultimate bending moment, it is difficult to improve the bearing capacity of the structure with a higher span, but the bearing capacity of this degree is enough to meet the current building needs. On the other hand, compared with the traditional reinforced concrete frame structure, the wear degree of the optimized frame structure decreases obviously in the process of experimental testing, and the wear situation tends to be stable with the extension of testing time. Due to the relatively large number of internal and external factors in the building, it is difficult to completely remove the damage of reinforced concrete. Therefore, it can be said that the research on Optimization of reinforced concrete frame structure based on parallel cloud computing can effectively reinforce the building structure frame, avoid the problems of frame structure direction displacement in the process of building, effectively guarantee the safety and stability of the building, and fully meet the research requirements.

The optimized frame structure is compared with the initial value, and the specific comparison is shown in Fig. 5. The left figure shows the model and stress analysis of the initial value, and the right figure shows the stress situation of the optimized model.

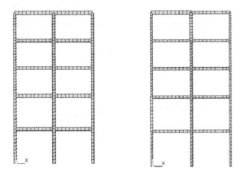

Fig. 5. Comparison of frame structure before and after optimization

Compared with the initial value, it has better optimization effect, which proves the feasibility and effectiveness of the algorithm in practical engineering.

4   Conclusions

The optimization effect of reinforced concrete frame structure based on parallel cloud computing is obviously improved. By standardizing the bearing capacity and load parameter information of reinforced concrete structure, the orthogonal intersection model of frame structure is improved to stabilize the bending moment of reinforced concrete frame structure. The optimization design of reinforced concrete frame structure is completed. Under the same environment, the bearing capacity of the optimized reinforced concrete frame structure is significantly improved, and its wear degree is significantly reduced, which fully meets the research requirements. In the structural system reinforcement analysis, we can further analyze the buildings with different seismic intensities and different structural forms, and obtain the corresponding relations, which can establish the analysis model more carefully and make the optimization analysis more accurate.

References

1. Melih, N.S., Gebrail, B., Xin-She, Y.: Metaheuristic Optimization of reinforced concrete footings. KSCE J. Civil Eng. **22**, 1–9 (2018)
2. Zongcai, D.: Flexural toughness and evaluation method of hybrid fiber reinforced ultra-highperformance concrete. J. Comp. Mater. **33**(6), 1274–1280 (2016)
3. Benavent-Climent, A., Ramirez-Marquez, A., Pujol, S.: Seismic strengthening of low-rise reinforced concrete frame structures with masonry infill walls: shaking-table test. Eng. Struct. **165**(jun.15), 142–151 (2018)
4. Aiping, Z., Congzhen, X., Tao, C., et al.: Experimental study on seismic performance of steel plate - concrete composite shear wall with shear span ratio of 1. J. Civil Eng. (10), 49–56 (2016)
5. Shuai, L., Gelan, Y.: Advanced Hybrid Information Processing, pp. 1–594. Springer International Publishing, New york (2019)
6. Jiangang, N., Jingjun, L., Yalu, Y., et al.: Experimental study on mechanical properties and optimum fiber content of plastic-steel fiber lightweight aggregate concrete. Silicate Bull. **35**(1), 87–91 (2016)
7. Bai, J., Cheng, F., Jin, S., et al.: Assessing and quantifying the earthquake response of reinforced concrete buckling-restrained brace frame structures. Bull. Earthq. Eng. **17**(7), 3847–3871 (2019)
8. Liu, S., Liu, D., Srivastava, G., et al.: Overview and methods of correlation filter algorithms in object tracking. Complex Intell. Syst. (2020). https://doi.org/10.1007/s40747-020-00161-4
9. Liu, J., Xiuli, D.: Mesoscopic numerical study on axial compression performance and size effect of reinforced concrete columns. J. Water Cons. **47**(2), 209–218 (2016)
10. Zheng, P., Shuai, L., Arun, S., Khan, M.: Visual attention feature (VAF): a novel strategy for visual tracking based on cloud platform in intelligent surveillance systems. J. Parallel Distrib. Comput. **120**, 182–194 (2018)
11. Liu, S., Lu, M., Li, H., et al.: Prediction of gene expression patterns with generalized linear regression model. Front. Genet. **10**, 120 (2019)
12. Afefy, H.M., Kassem, N.M., Mahmoud, M.H.: Retrofitting of faulty reinforced concrete frame structures using two strengthening techniques. Mag. Concr. Res. **71**(56), 309–324 (2019)

An Evaluation of the Intervention Effect of Autonomous English Learning Motivation Based on Knowledge Map

Zhou Zhi-Yu[✉] and Ruan Meng-li

Shandong Management University, Jinan 250357, China
jj356398632@163.com

Abstract. In the past, the evaluation method of the intervention effect of Autonomous English learning motivation was not fluent because of the lack of correlation between autonomous learning ability and motivation factors. Therefore, this paper proposes an evaluation of the intervention effect of Autonomous English learning motivation based on knowledge map. Based on the different degree of connection, the knowledge map of self-learning ability and motivation factors is constructed. Combined with motivation intervention, the evaluation index of intervention effect is determined, the weight of different indexes is calculated, the comprehensive score is calculated for the index assignment, and the evaluation of intervention effect is realized by combining the intervention effect evaluation grade table. The experimental results show that compared with the traditional methods, the designed method based on knowledge map has better fluency.

Keywords: Knowledge map · Autonomy · English learning · Intervention effect evaluation

1 Introduction

Knowledge map is a modern theory that combines the theories and methods of Applied Mathematics, graphics, information visualization technology, information science and other disciplines with the methods of bibliometrics citation analysis and co-occurrence analysis, and uses the visualized map to vividly display the core structure, development history, front fields and overall knowledge structure of the discipline to achieve the purpose of multi-disciplinary integration [1–3]. It can provide practical and valuable reference for discipline research. Knowledge map integrates all disciplines to facilitate the consistency of user search, find more accurate information for users, make a more comprehensive summary and provide more in-depth information [4].

Motivation is the internal state that directly promotes the activities of organisms to meet the needs of individuals, and it is the direct cause and internal motivation of behavior [5]. The level and strength of motivation determine the quality, level and effect of individual activities. According to the role of motivation in learning, learning motivation is defined as: learning motivation refers to the internal motivation that directly promotes students to carry out learning activities [6]. Learning motivation can

S. Liu and L. Xia (Eds.): ADHIP 2020, LNICST 347, pp. 518–529, 2021.
https://doi.org/10.1007/978-3-030-67871-5_46

explain why students study, how hard they work and what they are willing to learn. It is the core of the learning process, which is expressed in the form of learning intention, desire or interest, so as to stimulate students' enthusiasm for learning, so as to actively participate in learning activities and play a role in promoting learning. Generally speaking, the more correct the learning motivation, the stronger the learning requirements, and the higher the quality of learning. Students with strong motivation often have serious learning attitude and strong learning perseverance. The strength of students' dominant motivation will affect their academic performance. The relationship between learning motivation and learning effect is not only a one-way relationship, but also an interdependent two-way relationship [7]. Learning motivation can increase students' behavior to promote learning, but what students learn can in turn further increase learning motivation.

English learning motivation includes three aspects: attitude towards learning English, desire to learn English and efforts to learn English. Motivation is the internal motivation to promote English learning, and a psychological state of conscious initiative and enthusiasm of English learners in English learning activities [8].

At present, English learners have no long-term goals, their learning attitude is not positive, and their motivation for English learning is at a medium low level. Many learners learn English only for the purpose of examination or enrollment. This kind of instrumental learning motivation can't stimulate the learners' enthusiasm for learning English and keep their interest in learning for a long time. Once their performance is not ideal, their confidence in learning English will be hit and their enthusiasm for learning English will be reduced. Therefore, some scholars have intervened in different ways, and used the intervention effect evaluation method of Autonomous English learning motivation to evaluate the effect of Autonomous English learning after intervention. However, the method is not mature enough, because the relationship between autonomous learning ability and motivation factors is not close enough, making the evaluation method less fluent. At this time, we design a method to evaluate the effect of Autonomous English learning motivation intervention based on knowledge map to solve the above problems and promote English learners' autonomous learning.

2 Evaluation of the Intervention Effect of Autonomous English Learning Motivation

2.1 Establish the Relationship Between Motivation Factors and Autonomous Learning Ability

The relevance between autonomous English learning motivation and autonomous learning ability is based on the feedback of students' actual situation. A questionnaire survey was conducted among 220 sophomores in a university. The questionnaire is designed with reference to the relevant research at home and abroad. It consists of two parts: the basic information of the students and the multiple choice questions. The multiple-choice questions are all in the form of 5-point lektor scale, which is divided into five grades: "very inconsistent with my situation e" to "very consistent with my situation a".

Items 1–15 survey students' autonomous learning ability, covering five aspects: first, making learning plans; second, using learning strategies effectively; third, self-monitoring learning process; fourth, self-assessment of English learning results; fifth, understanding teachers' teaching objectives and requirements. Items 16–27 investigate students' learning motivation, which mainly includes four main motivational factors: first, internal interest; second, self-efficacy (expectation of English level); third, potency (value and meaning of English learning); fourth, motivational behavior (effort level) [9]. After eliminating the invalid questionnaires, such as complete or partial unselected, completely identical and so on, 197 valid questionnaires were tested for consistency. The test results showed that the reliability coefficient reached 0.85. The data collected were analyzed by SPSS15.0.

According to the results of the questionnaire survey, the specific situation of College Students' autonomous learning ability and motivation level is listed in the table below (Table 1).

Table 1. Scores of independent learning ability and learning motivation factors

	Ability/factor	Average value	Standard deviation
Self-learning ability	Study plan	3.5	0.83
	Learning strategy	3.52	0.77
	Self-monitoring	3.34	1.0
	self assessment	3.19	0.77
	Understand learning teaching goals and requirements	3.45	0.81
Motivation factor	Intrinsic interest	3.22	0.77
	Self-efficacy	3.73	0.86
	Potency	3.59	0.89
	Motivational behavior	3.17	0.85

It can be seen from the data in the table that there is little difference in students' autonomous learning ability. In fact, the single factor analysis of variance shows that the mean difference between them is not significant. The average score of all items in the table is more than 3, indicating that students have certain autonomy and have great expectations for English learning, but lack of intrinsic interest. Using statistics to calculate the correlation coefficient between self-learning ability and motivation level reached 0.65. Statistically speaking, there is a significant correlation between them. According to the above analysis, students with different levels of learning motivation have different performance in autonomous learning ability. As shown in Table 2.

Table 2. Differences of autonomous learning ability in different levels of learning motivation

Self-learning ability	High grouping		Low grouping		p值
	M	SD	M	SD	
Study plan	3.56	0.77	2.81	0.76	0.00
Learning strategy	3.76	0.82	3.21	0.59	0.02
Self-monitoring	3.59	0.88	3.04	0.56	0.02
self assessment	3.46	1.1	2.81	0.82	0.03
Understand learning teaching goals and requirements	3.89	0.58	2.85	0.66	0.00

The students whose average score of motivation level is more than 3.43 (including 3.43) are divided into high group and the students whose average score is less than 3.43 are divided into low group. From the data in the table, it can be seen that there are significant differences in various autonomous learning abilities between the high-level group and the low-level group. The students in the low-level group have low scores of autonomous learning ability, which indicates that their autonomous learning ability is not strong and they lack effective self-monitoring and evaluation ability. Regression analysis is made on the influence of motivation factors on autonomous learning. The results are shown in Table 3.

Table 3. Regression of motivation factors to autonomous learning ability

Independent variable (motivation factor)	Beta value	T value	P value
Intrinsic interest	0.23	2.4	0.02
Self-efficacy	0.21	2.26	0.03
potency	0.2	2.18	0.03
Motivational behavior	0.32	3.0	0.00

The results of regression analysis show that more than 80% of autonomous learning ability can be explained by the motivational factors of independent variables, and the influence of these four variables on the dependent variables is not repeated. T-test showed that four motivational factors, namely intrinsic interest, self-efficacy, potency and motivational behavior, were significant for regression. The results of multiple regression analysis in the table can be written into a standard equation: autonomous learning ability = 0.23 × internal interest + 0.21 × self-efficacy + 0.2 × potency + 0.23 × motivational behavior. Therefore, motivation factors will directly affect the ability of autonomous learning.

According to the analysis, it is determined that there is a correlation between the autonomous learning ability and the motivation factors. According to the correlation between the two, the evaluation index is determined by using the knowledge map.

2.2 Determination of Intervention Effect Evaluation Index Based on Knowledge Map

Knowledge map can show the relationship between knowledge development and structure through various graphics, describe knowledge resources and their carriers by using visualization technology, analyze, build, draw and display knowledge and their interrelations [10, 11].

Based on the correlation between the above-mentioned autonomous learning ability and motivation factors, a knowledge map of autonomous learning ability and motivation factors is constructed. As shown in the figure below.

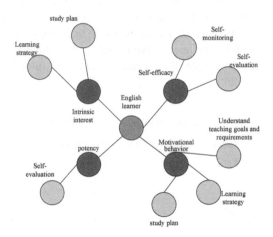

Fig. 1. Knowledge map of autonomous learning ability and motivation factors

On the basis of the knowledge map of autonomous learning ability and motivation factors, it intervenes the learners' English learning motivation from the early, middle and later stages of learning English. In the early stage of English learning, it mainly intervenes the learners' learning attitude and demand; in the middle stage of English learning, it stimulates the learners' emotion; in the later stage of English learning, it strengthens the learners' learning ability. The role of the above different stages and the relationship with learning motivation are shown in the figure below.

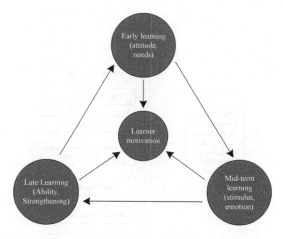

Fig. 2. Motivation intervention in different stages of English learning

Combined with the information in Fig. 1 and Fig. 2, the evaluation indicators of the intervention effect of Autonomous English learning motivation are determined. Interest in learning: Learners' interest in English learning can be improved through novel, uncertain and inconsistent events; curiosity: to stimulate learners' information searching behavior through questioning and other ways, and then to stimulate curiosity; concentration: to maintain students' interest in English learning by changing various teaching elements and avoid attention Loss of strength; learning objectives: to provide students with a teaching objective and formulate reasonable methods to achieve these objectives, linking short-term objectives with long-term objectives; motivation matching: to match teaching with students' various learning needs through the use of various strategies; familiarity: to use students' familiar language and experience in teaching, and put teaching in the existing situation To help learners connect new knowledge with existing experience; learning needs: to set teaching objectives of different degrees of difficulty, to help students establish a positive and expected attitude to success; learning ability: the ability of learners to complete tasks with appropriate difficulty levels; personal responsibility: to let students realize the success of learning in teaching, is their own earnest efforts to study Result. Satisfaction: let the students use the new knowledge or skills to solve the problems in the real situation, provide feedback, praise and motivate the students in time in the learning process, let the students feel the fairness and consistency of the evaluation standards, and make the students always maintain a positive attitude.

After the above analysis, the evaluation indicators of the intervention effect of Autonomous English learning motivation are determined as shown in the figure below (Fig. 3).

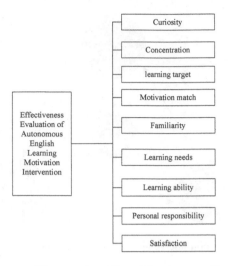

Fig. 3. Evaluation indicators of intervention effect of Autonomous English learning motivation

After determining the evaluation index of the intervention effect of Autonomous English learning motivation, the weight of the evaluation index of the intervention effect is calculated to determine the level of the subsequent intervention effect.

2.3   Calculation of Evaluation Index Weight

In view of the relationship between different indicators of evaluation, motivation factors and self-learning ability, the method of questionnaire survey with experts as the object is adopted to assign values to the indicators. The evaluation index weight is calculated by using the evaluation matrix of expert scores. The matrix is as follows:

$$g=\begin{Bmatrix} k_{11} & k_{12} & \cdots & k_{1i} \\ k_{21} & k_{22} & \cdots & k_{2i} \\ \cdots & \cdots & \cdots & \cdots \\ k_{j1} & k_{j2} & \cdots & k_{ji} \end{Bmatrix} \qquad (1)$$

The elements of each column of the judgment matrix are treated as one

$$G_{ji} = \frac{k_{ji}}{\sum\limits_{j=1}^{j} k_{ji}} \, (ji = 1, 2, \ldots, r) \qquad (2)$$

Sum the judgment matrix after normalization by rows:

$$G_j = \sum_{i=1}^{r} G_{ji} \qquad (3)$$

For the normalization processing, the obtained feature is the weight.

$$G_i = \frac{G_j}{\sum\limits_{j=1}^{r} G_j} \tag{4}$$

In the formula, there are different index scoring items k_{ji}, which ji are item number r and constant. In formula 2, they are rounded G_i to indicate the obtained characteristics and the calculated evaluation weight. According to the calculated weight quantitative evaluation index, calculate the index score value, quantitative evaluation index, and measure the intervention effect according to the quantitative score.

2.4 Quantitative Intervention Effect Evaluation Index

The weight of the intervention effect evaluation index determined in the previous step is unified to prepare for the quantitative intervention effect evaluation index. Considering the counting habit of the percentage system, multiply the weight of each index by 100 to become the standard value of the evaluation result. The final score of each sub index is:

$$J_i = \sum_{n=1}^{r} G_{in} k_{jin} \tag{5}$$

In the formula, G_{in} indicates the weight value i of the sub indicator n and the score value of the expert for the sub indicator. k_{jin} represents the quantitative j score value of the calculated sub index i. Combining the weight obtained by the above calculation process J_i with the quantitative score value i, the comprehensive evaluation value of intervention effect is calculated by the weighted superposition method. The calculation formula is as follows.

$$C = G_i \frac{\sum\limits_{i=1}^{r} J_i}{i} \tag{6}$$

The calculated comprehensive evaluation value of the intervention effect of Autonomous English learning motivation is corresponding to the intervention effect evaluation grade standard table, and the intervention effect grade standard is shown in Table 4. To evaluate the effect of Autonomous English learning motivation intervention.

Table 4. Evaluation level of Autonomous English learning motivation intervention effect

grade	Grading standards	Description	Level description
I	<1.5	The effect of the intervention is positive, and the learner has a longer learning motivation	Excellent
II	1.5–2.5	The effect of the intervention is a positive result, and the learner's motivation for a short period of time	Good
III	2.5–3.5	The effect of the intervention is that there is no obvious guidance result, and the learner maintains the original state	Medium
IV	3.5–4.5	The effect of the intervention was negative, and the learners became lax about English learning	Worse
V	>4.5	The effect of the intervention was a negative result, and the learner's learning intention approached zero	Very poor

According to the intervention effect evaluation grade standard in Table 4, the intervention effect evaluation of Autonomous English learning motivation can be realized. So far, the evaluation of the effect of Autonomous English learning motivation intervention based on knowledge map has been completed.

3   Experimental Study

3.1   Experiment Preparation

In UCI machine learning repository (https://archive.ics.uci.edu/ml/datasets.html) Select the experimental data.The goal of the experiment is 101 students from two classes in a high school. In the research, combined with the characteristics of senior high school students' English learning and the current situation of teaching, some measures are adopted to intervene in students' Autonomous English learning motivation. The intervention effect of Autonomous English learning motivation is evaluated by using the designed method based on knowledge map. At the same time, we use the traditional method to evaluate the effect of Autonomous English learning motivation intervention.

In the process of using different evaluation methods to evaluate the intervention effect, it is necessary to establish a connection between different evaluation indexes. The time consumed in this process is an important index affecting the fluency of evaluation methods. The longer the time is, the worse the fluency is, the shorter the time is, the better the fluency is. Therefore, fluency is chosen as a reference index to evaluate the intervention effect of different Autonomous English learning motivation.

In the experiment, the computer is used as an assistant to process the experimental data through the software analysis system such as word and excel.

3.2   Experimental Results and Analysis

Using the traditional evaluation method of the intervention effect of Autonomous English learning motivation and the designed evaluation method of the intervention effect of Autonomous English learning motivation based on knowledge map, the experiment results are as follows.

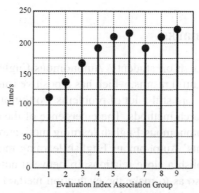

(a) Experimental results of traditional evaluation methods of intervention effect

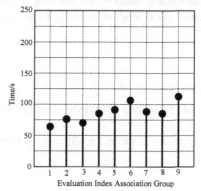

(b) Experimental results of intervention effect evaluation method designed

Fig. 4. Experimental results of different intervention effect evaluation methods

In the experiment, nine groups of experiments were carried out according to the evaluation indicators. The first group was set as no association between the indicators, the second group as the association between the two indicators, the third group as the association between the three indicators, and so on. The time of establishing the association between the indicators of different evaluation methods in each group of experiments was obtained.

The results in Fig. 4 show that the time of traditional intervention effect evaluation method is 115 s at least and 225 s at most. The average time is about 179.6 s after calculation. The results of the designed knowledge-based graph show that the minimum time required for establishing the association of indicators is 60s, the maximum is 113 s, and the average time is about 84.7 s after calculation. The longer it takes to establish the association of evaluation indicators, the worse the fluency of the evaluation method. From the above data, it can be seen that the traditional evaluation method takes about twice as long as the designed evaluation method based on knowledge map, which shows that the designed evaluation method based on knowledge map is more fluent than the traditional evaluation method.

4 Concluding Remarks

In recent years, the research on students' Autonomous English learning motivation is fast, and has achieved fruitful results, which has a positive role in English teaching and research. However, there are still a lot of problems worthy of discussion, whether it is the focus groups or research methods. The emergence of the evaluation method of the intervention effect of Autonomous English learning motivation is of great significance to the research of students' Autonomous English learning motivation. This paper uses knowledge map to establish a close relationship between autonomous learning ability and motivation factors, so as to make the evaluation method more fluent.In the future research, we should take students as the main body, and teachers should guide them properly. We can try to introduce autonomous learning platform into reading strategy teaching, cultivate students' good English reading habits and improve their English level.

5 Fund Projects

1. Key R & D plan of Shandong Province (public science and Technology) 2019GGX105013.
2. Social science planning research project of Shandong Province17CQXJ11.

References

1. Duan, P., Wang, Y., Xiong, S., et al.: Space projection and relation path based representation learning for construction of geography knowledge graph. J. Chin. Inf. Process. **32**(03), 26–33 (2018)
2. Jinmei, L.V., Shengtao, G.A.O.: Research on the quality evaluation of innovation and entrepreneurship education in colleges and universities—based on group G1 method. J. Anhui Univ. Sci. Technol. (Soc. Sci.) **20**(04), 85–89 (2018)
3. Xiong, S.: The performance assessment of blended learning from Kirkpatrick 's model perspective. J. Ningbo Inst. Educ. **21**(03), :96–99+125 (2019)
4. Shen, Z.: A research on the construction of practical teaching standards and quality evaluation for normal school students. J. Educ. Sci. Hunan Norm. Univ.ersity **18**(03), 111–117 (2019)
5. Zhang, Y., Zhang, M.: Competence evaluation and matching based on machine learning. Comput. Eng. Sci.ence **41**(02), 363–369 (2019)

6. Wang, W., Dong, Y., Hu, Y.: Research on evaluation factors of the students' learning behavioral engagement based on interpretation structure model. Math. Pract. Theory **49**(09), 107–116 (2019)
7. Liu, S., Liu, G., Zhou, H.: A robust parallel object tracking method for illumination variations. Mob. Netw. Appl. **24**(1), 5–17 (2019)
8. Dong, M., Jia, Z., Wang, J., et al.: Assessment of effect of situated learning on developing crisis intervention skills. J. Shanghai Jiaotong Univ. (Med. Sci.) **39**(05), 539–543 (2019)
9. Liu, S., Fu, W., He, L., Zhou, J., Ma, M.: Distribution of primary additional errors in fractal encoding method. Multimedia Tools Appl. **76**(4), 5787–5802 (2014). https://doi.org/10.1007/s11042-014-2408-1
10. Jiang, Y., Yuan, X.: Simulation research on accurate evaluation of credit degree in e-commerce transaction. Comput. Simul. **35**(07), 430–433 (2018)
11. Chen, X.: Marine transport efficiency evaluation of cross-border e-commerce logistics based on analytic hierarchy process. J. Coast. Res. **94**(01), 682–683 (2019)

Author Index

Printed in the United States
By Bookmasters